T0245270

CAMBRIDGE LIBRARY COLLECTION

Books of enduring scholarly value

Life Sciences

Until the nineteenth century, the various subjects now known as the life sciences were regarded either as arcane studies which had little impact on ordinary daily life, or as a genteel hobby for the leisured classes. The increasing academic rigour and systematisation brought to the study of botany, zoology and other disciplines, and their adoption in university curricula, are reflected in the books reissued in this series.

Life and Letters of Sir Joseph Dalton Hooker

Sir Joseph Dalton Hooker (1817–1911) was one of the most eminent botanists of the later nineteenth century. Educated at Glasgow, he developed his studies of plant life by examining specimens all over the world. After several successful scientific expeditions, first to the Antarctic and later to India, he was appointed to succeed his father as Director of the Botanical Gardens at Kew. Hooker was the first to hear of and support Charles Darwin's theory of natural selection, and over their long friendship the two scientists exchanged many letters. Another close friend was the scientist T. H. Huxley, and it was the latter's son, Leonard (1860–1933), who published this standard biography in 1918. The second volume details Hooker's management of Kew, his later travels, and the end of his long life.

Cambridge University Press has long been a pioneer in the reissuing of out-of-print titles from its own backlist, producing digital reprints of books that are still sought after by scholars and students but could not be reprinted economically using traditional technology. The Cambridge Library Collection extends this activity to a wider range of books which are still of importance to researchers and professionals, either for the source material they contain, or as landmarks in the history of their academic discipline.

Drawing from the world-renowned collections in the Cambridge University Library, and guided by the advice of experts in each subject area, Cambridge University Press is using state-of-the-art scanning machines in its own Printing House to capture the content of each book selected for inclusion. The files are processed to give a consistently clear, crisp image, and the books finished to the high quality standard for which the Press is recognised around the world. The latest print-on-demand technology ensures that the books will remain available indefinitely, and that orders for single or multiple copies can quickly be supplied.

The Cambridge Library Collection will bring back to life books of enduring scholarly value (including out-of-copyright works originally issued by other publishers) across a wide range of disciplines in the humanities and social sciences and in science and technology.

Life and Letters of Sir Joseph Dalton Hooker

VOLUME 2

EDITED BY LEONARD HUXLEY

CAMBRIDGE UNIVERSITY PRESS

Cambridge, New York, Melbourne, Madrid, Cape Town,
Singapore, São Paolo, Delhi, Tokyo, Mexico City

Published in the United States of America by Cambridge University Press, New York

www.cambridge.org
Information on this title: www.cambridge.org/9781108031011

© in this compilation Cambridge University Press 2011

This edition first published 1918
This digitally printed version 2011

ISBN 978-1-108-03101-1 Paperback

LIFE AND LETTERS OF
SIR JOSEPH DALTON HOOKER, O.M., G.C.S I.

Volume II.

Sir Joseph Dalton Hooker O.M., G.C.S.I.
from a portrait by Sir Hubert von Herkomer R.A. (1889)
by kind permission of the Linnean Society

LIFE AND LETTERS

OF

SIR JOSEPH DALTON HOOKER

O.M., G.C.S.I.

BASED ON MATERIALS COLLECTED AND ARRANGED BY LADY HOOKER

PORTRAITS AND ILLUSTRATIONS

BY LEONARD HUXLEY

AUTHOR OF 'LIFE AND LETTERS OF T. H. HUXLEY,' ETC.

VOLUME II

LONDON

JOHN MURRAY, ALBEMARLE STREET, W.

1918

CONTENTS

OF

THE SECOND VOLUME

APPENDICES

ILLUSTRATIONS

TO

THE SECOND VOLUME

LIFE OF
SIR JOSEPH DALTON HOOKER

CHAPTER XXVIII

ECONOMIC BOTANY AND THE NEW FLORAS

THE practical interests of Economic Botany constantly reappear in the correspondence of the sixties : such as reports on the Indian tea plantations (1868), the despatch of young tea plants and seeds to Jamaica (1863), an interest which led him to accept the dedication of a work on Tea by Mr. James Mac-Pherson, to whom he writes (November 2, 1870) :

> Such a book is very much wanted indeed, and will prove a great stimulus to the introduction of the Tea plant into many parts of the world to which we have sent the plant from Kew, and from whence I have enquiries for such a work.

Further appear the introduction of Ipecacuanha[1] and Mahogany from seed raised at Kew (1866 and 1867) to India, and the boyishly cheerful note to Dr. Anderson in the latter year :

> I am so jolly glad to have been the means of introducing Papyrus into India ; I really am proud of that.

Of special importance is the correspondence with Dr. Anderson of the Calcutta Botanic Garden on the introduction of Cinchona into India[2] at a time when the cost of quinine

---

[1] This was unsuccessful.

[2] Mr. (afterwards Sir) Clements Markham was the actual collector who pushed into the forests of Peru and Ecuador, and at great personal risk brought back young plants and seeds, which were raised in thousands at Kew for distribution. Anderson instituted the experiments which led to its successful cultivation in India.

1

to the Bengal Government was placed at £40,000. The first
reference is dated April 22, 1861 :

> I have not written since you left ; in fact I have been so
> anxious about the Cinchonas you so gallantly took out that
> I was indifferent to everything else in your way out till I
> should hear the result. ' Well done thou good and faithful
> servt.' Thwaites confirms your report of their well-being,
> and I do hope that McNicholl at Ceylon will rear them
> up in the way they should go.

A year later :

> I am truly glad to hear of the Cinchona success and
> sincerely hope my vaticination against Darjiling will fail—
> we know nothing at all by theory.

Indeed there was some trouble over the rival sites recommended for the plantations ; he recommends the avoidance
of partizanship, patient trial of each, for :

> Every day tells me that theory and practice have nothing
> to do with one another, and that in gardening operations
> they are wholly opposed.

And again, summing up Anderson's and his opponent's
views :

> I think you both presume inordinately on your several
> experiences, yours in Java, his in America, and that if
> Cinchona is to succeed it will be *in spite* of you both.

More plants were asked for in January 1863, but ' it is *impossible* to send them in winter. They would all be killed in the
Channel, and must wait till the sharp frosts are over.' By
April 1, 1863, ' The Cinchona growing at Calcutta is a wonder
—have you a photograph of it ? '

So it continues to ' go ahead fast and well. I do not believe
in an atom of difference between so-called *micrantha, nitida,*
and *peruviana.*'

Of a German, however, in charge of a West Indian station,
though able to write ' a splendid paper,' he complains;

How I wish he were a better Botanic Gardener—he has been instructed to propagate Cinchona in Trinidad and made a regular mess of it. A German scientific man is the most unpractical and impracticable pig in Christendom.

Meanwhile an attempt was being made to alter the nomenclature of Cinchonas ; of this he writes (February 14, 1863) :

Neither Bentham, my Father, nor Thomson nor I will have *Chinchona* at any price—true enough it is right in the *abstract*, but it is an innovation that will be forgotten and never followed. At any rate no non-scientific man has any right to dabble authoritatively in scientific nomenclature, and any scientific one is crazy to attempt it before securing the adhesion of a large class of men ; he should have consulted us, you, the French, the Germans and pharmacists before attempting to force a change down our throats. As it [is] names are means, not ends ; Cinchona is not only the long recognised equivalent for Count Chinchon's name, and what is of more importance, is the universally recognised name for the Genus. If we change it on grounds of derivation, so we must thousands of names in Botany, Zoology, Geography, and indeed in every-day language of life.

One of the Kew employés was sent to Dr. Anderson in the winter of 1863, a good gardener, but not likely to become a herbarium keeper or curator, to help at Darjiling cinchona plantations. A year later :

A first-rate man goes out to you in Scott ;[1] he is the author of first-rate papers on Hybridization, highly applauded by Darwin, and goes to India to get any appointment he can in Bot. Gardens, Tea or Cinchona. His only faults are his craze for science and a tendency to shirk work for science. In this respect he would suit you well.

[1] John Scott (1838–80), who had been working as a gardener in the Edinburgh Botanic Gardens. See C.D. iii. 300 and the interesting biographical note, M.L. i. 217. The latter book also contains Darwin's correspondence with him. Hooker's interest in Scott had been stirred by Darwin—whose letters of May 23, 1863, first suggesting the Indian appointment, and of May 22, 1864, when it was settled, are given in M.L. ii. 319 and 331.

The general success of the Cinchona plantations is reported to the sympathetic ear of Sir Henry Barkly[1] at Mauritius (June 17, 1867):

The Cinchona is at last established in Jamaica and in a fair way of being successfully established: 200 acres have been enclosed for the cultivation and 2500 plants have sprung from the seeds I transmitted and which Thwaites at Ceylon ripened.

And further on (July 6, 1868):

Cinchona continues to thrive in India and in most of the Colonies that are warm enough. I have obtained permission and sent out a good gardener for 5 years to St. Helena, of whom I heard an excellent account to-day. His first attentions will be given to Tobacco and Cinchona. The former will be, I suspect, the more profitable produce of the two, the want of really good Tobacco is so great. The London Merchants complain bitterly of the dearth of good leaf for cigar manufacturing in England. Letters from Bahia, received two days ago, tell me that they export from that Port 10,000 *tons* annually, but of so low a quality as not to command a market in England !

On the other hand, though he sends Cinchona plants with others to Sir H. Barkly in 1874 for cultivation in the Gardens in Cape Colony, he repeats the warning he had given to Mr. Bolus[2] six years before : ' There is not a ghost of a chance of Cinchona succeeding in S. Africa.' And ' I cannot fancy what you will do with the Cinchonas, for which I fear you are too cold and dry.'   But though

---

[1] Sir Henry Barkly (1815–98) was a very successful colonial governor, British Guiana 1849–53, Jamaica 1853–6, Victoria 1853–63, Mauritius 1863–70, Cape Colony 1870–7. He was elected F.R.S. 1864, his scientific interests being principally in botany.

[2] Dr. Harry Bolus (1834–1911), botanist and collector, went out to the Cape in 1850, accumulated a large fortune there, and was a liberal patron of botany and education. He founded the Bolus professorship of botany in the South African College at Cape Town, and left a large sum for scholarships, &c., and his valuable herbarium and library to the College. He first corresponded with the Royal Gardens, Kew, in 1867, and continued this during his whole life, presenting large collections of his duplicates to Kew. He published many works on the South African Flora; principally on the heaths and orchids. Elected F.L.S. in 1873, and Hon. D.Sc. of the South African University later.

I cannot suppose that you will succeed with the Cinchonas, I hope that you will with *Eucalyptus citriodora*, a charming plant, the odour of whose leaves far supersedes the ' Lemon Verbena.'

Cinchona was also introduced successfully into Jamaica, but the full triumph of Cinchona in India appears in 1893, when Sir George King, sending his latest Report on its cultivation, writes (September 24) :

We have at last compassed the end which Govt. set before itself in introducing Cinchona into India—an enterprise with the initiation of which you had a great deal to do—viz., to put Quinine within the reach of the poorest native.

Now anybody in Bengal who possesses a farthing (the equivalent of a pice) can buy for himself at any post-office in Bengal a dose of 5 grains of perfectly unadulterated Quinine !

It may interest you to see what these pice packets are like, so I enclose a few. They can be had printed in any Indian vernacular. The scheme was begun last January ; and up to the end of August 368½ pounds of Quinine had been sold in this way.

The Colombian barks were propagated in India ; cork oaks in the Punjab ; seed of better kinds of tobacco was sent to Natal ; Liberian coffee that was first grown at Kew in 1872, became a flourishing crop alike in the East and the West Indies. To Dominica it promised special success, as being immune to the ' white fly ' which destroyed ordinary coffee. Chocolate also was introduced into Ceylon. The *Elaeis guineensis*, source of palm oil, was taken to Labuan ; experiments were made with a new tanning material, the Atgarrobo of Chile, and various fodder grasses were brought to new centres. The plant most widely in demand from Kew in the late seventies was the Eucalyptus, enemy of malaria ; but perhaps the most valuable achievement of Kew was the transportation of the rubber plant from the dangerous forests of the Amazon and the Orinoco to our own healthier colonies. In 1873 Hooker persuaded the Government to send an expedition to obtain the seeds of the *Hevea brasiliensis*—the Para rubber tree. From the seeds

collected, about a dozen plants were raised at Kew and sent to
Calcutta, but all died.   Then in 1876 Mr. H. A. Wickham was
sent out by the Kew authorities.   He found the best trees
growing not in the swamps beside the rivers, but upon the up-
lands, and therefore insisted on *Hevea* being treated as a forest
tree, planted not more than forty to the acre.   As in the case of
the Cinchona, a government jealous of its monopoly might have
raised difficulties had it been certain that the seeds collected
could have started a rival industry ; but the previous experiment had failed, and Mr. Wickham's 70,000 seeds, specially
designated for delivery to Her Britannic Majesty's Royal
Gardens at Kew, passed unchallenged.

This time the experiment was successful.   From the 70,000
seeds, some 2800 plants were raised and sent to Ceylon, where
their cultivation was studied and new seeds in turn sent out
by Dr. Thwaites to Fiji, Queensland and Sydney, Jamaica
and Trinidad, Java and Zanzibar, to be the foundation of the
new rubber industry.

Thirty years later, when at length the rubber plantations
had become a valuable national asset, Mr. Wickham wrote to
Hooker as follows :

August 10, 1906.

Will you permit me to congratulate you on the now, at
last, after so many delays, development in systematic cultivation of the Hevea (Para) Indian Rubber ; remembering, as
I do, your foresight and initiative in securing the free hand
enabling me to bring away the original stock on which it is
founded, from the forests of Alto-Amazonas.

Hooker foresaw the future of rubber from the first.   Writing
to Lady Hooker from Trichinopoly in December 1914, Captain
J. S. Hooker tells how he met an ex-tea planter, a bit of a
botanist, who had several times been to Kew in the old days.

He told me that if he had followed ' Lion's ' [1] advice when
he first came out in, I think, '75, as a tea planter, he would
have been a rich man now.   ' Lion's ' parting words were,
' If you take my advice you will go in for rubber.'   Fancy

---

[1] ' Lion ' was a nickname for Sir Joseph (see p. 367).

his having foreseen the possibilities of rubber as long ago
as that !

In 1868 he strongly encourages Sir Henry Barkly in his
efforts for the salvation of the forests in Mauritius, where, as
in New Zealand and the Western Cape districts, a decrease of
rainfall and general humidity appeared to follow forest destruction by axe and fire. India had found a remedy by
inaugurating a staff of well-paid forest officers, who received
two years' training in the forest schools of Germany and
France ; but ' our arbitrary Indian measures would not suit
a Colony.'

July 6, 1868.

Even in England we are suffering from over drainage,
and the desiccation of the air and extremes of cold rainy
seasons and protracted droughts are no doubt due to this.
At Kew, where thirty years ago very good collections of
Mosses and Hepaticae were to be made in the wood, there
are now not a dozen species, and the underground streams
and back springs of Richmond Hill being diverted into drains,
the trees suffer frightfully and die by scores. . . . I am now
introducing watermains and standcocks all over the grounds
and having reservoirs built on Richmond Hill for the supply
of the Gardens with water which we pump from the Thames
up to the reservoir.

Later, he reports progress with regard to New Zealand.

To Sir H. Barkly

July 6, 1874.

The Colonial Govt. have sent me £100 to be expended
on boxes and carriage of forest plants which Kew is to
supply to the Colony during the summer. I am very glad
of all this though, as it will tend to impress the Govt. with
the practical value of Kew to the State, of which the last
Govt. were absolutely ignorant, and showed no wish to be
instructed.

It was the business of Kew to maintain a correspondence
with other great Botanical Gardens ; but Hooker's own friendship with many of the men at the head of these gave the corres-

pondence a personal turn, covering a wider field of interests
than official communications. Such, for instance, was his
correspondence with his old friend Dr. Anderson, who took
charge of the Calcutta Garden in 1860, the letters amounting
to more than one a month for nine years.

The Anderson letters show the exchange of plants proceeding and the sending of drawings, especially of Orchids,
to be copied at Kew; the safe arrival of three Himalayan
Magnolias; the loss of plants in transport and the fatal damage
to Ward's cases, especially if transhipped ['all steamships hate
Ward's cases,' he assures Mr. Bolus, February 24, 1868]; the
superior facilities possessed by the great nurserymen for reviving and ' growing on ' the plants which reach them, so that
their collectors are credited with sending better materials than
the Gardens and their collectors :

> The real truth is, we cultivate orchids under *very great
> difficulties*, and cannot hold a Dendrobe [1] to a Nurseryman.
> Since my expedition into Sikkim *not one Alpine Sikkim*
> plant has been introduced.   You know I dried all seeds my-
> self and sent them off at once by post straight to England.
> We much want that class of Darjiling plants that are so
> common and gay about the station.   Do make an effort.
> I then introduced a great many, but they have been lost
> since.

Further, time after time he begs that ' the plan, so successful with me,' be adopted ' of sending a few seeds of the rarer
Sikkim things in letters by post *at once* and repeatedly,' and
Alpines collected ' with your own hands by pinches not by
pecks through natives,' who cannot be trusted to see that
they are properly ripe and dry.   For the miscellaneous collections of seeds come up very badly both at Kew and elsewhere.
If seeds from England also fail, let them be bought from
Vilmorin of Paris; they will have ripened better in his southern
gardens.

On the other hand he no longer wants miscellaneous collections of Indian plants sent for the Herbarium; they do not

---

[1] The Dendrobium is a very handsome genus of orchid.

repay the trouble of collating. There were abundant duplicates of almost everything :

> Half the day Thomson and I spend over the huge supple-
> mental Indian collections, most of which are mere lumber,
> and we are burning cartloads of specimens. . . . For now
> 12 years we have been groaning over collections from
> India, and we still have Falconer's and Wight's to do.

His active interest in the Calcutta Gardens had continued
unabated since his first visit in 1847. From the first he would
have liked to see them moved to a more convenient position,
say at Alipur. Opportunity of pressing the point came in
1867. At the end of the year Anderson wrote reporting the
destruction wrought by a terrific cyclone ; if the Gardens be
reconstituted, it should be nearer Calcutta. To have the
Botanic Garden where it would be accessible to students of the
Medical College and to the public, would be an immense boon,
Hooker knew, and he replies :

> I immediately wrote a leader for G[ardeners'] C[hronicle]
> in which, with my usual stupidity, I put the Garden on the
> wrong side of the river ! It is a constitutional disease with
> me not to know right hand from left till I stop to think.
> I have consulted Thomson, Sir Lawrence Peel and others,
> and all think the principal Botanical Establishment, Library,
> Herbarium and a good type-named Garden Collection should
> be at or near Calcutta, nearer than Garden Reach, and a
> noble large general Garden perhaps at Darjiling or elsewhere
> in hills on rail to Calcutta.
> *February* 19, 1868. I continue to sympathise most deeply
> in the matter of the Calcutta Garden, and will co-operate
> gladly to the extent of anything short of having no Botanical
> Establishment at Calcutta.
> You ask if I and Thomson ' will urge you to remove the
> site.' We will gladly do anything in reason, but you must
> mature a plan first. I suppose that you could not come
> home for 6 weeks and discuss it here ? Govt. deputing
> you ? It is a matter of vital importance. If you could come
> at the British Association time we could do a stroke of
> work.

After reporting official sympathy with this view in his next letter, he urges on June 4 :

> I am most anxious about the future of your Garden and take care to ventilate it everywhere. Kidderpore all agree is a capital idea [a site was available there]—Alipore way was always my view since 1847. Do not frighten the Govt. by too great demands. My Father's plan always was to ask for so much of one thing at a time as could be done and *form a complete affair by itself ;* the next year another, and so on. It would be most advisable that you came home for a short leave to take hints from European Gardens. This would commit the Govt. to move.

In 1867 he is ' heartily glad to find that there is another Student of Botany in South Africa—a Colony which I think boasts of more than any other.' This was Mr. Harry Bolus, already mentioned, with whom he exchanged plants and seeds. But there was a limit to the powers of Kew. Collating doubtful species took time, and no one at Kew was so familiar with Cape botany as to distinguish common from scarce plants or to name off hand. Therefore let his correspondent mark those specimens only of which he has any doubt—not sending more than twenty or thirty at a time.

' You must expect now and then a difference of opinion,' he writes, ' as to the species : we work from dried specimens, you from fresh, and we have each *much* to learn from one another.' (September 9, 1867.)

On his father's death in 1865, Hooker had taken up the correspondence with Sir Henry Barkly, then Governor of Mauritius, where there was a fine Botanical Garden. Sir Henry was a keen botanist, and Lady Barkly a collector of ferns. A packet of ferns which she had sent to be named had to stand over awhile for ' we have now no one at Kew capable of naming Ferns.' Early next year, however, as the post of Assistant Director had been abolished on Hooker's accession to the Directorship, an additional assistant in the Herbarium was sanctioned, and ' I have my eye on a man who will take Ferns in hand.' Meanwhile Lady Barkly's ' Pteridomania ' would be ' remembered when we have a distribution of duplicates at Kew.'

Their personal interest in his own science was stimulated
by many practical touches. Hooker buys a set of rare Samoan
ferns and sends them out that Sir Henry may have the refusal
of them ; sees about the naming of Reunion ferns for Lady
Barkly, with the promise of examining one rare specimen
himself ; discusses a knotty point of nomenclature with her ;
sends cultural hints from Kew experience of growing the delicate *Hymenophyllum* group under bell glasses coloured green,
or in artificially arranged shade at Calcutta ; sends out books
that are wanted, and introduces an American botanist who
will correspond with him on ferns.

More officially he busies himself to confirm the appointment
of a first-rate curator of the Mauritius Garden whom his father
had chosen, Dr. Meller, one of whose chief interests should be
to make known the rich vegetation of Madagascar. For Dr.
Meller he picks out rare books at sales, and helpfully adds :
' Always ask me to do anything of this kind, as I can generally
hear of cheap copies.'

Most important of the various transactions with Sir H.
Barkly was that a little before Sir William Hooker's death, the
Mauritius Herbarium had been sent to Kew for collation with
the collections there.

In this connexion a couple of letters may be quoted. The
first touches on the idiosyncrasy of the assistant who had
been dealing with the ferns sent for identification.

June 29, 1866. I quite understand what you observe
of his tendency to over-work—it is inevitable with men
who, having long closely studied local Floras and subdivided them to the (very) root, are suddenly confronted by
large collections and extensive suites of specimens from
many localities ; a reaction sets in, and like all neophytes,
they are apt to be carried too far [when going from critical
analysis to careless or superficial synthesis, March 23, 1867].
I am always fearful of assuming the position of scientific
mentor over my subordinates, at first especially, but shall
lose no opportunity of keeping him straight before he finds
his way for himself, which he is sure to do in time. . . . Any
hints you give me for him, or his nomenclature, will be

immensely valuable, both for their own sakes and as showing him that correspondents are not mere lookers on.

At the end of the next month, the Mauritian Herbarium was to go back, as soon as the assistant had revised the lists :

He tells me that he catalogued them all, took out specimens of all that were wanting with us and kept corresponding numbers, so that any query arising in Mauritius can be answered at once by a reference to us. These collections should be deposited at the Bot: Garden; where alone they can be made useful, and to which establishment they are essential.

But Sir Henry was disappointed and dissatisfied with the Herbarium when it reached Mauritius. He had expected an entire critical revision of its contents. Hooker had to explain that this was a much vaster work than either Sir Henry imagined or Kew in reality had time or means to undertake. The simple collation of the materials with those at Kew, a point of great value for future reference, had involved many weeks' labour for Professor Oliver at a time when he was overburdened with work owing to the death of Sir William and the illness of Dr. Hooker.

With the return of T. Thomson from India in May 1861, there was a renewed prospect of finishing off the arrangement of the Indian materials and publishing a complete Flora Indica.

As regards the former, the work was greatly prolonged. Dr. Thomson himself was broken in health, and though after paying another visit to India, forbidden by his doctor, he left Reigate and definitely settled at Kew in 1863, progress was slow. When the distribution of the existing Indian collections was finished, Dr. Wight's Herbarium of Peninsular India also came to Kew, for distribution, so that the catalogues and material preliminary to the main enterprise were not finally ready till 1870. The first part of Vol. I. appeared in 1872.

As to the form the Indian Flora was now to take, it was that of the Colonial Floras which were being put in hand by the local Governments.

Thus the matter is broached in a letter to Anderson, August 18, 1861 :

The Australian Govts. have taken up the Flora in earnest and will pay Bentham well to do a Flora Australasiae. I wonder when the Indian Govts. will. I have been really thinking that if the Indian Govts. (Calcutta, Bombay and Madras) would club to make it worth my while I would yet do ' Flora Indica ' and give up examinations and all my other emoluments but Kew for 8 years. I should do it in English, like Flora Hong Kong in all respects. I think it would all come into 8 volumes of 1200–1400 species in each, and if Bengal would grant £100 per volume and Madras and Bombay £50 each, and they would together take 100 copies from the publisher, at not more than 20s. per volume, I would undertake the work and devote my faculties to it till finished. I have now been 14 years working at the Indian Flora continually, and I must confess I feel loth to leave the work to others now that the way is all cleared by myself.

His hope was that Anderson might arrange for the proposal to come from the Indian Government, and he writes (April 21, 1862) :

I shall be very anxious to hear what terms you shall have made with Laing about Flora Indica. I really cannot make up my own mind as to Latin or English and shall only be too glad to have that settled for me ; *du reste* we are agreed on plan &c.

And in July :

My impression is that for a Flora of all India *Latin* would be best, for departmental or Presidential English.

The matter dragged, as official matters do, and he was vexed by the delay.

August 4, 1862.

Sir W. Denison has just written asking me what steps are taken in Calcutta about ' Flora Indica.' I have answered, sending extract of yours of March 10th, saying that you had prepared a plan which Laing *had promised to sanction*, and that I was to have option of refusing it—which was all very fine, but never a word have I heard since from you or anyone else. I am heartily sick of you Indians and your

talkee talkee of Flora Indica. I wish I had adhered to my
resolution of listening to no more proposals. . . . I really
am sick of proposals, and feel rather indignant about the
whole thing. I have had correspondence enough to make a
bonfire of about it, first and last. Denison is in earnest,
however, and is a very fine fellow, and I should not like to
give him umbrage by appearing ungracious about what he
has really taken an active interest in.

Decision still tarried : ' I do wish,' he writes on October 27,

you would put me out of my pain about ' Flora Indica.' I
would be thankful to be told there is no chance of its being
undertaken rather than this perennial uncertainty.

By the new year another delay arose.

I am booked for Flora of both British N. America and New
Zealand, so good-bye to Flora Indica for the present. Here
are Australia, New Zealand, Brit. N. America, the Cape and
West Indies *all writing Colonial Floras,* and India NOWHERE.
If you still think of it I would only undertake it with Oliver
and set him to work while I get these little works off my
hands. New Zealand will not cost me much trouble as by
good luck I am now well through revising the Flora for my
own satisfaction, and so it will be all *writing out* chiefly.
(January 4, 1863.)

Still negotiations proceeded. The India Council at home
took up the matter, though, as he tells Anderson on September
10, 1863,

the initiative had far better come from Calcutta. As it is
I do suppose that it will be more your affair than any one
else's ; true enough we may begin it here, but we are getting
old, and the work can never be finished by me, I fear. As it
is my hands are very full without it. Thomson, however,
is most anxious to begin.

But the initiative was not to come from Calcutta. The
official responsible, ' to my knowledge a very " pernicketty "
fellow to originate a thing with,' was obdurate, despite Anderson's ' gallant fight.' News of this defeat reached Kew by
the same post as an intimation that, prompted by Sir Charles

George Bentham and J. D. Hooker, 1870.

Wood, who forestalled Denison's offer in November to see things through in Calcutta, the India Council at home were beginning to move. And he tells Anderson (September 19) :

> I am sure you have done all in your power and well done too. I am not at all disappointed and will do my 'little possible' still ; under whatever shape the work is sanctioned by whatever department of Govt. I am your man. . . . As for the personal pay, I feel past *paying* for and *praying* for . . . but Thomson will no doubt take his share (of the work) and I do suppose that T. Anderson, late Director of the B. Bot. Gard. Calcutta, will in 1870 carry on the work. . . . I find the N.Z. Flora so onerous and laborious, though I have *thrice* worked it all out, that I do dread the Flora Indica.

This was looking far ahead. But his fears as to its completion were sadly justified, more especially by Thomson's illness and Hooker's accession to all his father's duties and such unlooked-for tasks as the completion in 1868 of the ' Genera of Cape Plants ' [1] on Dr. Harvey's sudden death, for Dr. Sonder, who was nominally collaborating, proved a broken reed.

For a long time, so far as Hooker was concerned, amid the endless pressure of duties at Kew in addition to work at the laborious Genera Plantarum, he could only hold a watching brief for the Flora Indica. A typical note is dated December 30, 1864 :

> I am working desperately hard at Herbarium and Garden work, Genera Plantarum and Cryptogamic portion of the New Zealand Flora. I have also undertaken to finish Boott's Carices and to publish 200 plates thereof. The whole of the collections have come to Kew. Flora Indica makes no progress.

[1] Though Harvey's *Cape Flora* had to be abandoned, Hooker, in default of other aid, finished the *Genera* himself before July 7, when he wrote to Mr. Bolus at the Cape. ' It is true,' he notes a fortnight later, 'that each little hiatus was little, but they could only be supplied by a full consideration of collateral subjects.' But it was a laborious task for one already so busy who was not personally familiar with the Cape Flora. He had added ' Sketches of the Arrangement of the Classes and Orders that may assist the Students, and the very improved Introduction to Botany from Harvey's *Cape Flora*.'

The Flora of Tropical Africa is ordered, and Oliver and I have to undertake it ! We shall not attempt a complete Flora, but sort of sketch of each genus as far as its species are well known, easily discriminated and worth describing if new, and so on.

In short (October 12)

we are all worked within an inch of our lives, and as my own family grows up and my Father advances in years (he is now 79) my daily cares increase in every way, so that I am at times utterly stranded with work.

Nine years had increased the note of pessimism since Colonel Munro's departure in 1855, and the farewell letter that looked forward eagerly to his return :

How soon will that be ? I shall hope to have worked out the Indian collections down to Gramineae [Col. Munro's speciality] by that time, when you really must relieve guard, or I shall lay down my musquet—but indeed I do hope that you will have laid by your real one before that, and have left active soldiering to younger men, who have·not the stores of intellectual matter to dispense that you have. We Botanists have some property in you and do not wish to lose it.

But in 1864 the hoped-for progress has not been accomplished ; and he repeats to Anderson :

I begin to look to your return before any material progress can be made in so laborious and extensive an undertaking.

Again, May 20, 1868 :

I wish indeed you could scheme a few months in England to talk over matters, and still more that I could scheme a cold weather at Calcutta to help you ! In two years you might make a good stroke into the Flora Indica, which Thomson will never do a stroke of—and as for me, my share is done in the 7 years of hard work I had in naming and arranging the whole Indian collections of ourselves, Jacquemont, Griffith, Falconer, Helfer, Wight and all

others, and incorporating with the Wallichian, &c. &c., at
Kew. That and the Precursores must stand as my contribution. There is now no difficulty in taking any genus down
and describing the species—all that remains is to collate
with the original Wallichian Herbarium at Linn. Soc.

In 1869, the sixth year since the work had been officially
sanctioned (November 10, 1863), the cry was still for *workers*.
The Botany of India rested in a ' chaotic and disgraceful state.'
Even the Calcutta Botanic Gardens remained scarcely accessible
to students in Calcutta. Fruitless search for J. L. Stewart's
' Punjab Plants,' printed by the Punjab Government, of which
he had heard by accident, revealed the fact that

the local Governments of India habitually print purely
scientific works on Botany which they neither advertise,
nor publish, nor distribute to Botanists, and which, as in
this case of the Punjab Flora, are quite inaccessible to working Botanists, even when they hear of them. [In short] it
is really a pity that steps are not taken to centralize and
utilize the Scientific efforts of the Indian Govt. Indian
Botany is the *bête noire* of Botanists. (To Sir Mountstuart
Elphinstone Grant Duff, November 3, 1869.)

He had already suggested that Dr. Cleghorn, a retired
civilian, should be employed in writing books on Indian
Forestry.

His little work on the forests of Southern India is a most
excellent one, and should be followed by others on the
forests of the N.W. Provinces, Bengal, &c. It is a sad pity
that the experience of such men, who were the organisers
of the Forest System, should go to the grave with themselves. (To M. E. Grant Duff, February 6, 1869.)

CHAPTER XXIX

ALTHOUGH the last five years of the Assistant Directorship were a period of great pressure administratively, it was also a productive period in scientific work.

Chief among Hooker's publications were 'The Outlines of the Distribution of Arctic Plants,' which after being first read at the Linnean, June 21, 1860, was completed for publication in the transactions for 1862 ; a series of publications on the flora of the Cameroons, based on Gustav Mann's collections ; [1] part of the Handbook of the New Zealand Flora ; the famous Essay on Welwitschia ; the Botany of Syria and Palestine for Smith's Dictionary of the Bible, and the first parts of the Genera Plantarum.

Of these, the monumental Genera Plantarum deserves first mention, for it marked an epoch in botany. With the advance of knowledge, previous systematic works of the kind were no longer adequate. These had been based on examination of a relatively small number of plants, and were quite inadequate in face of the vast numbers of plants that came to Kew from every part of the world. A great summary was more than ever needful.

Hooker did well by inspiring Bentham to join in this monumental task at the very moment when he was inclined to retire from botanical work. Both had long felt the need of a complete summary of botanical diagnosis, but realised that it was

[1] One of these was an enumeration of the Mountain Flowering Plants and Ferns of the whole region for Burton's *Abeokuta and the Cameroons Mountains*, 1863.

beyond the powers of any one man to undertake. About 1857 they found that they had this idea in common. Thereupon plans and guiding ideas were fully discussed. Work began with the sixties, and by 1862 the first part of Vol. I. appeared. This was completed in 1865. Of Vol. II. the first half appeared in 1873, the second in 1876 ; of Vol. III:, similarly in 1880 and 1883, the work, as long as it was arduous, thus covering nearly a quarter of a century.

The aim of the work was not so much, like that of so many others, to produce a complete new system, as to lay the foundation for this by the accurate definition of the smaller groups. Systematic botany was taken not as an end in itself, but as a means of illustrating the laws of evolution and the dispersal of species, and the relation of physical changes to these laws. The authors set out to give a revised definition of every genus of flowering plants, a view of its constituent species, geographical distribution and synonymy, with references and notes. ' It is difficult,' Hooker once wrote to Bentham, ' to keep one's wits sharp in revising such pregnant matters.' The especial value lay in the fact of a personal re-examination of thousands of specimens, living or dead, whenever practicable, for between them the authors had an extent of knowledge and a command of materials never previously attained. At the same time as he analysed his materials for the Genera, Bentham took the opportunity to discuss fully some of the more important orders in the Linnean Society's Journal.

The general framework upon which Bentham and Hooker built their work was not a new one. It was adapted, with advantageous modifications, from the system set forth by De Candolle. This they chose as the most satisfactory of the many with which the path of botanical science had been strewn with increasing frequency from 1789 to 1857. Botanists were constantly striving after a natural system of classification as opposed to the artificial, non-natural system of Linnaeus, which long held the field by reason of its utility in identifying plants.

A natural system, said Ray, was not to bring together dissimilar species, nor to separate those which are really allied.

But his dictum gave no clue to the principle on which this grouping should be made. What was the true test of affinity ? It was something more than the casual resemblances which in early days led to the division of plants into trees, shrubs, and herbs.

As knowledge advanced, Linnaeus, with his gift of lucid discrimination and concise terminology, was able to mark off species clearly by their structure and group them. Here was a firm step for future advance to a natural system ; but that advance was stopped by the very success of his non-natural identification system.

He applied the strict principle of formal logic, whereby a species is defined as possessing the attributes common to a wider class (genus) together with the attributes peculiar to itself. Hence his scheme of the two names, one generic, the other specific, which labelled and ' placed ' a species, the ' barbarous binomials ' of a later sneer, which ignored Linnaeus' logical mind and the orderly basis he laid for future workers.

De Jussieu revived and filled out a conception which had already been partly applied. Nature had given plants as they germinated, either one seed-leaf or two or none. Here, then, were the three primary groups of De Jussieu's system (1789), monocotyledons and dicotyledons (together the flowering plants) and acotyledons (the cryptogams, mainly). He completed his subsidiary grouping by dividing the flowering plants into fifteen classes, somewhat artificially arranged, and these again into 100 natural orders, each made up of a group of genera with characters in common.

This system De Candolle recast. De Jussieu's classes were scarcely satisfactory ; the addition of whole new floras, such as those of the Cape and Australia, meant much reorganisation. The great virtue of De Candolle's system was that, in the main, it was established on a morphological basis. True that he employed physiological characters as well for some of his definitions, but he recognised the comparatively small value of these in classification, unlike Lindley, none of whose classificatory schemes held good, for the most diverse plants may show similar physiological adaptations to their conditions, e.g. when they turn parasites. Moreover, he showed that his morphological basis held good even where obscured by structural abortion or degeneration or union of parts. Pushed to its conclusion, indeed, this implied that existing species were not originally such as we know them. To build on such a foundation was not irreconcilable with Darwinian developments, though, like all other pre-Darwinian systems, it was based on the belief on the fixity of species, and so had missed the real clue to nature's order.

Had the order of events been changed by ten years, and the planning of the Genera Plantarum followed instead of preceding the ' Origin,' it would have been arranged to show as far as possible a grouping by lines of descent. But the original scheme had been worked out before the ' Origin ' appeared, and it was not till nearly six years afterwards that Bentham confessed himself a complete convert. A new scheme, so far as the two authors were in agreement as to affinity by descent, would have meant a new survey of the whole botanical field and a thorough re-working out of the evolutionary idea as applied to botany. Leaving these, then, to the ripening effect of time, they proceeded with their original plan, only with such clear recognition of natural affinity by kinship that it became, if not a Darwinian exposition, at least an arsenal of material for such an exposition.

Bentham's appreciation of the Candollean system was perhaps intensified by the fact that he had been brought up on it and had worked with De Candolle himself ; but this implied no disposition to follow De Candolle slavishly. His system was used as a basis, not a complete scheme ; the great groups, with special reference to the Gymnosperms, were more evenly balanced ; a new series, the discifloral, was introduced under his first group ; and elsewhere we see—here, a new sub-class introduced : there, the whole series of natural orders re-cast, the morphological grounds of classification being extended so as to include differences of internal structure.

The arrangement of the work, if assumed to proceed in each great division from the simpler to the more complex orders, seemed to support the theories as to the primitive type of Angiosperms advanced by Dr. Arber and Mr. Parkin in 1907.[1]

Dr. Arber accordingly wrote to enquire if the sequence of orders and families adopted, to give a simple example, in the ' British Flora ' of Bentham and Hooker, was in accordance with some scheme of beginning with the most primitive. What precisely was the principle involved ? The reply was as follows :

To Dr. E. Newell Arber [2]

14 South Parade : May 13, 1907.

With regard to your queries respecting the primitive type of Angiospermous ' plants,' that subject has never been far from my mind for upwards of half a century, during which period I have failed to grasp a feature in the Morphology, Physiology or Geographical distribution of Angiosperms, that gave much color to whatever speculations I may have indulged in respecting it.

I do not share Engler's views as expressed in his classification and writings. The classification is neither better nor worse in the abstract than De Candolle's (so-called), and is far more troublesome to apply for practical purposes. I hold to Robert Brown's view of the orders being reticulately not lineally related.

The Cohorts of the Genera Plantarum were the result of long study and anxious deliberation on Mr. Bentham's and my part ; they are in a measure compromises, intended to show the relationship of the orders and at the same time enable users of the work to recognise them (and the plants belonging to them) by our descriptions.

You ask why ' in the British Flora of Mr. Bentham and myself I begin Dicots with Ranunculaceae ' ! Premising that I had no part in the authorship of the work, I can only assume that Mr. Bentham, having regard to the object of the work which he sedulously puts forward, adopted what

---

[1] ' On the Origin of Angiosperms,' read at the Linnean Society in March, and published in the July Proceedings.

[2] Edward Alexander Newell Arber (b. 1870), M.A., Sc.D. Cantab., F.R.S., F.L.S., Hon. Member of the New Zealand Institute ; University Demonstrator in Palæobotany, Trinity College, Cambridge.

he considered the sequence best adapted for his purpose—
that is the so-called Candollean. I am quite sure he had
no hypothetical view.

Lastly with regard to the primitive Type of Angiosperms.
I am disposed to think that apart from Geological Evidence,
the channels along which this is to be sought have not been
explored, if found.

An excellent description of the Genera Plantarum is given
by Professor F. O. Bower, which I quote from ' Makers of
British Botany,' pp. 313-14.

It consists of a codification of the Latin diagnoses of
all the genera of Flowering Plants. It is essentially a work
for the technical botanist, but for him it is indispensable.
Of the known species of plants many show such a close
similarity of their characters that their kinship is recognised
by grouping them into genera. In order that these genera
may be accurately defined it is necessary to have a précis
of the characters which their species have in common. This
must be so drawn that it shall also serve for purposes of
diagnosis from allied genera. Such drafting requires not
only a keen appreciation of fact, but also the verbal clearness
and accuracy of the conveyancing barrister. The facts
could only be obtained by access to a reliable and rich Herbarium. Bentham and Hooker, working together at Kew,
satisfied these drastic requirements more fully than any
botanists of their time. The only real predecessors of this
monumental work were the *Genera Plantarum* of Linnaeus
(1737-1764) and of Jussieu (1789), to which may be added
that of Endlicher (1836-40). But all these were written
when the number of known genera and species was smaller.
The difficulty of the task of Bentham and Hooker was
greatly enhanced by their wider knowledge. But their
Genera Plantarum is on that account a nearer approach to
finality. Hitherto its supremacy has not been challenged.

Notable in another way was the monograph on the strange
plant *Welwitschia mirabilis,* named after Dr. Welwitsch, who
had discovered it in Angola. Hooker did not do much in the
way of microscopic botany, but what he did was fifteen years
ahead of contemporary work, and remained of permanent value.

The monograph on *Welwitschia,* patiently working out its morphology, development, and histology, still holds its place, though recently many papers on it have been written under the direction of the late Dr. Pearson [1] of the Cape, and new light has been thrown on it by subsequent botanical generalisations.

The determination of this highly anomalous plant was a matter of great labour and prolonged microscopical examination directed by unrivalled botanical knowledge. ' I expect it is going to be your Barnacles,' wrote Darwin with a jesting glance at his own long drawn labours with the microscope on that genus ; and Hooker himself regarded this as his greatest triumph of the kind.

' I brought my remarkable plant before Linn. Soc. last Thursday (he tells Darwin, January 19, 1862) with some effect— it was thought quite as curious as I represented.'

And the following day he writes to Huxley :

Then this blessed Angola plant has proved even more wonderful than I expected—*figurez vous* a Dicot. embryo, expanding like a dream into a huge broad woody brown disc 8 years old and of texture and surface like an overdone loaf, 5 *feet diam.* by 1½ high above the ground, and never growing higher, and whose two *cotyledons* become the two and only two leaves the plant ever has, and these each a good fathom long. From the edges of this disc, above the two leaves, rise branched annual panicles, bearing cones something like Pine cones, which contain either all female flowers, or all hermaphrodite flowers ; the hermaph. flowers consist of one *naked* ovule absolutely the same as of *Ephedra,* in the organic axis of the flower, surrounded by six stamens and a four-leaved perigone. The ♀ flower is quite different ! Lastly,

[1] H. Harold W. Pearson (1870–1916). He was educated privately, and after holding a teaching post at Eastbourne, he entered Cambridge in 1893 ; was Foundation Scholar of Christ's College in 1896, Darwin Prizeman and Frank Smart Student of Botany at Gonville and Caius College 1898. Visited Ceylon as Wort's Travelling Scholar 1897–8, and gained the Walsingham Gold Medal in 1899. B.A. 1896; M.A. 1900, and Sc.D. 1907. Assistant for India, Royal Gardens, Kew, 1899–1901. Assistant to the Director 1901–3. Appointed Harry Bolus Professor of Botany, S. African College, 1903, he travelled a good deal, especially in Namaqua Land; and contributed various botanical and geographical papers. Through his ceaseless exertions an unrivalled Botanical Garden has been formed at the Cape. F.R.S. 1916.

fancy my joy at discovering the key to the development of this hypertrophical embryo taking to become a plant after the fashion it does : and at my being able to show that though neither Dicot, Monocot, nor Gymnosperm in flower or Exogen or Endogen in structure of axis, wood or bark (its cambium ring is facetious in the extreme), it is still undoubtedly a member of the family Gnetaceae amongst Gymnosperms, as the structure of the ovule and development of the seed and embryo clearly show. It is out of all question the most wonderful plant ever brought to this country—and the very ugliest. It re-opens the whole question of Gymnosperms as a class, will (in the eyes of most) raise these, as I always said they would be raised (by its hermaph. state and perianth) to equivalence in these respects with Angiosperms, assuming (which I do not) that such unisexuality is a sign of low type in Phaenogams, strikes at the root of Brown's placentation theory and of that which ranks the radicle of embryo as an internode : and is a strong argument in favour of a new French doctrine that the Gymnospermous ovule is all a delusion and a snare.

There then—having bepraised myself I will turn the cock on you. I am very much obliged for the Edinburgh paper slip, which is very gratifying ; the outline seems capital, and I do not wonder that you found sinners enough in ' Saintly Edinburgh ' to go and hear [it].[1]

By August 4 he could tell Dr. Anderson that he had spent fully seventy hours already over the microscope, and yet had all the wood and leaf anatomy to do ; and on the 20th arouses Darwin's admiring envy at such a feat by having

sat 5 hours together at microscope at least 6 times lately, besides all the odd days and hours I have spent over it ; and am very far from finished yet. Every part is so curious.

He was deep in all the ' horrid complexity of Gymnospermous embryology.' At this moment he was fortunate

[1] On January 4 and 7, 1862, Huxley lectured in Edinburgh ' On the Relation of Man to the Lower Animals.' A furious outcry followed in the local religious papers. See *Life and Letters of T. H. Huxley*, i. 278 *seq.*

enough to receive five splendid specimens from a Mr. Monteiro of Loando, who ' like a trump ' sent down the coast at his request to get them. And during his absence from home in September, still ' staggered with the intricacy of Welwitschia,' much help was given by Professor Oliver, ' who is a real blessing,' and had been examining the tissues where he had left off, making ' some charming drawings that will save me a world of trouble.' (To C. D., September 16, 1862.) The completed monograph was read at the Linnean in December, and published in the Transactions for 1863.

The inevitable sense of staleness after a protracted piece of work appears from a letter to Darwin of October (12 ?).

My wife went to Cambridge and enjoyed it ; I stayed at home ! (and enjoyed it), working away at Welwitschia every day and almost every night. I entirely agree with you by the way, that after long working at a subject, and after making something of it, one invariably finds that it all seems dull, flat, stale and unprofitable—this feeling, however, you will observe only comes (most mercifully) after you *really* have made out something worth knowing. I feel as if everybody must know more of Welwitschia than I do, and yet I cannot but believe I have ill or well expounded and faithfully recorded a heap of the most curious facts regarding a single plant that have been brought to light for many years. The whole thing is, however, a dry record of singular structures, and sinks down to the level of the dullest descriptive account of dead matter beside your jolly dancing facts anent orchid-life and bee-life. I have looked at an Orchid or two since reading the Orchid book, and feel that I never could have made out one of your points, even had I limitless leisure, zeal and material. I am a dull dog, a very dull dog. I may content myself with the *per contra* reflection that you could not (be dull enough to) write a ' Genera Plantarum,' which is just about what I am best fitted for. I feel I have a call that way and you the other.

The Arctic Essay was one of those where his own work had ranged far into rewarding fields under the stimulus of Charles Darwin's questioning, and after patient marshalling of the facts and analysis of their meaning, brought back new support for Darwin's ideas, wherein was to be found the only intelligible explanation of the problem.

After enumerating the 762 known Arctic flowering plants with their localities (examining specimens in every possible case), and distinguishing the five Arctic areas characterised by marked differences in vegetation, he traced the distribution of the Arctic plants and their close allies into the temperate and alpine regions of both hemispheres, and showed how this distribution was accounted for by slow changes of climate during and since the Glacial period.

The five botanical areas differed greatly in the abundance of their flora, while many types were restricted to a north and south range in their own area. Richest of all was the Scandinavian section of the European area, containing three-quarters of the whole Arctic flora, three-fifths of the species and nearly all the genera. Hooker had already pointed out in the Tasmanian Flora that this Scandinavian flora alone of all groups was present in every latitude of the globe. The fuller the investigation, the more clearly all pointed to a southward migration of plants as the Glacial cold devastated the northern lands—and a subsequent return northwards at the end of the Glacial period, though in each area certain species had changed during long isolation—so as to be botanically defined as closely allied representative species—and again in each area the march north had been accompanied by hardy plants from the southern lands temporarily occupied, giving a slightly different character to each area.

Greenland presented a crucial case. Its flora was scanty ; it possessed scarcely any American species, though so near to America ; and yet, although European in character, it lacked some of the very common Scandinavian types, which ranged far north elsewhere. Its poverty was due not to climate, but to a large abstraction of Arctic types from some other cause. In effect, the plants retreating before the cold found their retreat cut off at the end of the peninsula. Many, having no further refuge, perished. The survivors spread north again with the milder climate, but the sea still sundered Greenland

from America, and they were not overtaken and reinforced
by the migrants on the American shore.

These conclusions then, drawn from much laborious comparison of species and tabulation of statistics, could only be
accounted for by admitting Darwin's hypothesis of the southward migration of northern forms—an hypothesis begun by
Edward Forbes and extended by Darwin to transtropical
migration. Nevertheless, Hooker felt doubts as to the extent
of the world-wide cooling invoked by Darwin to account for this
transtropical migration ; and the amount of equatorial cooling
needed would, he considered, have killed off all the purely
tropical vegetation such as we know.

The same Darwinian interest extended to his technical
work on Mann's Cameroon plants, so interesting as connecting
the Cape and Europe. This was to lead to a discussion of the
cold period ' *quoad* tropical African Mountains and Flora,' and
letters to Darwin are full of information as to what northern
plants were preserved in the cooler tracts of these tropical
mountains.

I do not know what to think of Tropical plants during
the cold period [he writes on March 17]. As to their living
through it, it is an *impossibility*. I quite go along with you
in suggesting as many Tertiary or Secondary cold periods of
migration as you please. But that such an order as Dipterocarpeae, whose species are all ultra-tropical, all trees, containing many diverse genera and species, should have survived
a cold period, or have been developed since, are equally pre-
posterous surmises in the *present state of science*.

Darwin in return repeated the claim of the ' Origin ' (ch.
xi.) for no more tropical cooling than Hooker himself had found
in the Himalayan zone where tropical and temperate flora
commingled, and confidently believed it would be found that
the ultra-tropical plants mentioned could adapt themselves to
this amount of cooling in conjunction with other changes in
physical conditions, such as moisture.

Against this, however, he still held out, writing on
March 18 :

I wish I could see any way of ' ingenious wriggling ' that would remove the crushing evidence in the shape of tropical forms—against tropical cold. You have no idea of the magnitude of such a case as the Dipterocarpeae, a Nat. Ord., not a mere genus, of 10 genera and 112 species all from Ceylon, the Malayan Peninsula and Islands—and of which a good 100 more species and many more genera are still to come from Borneo, Sumatra &c. All are woody, and far the larger proportion are large timber trees—not one ascends at all to any height—and analogous species to living are found in tertiary coal-beds of Labuan &c.

Darwin's appreciation of this Essay is recorded in his letter of February 25. (M.L. i. 465 *et seq.*) ' Such papers,' he exclaims, ' are the real engine to compel people to reflect on modification of species ' : and ' What a splendid new and original evidence and case is that of Greenland.'

To this Hooker replies on the 27th :

I am greatly pleased and indeed relieved by your letter, for no one but Oliver (who can judge) has pronounced any opinion on my Greenland paper, and I find that one is so easily deceived as to the value of such researches that I was anything but sanguine of your approval.

In a subsequent letter (March 3) he refers to certain corrections which had not been put into the final proofs—errors which required the eye of Darwin to detect—and replies to several questions raised by Darwin.

I am really sorry about the blunders in my Arctic paper (and, in anticipation, for the others you will find), but it is of mighty little consequence, you being the only one who has found it out ; it is well this should be so, I should never have written such papers but for you ; and the evulgation of your views is the purest pleasure I derive from them.

I am staggered equally with you by the idea that Greenland ought to have been depopulated during the Glacial period ; but if so, how is it that its *temperate* flora is no richer than its arctic—if it had been populated by migration since the Glacial Epoch, surely some species suited to the south end would have got over there—there are plenty such in Iceland ; then again the absence of *Caltha* anywhere in Greenland, and other plants that *swarm* elsewhere all round the circle, is as fatal as any indirect evidence can be to the population of the whole by chance migration. If you intend to ask me when we meet how I account for richness of Lapland Flora, I will take care to flee your presence. I am utterly at sea when I attempt to jog out of the quiet *locus standens* of Lapland being the focus for lattermost migration. I grant that the idea may be utterly false, of its being the centre. I have some vague notion that the pre-glacial focus of Scandinavian plants was a *terra polaris* that United Greenland, Iceland and Scandinavia (not perhaps in latitude, but somehow) ; what it may have embraced to the North of America and Asia I neither know nor care ; for it is quite clear that there have been very great modern changes of level amongst the Polar American Islands, which I suppose are rising. I only call this vegetation Scandinavian because it is now represented best in Scandinavia, and this partly because of present climate of Scandinavia and partly because of its mountains having afforded a favoring climate to said plants during post-glacial warm period. I cannot too strongly impress the fact that Greenland is unaccountably poor in plants ; its comparatively equable (for an arctic) climate is singularly favorable for a northern Flora. In summer the line of perpetual snow in Disko is about 4000 *feet* I am told. Just look again at the list of Arctic species at p. 272 found in Europe and America but *not in Greenland*. I have not a shadow of doubt about wholesale extinction in East N. America.

The criticism of naturalists able to appreciate the value of the botanical argument was the only criticism he considered worth having. Thus on August 20 he could tell Darwin, ' I am hugely pleased with Asa Gray's review of my Arctic Essay.' [1]

On the other hand, a review by Dr. Dawson,[2] a geologist with inadequate knowledge of botany, attacked the Essay specially on geological grounds, and accused Hooker of

[1] *American Journal of Science and Arts*, xxxiv., and in Gray's *Scientific Papers*, i. 122.
[2] Sir J. William Dawson, C.M.G., F.R.S. (1820–99), was born in Nova Scotia, and studied in Edinburgh 1841–2. He was President of the McGill University, Montreal, from 1855–93.

asserting a subsidence of Arctic America, which never entered into my head. . . . Indeed I need hardly say that I set out on Biological grounds, and hold myself as independent of theories of subsidence as you do of the opinions of Physicists on heat of Globe! (November 2, 1862.)

In fact he had been over the geological ground twice, with Lyell, and again with Hector. Dawson's review, as he tells Darwin, he treated with scant respect, and in the course of discussing his geological argument roundly told the writer

that it was impossible to entertain a strong opinion against the Darwinian hypothesis without its giving rise to a mental twist when viewing matters in which that hypothesis was or might be involved. I told him I felt that this was so with me when I opposed you, and that all minds are subject to such obliquities ! the Lord help me, and this to an LL.D. and Principal of a College ! [1]

As a curious anthropological pendant to the whole question he notes the following to Darwin (November 2) :

By the way, do you see the *Athenœum* notice of L. Bonaparte's Basque and Finnish language—is it not possible that the Basques are Finns left behind after the Glacial period, like the Arctic plants ! I have often thought this theory would explain the Mexican and Chinese national affinities.

At the end of 1862 the scientific world was anxiously awaiting the appearance of Lyell's book, 'The Antiquity of Man,' which was to proclaim definitely his acceptance of the mutability of species—rare instance of a man past sixty being converted from the opinion of a lifetime. The book appeared in March 1863. The situation of the moment and the unceasing expansion of Darwin's research work are Hooker's theme in the following.

To Brian Hodgson

December 6, 1862.

You ask about Lyell. I saw him the other day, still polishing away at his work on age of man, which he told me

[1] See the letter of November 9, 1862, in M.L. i. 209.

would not be out *before* Christmas, which means, not till an indefinite period after it. He will have a pretty job to reconcile all his old Geology and Biology to the new state of things brought about by the discoveries relative to the early condition of man, and the Darwinian controversy, theory, heresy, truth, or whatever else it be hight. Lyell accepts both and will be pitched into accordingly ; he has the ear of the public, however, and the sale of his work will be *prodigious*. It will be followed by a very clever and most amusing one by Huxley, on the relations of men to the lower animals, of which I have seen some sheets ; it is amazingly clever. This polemical Philosopher is resting on his spear at present, and giving Owen a little time to commit himself again. I heard a fraction of Owen's paper on the *Grypho-saurus* at the R.S. ; it was very interesting but too verbose and minute, reading out all the measurements of minute parts to inches and lines, etc. The general opinion was that Owen demonstrated its ornithic affinity and proved it to be a bird with the tail-feathers set on a jointed tail instead of the truculent hump that most birds have, but some say that there are peculiar bones or organs amongst the bones that may yet prove it to be Reptilian. The most curious part of its history is its confirmation of Darwin's much disputed dogma, the 'imperfection of the geological record.' This animal is only now found in the identical quarries that have been worked for all the lithographic stones used all over Europe, ever since lithography was an art !

Darwin still works away at his experiments and his theory, and startles us by the surprising discoveries he now makes in Botany ; his work on the fertilisation of orchids is quite unique—there is nothing in the whole range of Botanical Literature to compare with it, and this, with his other works, ' Journal,' ' Coral Reefs,' ' Volcanic Islands,' ' Geology of Beagle,' ' Anatomy, etc., of Cirripedes ' and ' Origin,' raise him without doubt to the position of the first Naturalist in Europe, indeed I question if he will not be regarded as great as any that ever lived ; his powers of observation, memory and judgement seem prodigious, his industry indefatigable and his sagacity in planning experiments, fertility of resources and care in conducting them are unrivalled, and all this with health so detestable that his

life is a curse to him and more than half his days and weeks
are spent in inaction—in forced idleness of mind and body.

The following is apropos of Huxley's book above mentioned
on ' The Relation of Man to the Lower Animals.'

To T. H. Huxley

Kew : Friday.

I am making a précis of our poor German collector,
G. Mann's, West African letters, to contradict Burton's
assertions, and have come across a passage that will amuse
you. Talking of the Gaboon natives, he says, ' They
generally were touching my beard and hair, lifting my hat to
see if the whole head was covered with the same hair, and
found it as they said, very strange that I had hair *like the
Monkeys and not like mankind.*'

So you see there are two opinions as to the value of the
similarity between men and monkeys. I do not think this
would have struck any but a nigger looking from a Nigger's
point of view. I wonder what the Monkeys find.

As to Lyell's book itself, he agreed with Darwin's verdict
as to the excellence of the Glacial chapters, the force of the
aggregation of evidence as to the origin of Man, and the skill in
picking out salient points in the argument for change of species,
combined with disappointment at the timidity which prevented
him from giving any judgment of his own on the materials set
forth.

In a letter to Darwin of March 15, 1863, he writes :

I have been having a long correspondence with Lyell, and
have given him quite as *deflagrating* a yarn as I sent you,
and likened him to the Theologians ! adding, that I had always
hitherto classed him as the sole sexagenarian philosopher
who could change his opinion on good ground. He proposes
some alterations of the two obnoxious passages, which will
at any rate *do justice* to the hypothesis as he states it, which
the former ones did not. Lyell dwells, and with reason,
on the fact that he makes as many converts whether he
withholds or gives his own opinion. I tell him *perhaps
more*, as people like to draw their own inferences, but that
is not the particular point we as his friends now look to.

I have finished Lyell and am enchanted with the Glacial Chapters, language, and the whole treatment of the Origin and Development subjects (with above qualifications) : it is certainly a grand book on the whole, and well worthy of Lyell's scientific reputation. He never rises to the magnificence of Huxley's language, nor to the sublimity of some of the passages in H.'s little book on the Position of Man, which you can read 1000 times with fresh delight.

Of his own work, indeed, as compared with Darwin's— whom he once apostrophised as ' you *facile princeps* of observers '—he always felt and spoke with humility. Thus he writes on October 2, 1862 :

The dismal fact you quote of hybrid transitions between *Verbascum Thapsus* and *nigrum* (or whichever two it was) and its bearing on my practice of lumping species through intermediate specimens, is a very horrible one ; and would open my eyes to my own blindness if nothing else could. I have long been prepared for such a case, though I once wrote much against its probability. I feel tolerably sure I must have encountered many such, but have not had the tact to discern them, when under my nose, and I hence feel as if all my vast experience in the field has been thrown away. Your Orchid Book has pretty well convinced me that such cases must be abundant, and they only tend further to disturb our ideas of physiological versus structural species. Perhaps my intermediates between *Habenaria chlorantha* and *bifolia* (of which I retain a lively recollection) were of this hybrid nature. Certain it is, that I had only to look for Hybrid Orchids in Switzerland to find two different sorts, and numerous specimens of one of them.

Besides correspondence touching Darwin's immediate interests in the study of cross-fertilisation and in climbing plants, many specimens of which he sent to Down for experiments, topics discussed with Darwin include the relations between Islands and Continents, the parallel between Alps and Himalayas, Variation and Environment, the latter leading to a curious application of Natural Selection to Sociology.

May 13, 1863.

I have perfect faith in your doctrine of absence of competition favoring retention of continental forms on Islands, though how the devil one is to reconcile that with the extraordinary modifications of other continental forms on same Islands passes my comprehension, except what you won't admit— that they were common to continent and island before disjunction of latter, and the modification is of the *continental* forms, the insular being the old original type. This is turning the tables over you with a vengeance, but I will work it out in spite of you. Go to—weep and howl! The Ferns of Ascension and St. Helena are totally different from one another and from Cameroons ; this is, or ought to be, a death-blow to all aerial migration, for Ferns are notoriously widely dispersed and dispersable. I wish I had never wasted a thought on the stupid subject.

May 24, 1862.

Thanks for your exposition of your island views,[1] I think I understand them precisely, my difficulty in accepting them arises from the want of apparent accordance between the plants common to island and continent, and what I should have expected to be common. In other words, migration is inadequate to explain the presence of what is common to both and the absence of what is absent in one. I am far from believing in ancient commotion, all I hold is that in the present state of science it is to me the least difficult hypothesis, though a very bad one. Cameroons Mountains have shaken my faith in our having any clue to ancient or modern migration as yet. We want some new hypothesis, as novel as Nat. Selection, or Glacial Cold, and as stupendous as Continental Connection.

Samaden, Engadine Valley: July 10, 1862.

This, the valley of the Inn, appears to me to combine the beauty of the Tyrol with the savage grandeur of Switzerland in a remarkable degree. In science I have seen little but Heer's fossils, he shewed me a leaf apparently Dicotyledonous from the Lower Lias in Jura, which please tell Lyell of. He has a wonderful collection of fossil insects and crustaceae

---

[1] See Darwin's letter, M.L. i. 241. ' With respect to Island Floras, if I understand rightly, we differ almost solely how plants first got there.'

from the same, beside which the fossil plants are as nothing, in point of absolute value of characters for systematic determination. I am as always impressed with the identity of physical features and wonderful analogy of *biological* between Alps and Himalayas, the former we can suppose we understand, because physical causes are the same everywhere and the sequence of these is probably the same in Alps and India: The representation of allied species too we can now (thanks to you) account for largely, but the repetition of forms in plants and animals in no way allied is always a puzzle, especially when accompanied by startling contrasts between allied forms. These latter can best no doubt be accounted for by the indirect action of physical causes (i.e. Nat. Selection) and I think there are already many reliable facts to be quoted in illustration of this, and that after the course of alteratives you have administered I could write a suggestive chapter, comparing the vegetation of Alps, Andes and Himalayas in my (never to be begun) book on Plants.

I cannot yet give up my dream of meeting you in Switzerland one day ; if you ever did come here, and I could see you for 5 minutes a day, I should be the happiest man alive. These rocks, plants and insects teem with thoughts of you and reminiscences of your writings.

Your Orchid book, which I have not read through, has suggested to me that insects &c. may have a wonderful deal more to do with checking migration than climate or geographics, and that the absence of whole genera may thus one day be accounted for by absence of genera of insects : in short that the Cat and Clover story is capable of immediate expansion by any one having sufficient knowledge of Plants, Insects and Geography.

Thursday, July (24 ?) 1862.

I was delighted with Heer, and went over all his collections, which are grand and good ; they serve to convince me that the Miocene vegetation was Himalayan, not American as H. supposed. Heer's error was very natural, for no one knows from any published works what the real nature of the Himalayan vegetation is:

Darwin's answer to the following is given in M.L. i. 197.

March 17, 1862.

I am greatly puzzled just now in my mind by a very prevalent difference between animals and vegetables : inasmuch as the individual animal is certainly changed materially by external conditions, the latter (I think) never except in such a coarse way as stunting or enlarging—and this is because in animals there is a direct relation between stimulated function and consequent change in organs concerned in that function ; e.g. no increase of cold on the spot, or change of individual plant from hot to cold, will induce said individual plant to get more woolly covering, but I suppose that a series of cold seasons would bring about such a change in an *individual* quadruped, just as rowing will harden hands &c. The cases are not parallel, because the parts of plants that could be so changed are annually lost, and the only conceivable parallel is afforded by bark : would a cycle of cold seasons cause the bark of a tree to thicken more than it otherwise would ? I cannot suppose that the buds of the individual would get thicker, or more scales, or more resinous scales ; or that its successive leaves can become annually more hairy : except indeed we assume the annual death of a large proportion of the buds, and that those alone are preserved that have most ' woolly ' leaves— when no doubt the woolly tendency would be inherited by the successive phytons of that bud, as by successive generations from seeds.

Be all this as it may, in neither plant nor animal would the induced character be of necessity inherited by the offspring by seed of the individual, to any greater extent than if it had not been changed—at least so far as the animal is concerned ; though with regard to the plant it might be, the seed being that of the phyton, not of the whole tree, or average tree. Thus a wild complication is introduced into the whole subject that perplexes me greatly.

Berkeley's article on acclimatization is very unclear I think (see last Saturday's *Gardeners' Chronicle*).

I cannot conceive what you say, that climate could have effected even such a single character as a hooked seed. You know I have a morbid horror of two laws in nature for obtaining the same end ; hence I incline to attribute the smallest variation to the inherent tendency to vary ; a

principle wholly independent of physical conditions—but
whose effects on the race are absolutely dependent on
physical conditions for their conservation.

Huxley is rather disposed to think you have overlooked
' Saltus,' but I am not sure that he is right. Saltus quoad
individual, is not saltus quoad *species*, as I pointed out in
the Begonia case, though perhaps that was rather special
pleading in the present state of science.

The exchange of letters continued while Hooker paid a
few days' visit to a big country house. Observation of the
life there led to an effusion on High Life by ' the future author
of " Aristocracy " or " Darwin in all in all." '

Kew : Sunday (March 20, 1862).

MY DEAR DARWIN,—I returned last night and found
Bates' letters which I send herewith, I have no time to compare them. I hope I have not abused you unmercifully in
my letter to Bates—you must take your chance !

I had a very profitable stay at X——, considering all
things, and came away with food for much reflection. I
could not make up my mind to stay over Sunday though
kindly pressed with real English hospitality. Some of the
family are very nice, all the ladies particularly so, the servants
perfection (*such* Nat. selection of flunkies), the food good
and plenty, the country beautiful—the weather detestable
and the habits and hours of the house quite intolerable.
It would take a letter from you every morning to have supported me under such a system of killing time and outraging
the stomach. However it does one good to go to such
places rarely, gives one much food for reflection, and will add
a chapter to my posthumous work ' On the principles which
regulate the development of an aristocracy.' The principal
part of this work will consist of 4 chapters, each headed
with a B, viz. Blood, Blunt, Brains, Beauty. These are
all good things, of use to the organism possessing them,
and hence sought after by all human organisms, and their
accumulation, by natural selection, must culminate in an
aristocracy, or there is no truth in Darwinism. The better
these are blended, the better will be your aristocracy, the
more they are separated the worse, and it is hard to say
which is worst *per se*, or which is best when all are mixed.

You have the aristocracy purely of B1 in Germany ; of B2 in America ; of B3 in France ; of B4 everywhere, but of 4Bs in England only : where indeed we have 4Bs in the highest nobility.   I met nothing beyond B1 and B2 at X—— however, perhaps with ever so small an element of the two others I might have been induced to stay Sunday, for I do maintain that the union of all must be irresistible in every degree and condition of life, from Fuegia to London.

I have no time to answer your kind long letter. There must be, as you say, something effective in the alteration of the reproductive system under variation, not necessarily *induced* by domestication but accompanying some variety artificially selected. I cannot however forget that it is through *marriage* alone that the 4 B's are usually recruited in after life, and so there may be something in what you say ! ! ! that's my philosophy—make the best of it till we meet.

To C. Darwin

June 29, 1863.

I went to the Guards Ball the other night, and was deeply interested—of course I know so few people that I had abundant time and opportunity to roam about, and observe, and listen—admire and despise—the contrasts of old and young were ghastly—my God, there were hideous old women in bride's robes enough to keep you in nightmares for a month of Sundays, and lovely girls enough to fill all the paradises of all the Turks. The intellectual cut and exceeding handsomeness of both men and women was very satisfactory in the main, as was the cleanliness and general health of the whole stock of high-bred humanity. To compare them with an equal number of the lower classes suggested many reflections, and strengthened me in my dogma that Brains + Beauty = Breeding + Wealth. I should extremely like to go to a similar selection in America, France or Austria ; my impression is that the comparison would be ludicrous.

The same view is pursued in the matter of Democracy in America, prompted in part by reading De Tocqueville, in part by the stir of the American Civil War. His own sympathies at the time may be described as not so much positively in favour

of the South as negatively against the North, resenting as he
did the unfairness of Northern criticisms of England, and the
overbearing and loud-mouthed tone of meetings held even in
cultured Boston, while he deplored the blinding and undigni-
fied effect produced on the tone and temper of such a man as
his friend Asa Gray—' I mean of course in his capacity as a
citizen, for I have the same high opinion of him as a man, as
ever,' he tells Darwin, with whom as well as Hooker, Gray
maintained a correspondence. When Gray spoke of the
two nations as naturally destined to be on the best of terms,
he reflected on the inevitable contention in the struggle for
life between two great organisms at once so like and so bent on
the same ends. In writing to Gray his only allusion to the
war ' was to the effect that it would clear off the mass of scum
under which, I considered, his nation groaned—this I intended
as the only conceivable good that could come out of such a
political contest '—and Gray had taken this as applying solely
to the opposite side. ' You and I,' he tells Darwin, ' have
always differed a good deal about America,' and continues
(March 10, 1862) :

> Our aristocracy may have been (and has been) a great
> drawback to civilisation, but on the other hand it has had
> its advantages, has kept in check the uneducated and
> unreflecting, and has forced those who had intellect enough
> to rise to their own level, to use it all in the struggle. There
> is a deal in *breeding*, and I do not think that any but high bred
> gentlemen are safe guides in emergencies such as these. The
> moral effect of Lord Russell's despatches on the English
> mind has been quite astounding, and I do not think you can
> point out a dozen men in public life, but of less breeding
> and culture (I do not mean by this *aristocratic training*, a
> specific thing) who would have been *safe* to have behaved
> with equal prudence, dignity and consideration, and yet Gray
> calls *this* the pressure of a mob ! If there is anything at all
> in force of circumstances and Natural Selection, it must
> arrive that the best trained, bred and ablest man will be
> found in the higher walks of life—true he will be rare, but
> then he will be obvious and easily selected by a discriminating
> public. When got too he is removed above a multitude of

temptations and conditions that prove the ruin of 9/10ths of
the rising statesmen of a lower class of life. Your ' Origin' has
done more to enhance the value of an aristocracy in my eyes
than any social, political or other argument. Now I never
allude to politics in writing to Gray—it is useless I know,
and furthermore wherever we did agree, it would perhaps
most often be on totally different grounds, and this leads to
endless misunderstandings.

What folly he talks of 2 such nations as England and
America ever being on the best of terms. What is there in
the whole history of the human race to quote for such a state
of things as ' best of terms ' between two nations of the
same blood and bone, and with the same aims and prospects ?
Nothing but the power of despising us, or we them, ever can
or ever will bring one of us to look amicably on the other.
It is not in the bounds of possibility that two nations so
powerful, so ambitious, so *like* should love one another, and
it will be a bad day for one or both when they do.   A. Gray
knows no more of the philosophy of the ' struggle for life '
than the Bishop of Oxford does.   You might as well talk
of High Church loving Low Church, God knows they are each
powerful enough and *like* enough to form one body religious
with a common aim and object, *if* they would sink differences
and agree *each to be nothing*, or one to be everything and the
other nothing.

Kew, Sunday (Dec. 1862).

I am actually reading de Tocqueville's Democracy in
America ; it appears to me a most able book, though I do not
at all agree with it (bigger fool you, you may say, and double
big fool I am to say so), but I cannot help it. He assumes
that D. in America was a success. Now I never regarded
America as having cohesion enough to be pronounced either
a success or a failure : there has been hitherto far too much
freedom of motion there, too little ' struggle for existence '
to develop any settled Govt. at all, and it is impossible
to predict what shape the existing (*introduced*) form of Govt.
would take in 100 years, even if this war had not stepped in
to confound all calculations. Democracy has *persisted* in
America, because there has been no cause for its overthrow,
just as Monarchies might persist indefinitely (though they
persist under much greater disadvantages). Specialisation

I conceive to be a dominant law governing everything, and I cannot see how either a Democracy or Republican form of Govt. can resist the effects of Natural Selection. In short, I regard a pure Democracy as visionary as a country peopled by one invariable species. This with me is no question of what is good or bad, but of what must ever be, and I do hold that a Govt. must always eventually get into the hands of an individual, or a family, or a class, or there is no truth in Natural Selection. Q.E.D. as you say.

To Charles Darwin

January 6, 1863.

I have finished De Tocqueville's Democracy in America and cannot help thinking how differently he might have written had he read the ' Origin' and applied it ; all his fallacies are attributable to ignorance of its principles, specially his want of perception that the versatility and variety of resources each Yankee possesses (to which he attributes all their excellences more or less) is simply the result of want of competition, and that when the land is filled with people this superiority will vanish, each will be good at his speciality only and the evil effects of Republicanism will burst out all over the people and communities. I do not believe any nation can last for ever, either under a Republic or Monarchy (both being bad). . . .

Then too all De Tocqueville's comparative vaticinations are frustrated by the growth of England's colonies; which he (Frenchman-like) utterly ignores.

January 24, 1863.

How dreadful the New York papers are; we see them here and I read and moralize over them by the hour. I believe that a Republican is the worst form of Govt. that can be *given* to a people, but perhaps the best they can make for themselves ; the mistake is to suppose that the Americans made it for themselves—they never did so, they accepted it from the hands of the few great men of that day, and so long as there was no struggle for existence it was never put to the test ; when the struggle came they found out that what they accepted as a working theory had not taken root enough in the hearts of the people to be upheld at any price. Really there is no bright spot in this sad, sad world but in shops that sell Wedgwood ware, which I have been haunting with some success. As I know that you will listen to nothing from me after this I will shut up.

The same subject is continued in letters to Asa Gray after the war.

I continue to read the *Nation* regularly and with great interest. I am so glad that it is the Tories who are going to take up the Alabama case. Though a whig myself (if anything), I always believed that the Tories and Aristocracy generally had better and wiser ideas during the American war than Goldwin Smith, &c. gave them credit for. I have no hesitation in thinking that the honor of our uppermost Tory classes is of a higher order than of the middle, just as their vices are more conspicuous. They can *afford* to be more high-minded, just as they can *afford* to commit sins that damn a lower class, and upon the whole I expect they are less vicious than the middle class and infinitely less than the lower. Indeed upon scientific grounds I have stated before (Natural selection, and continued success only attending honesty) I think it should be so.

Did you ever read that painful book, Malthus on Population ? I did the other day, and was painfully impressed by it. I had supposed he was a sort of materialist, who *advised* the checking of the population by restrictive means, and was surprised to find nothing of the sort, and a rather fine exordium at the end on a future state and the benefits of Christianity ! His arguments seem incontrovertible to me.

To A. Gray

March 22, 1867.

I was amused with the *Boston Advertiser* pitching into the *Pall Mall* as representing ' the Governing Class.' I suppose the fashionable name misleads them with regard to the point at issue. The *P. M.* is logically right and *B. A.* clearly wrong ; the error is in the *B. A.* assuming that we are a ' free nation.' We are nothing of the sort, and *the masses do not wish to be so.* They are engaged in the struggle for existence and care nought for freedom or politics—so we have bribery and corruption in all our elections—even amongst the lower *educated* classes—rampant. Take away bribery—and other extraneous motives for voting—and scarce any would vote but the middle and upper middle classes. The more I compare your and our papers, the more I see that you have no representatives of our lower middle and lowest classes, except perhaps in New York—of our *masses*, in short—any more than you have representatives of our Aristocracy. You represent our middle and upper middle classes, who wield all the power with us as it is. We are in no way comparable as a people ; our political virtues and vices are quite different. Upon the whole you are the gainers ; but it will not last. You will one day have a poverty smitten *residuum* that will yearly increase in the same ratio as wealth at the other end— a class who won't be educated, and who will vote for equal distribution of property and of all God's gifts, for no ' meum ' and ' tuum,' but for ' God for us all,' and that god their bellies. Power and wealth will lapse into the hands of the strong with you, and laws will keep it there.

I also notice the *Nation* pitching into the Londoners for the state of London during the snow, and citing it as a proof of the lamentable inability of the English to improve &c. ; and on same day in the New York paper were frightful letters on the state of the river and ferries of New York, where people were kept all day and could not cross. So we go on every day ; it is the Beam and the Mote,—and so it will be to the end of the Chapter.

The Herbarium affair is now settled,[1] and I expect the money next week, not before it is wanted for my wants and position here, which I must have abandoned, were it not settled. I could not live on here without a complete alteration of all my household affairs, on my present income, besides which I must have given up all my functions as head of my own and Henslow's family, and mover of Botanists, &c.

To A. Gray

March 12, 1868.

Anent politics I have nothing to say. On both sides of the water we seem to suffer under the inevitable evils of our respective forms of Government, and all I believe is that if you had our form you would be ten times worse than you

---

[1] I.e. the purchase by Government of Sir William's collections and library (see p. 48).

are, and if we had yours ditto ditto. I suppose that amongst civilized peoples not engaged in warfares that distract their attention from home affairs, the Government is pretty much what the masses like—a part of themselves in fact—and I do not believe in any abstract good or bad form of Government. If we like an Aristocratic Govt., it is because we like that form of the haphazard that settles the Govt. on birth. You, on the contrary, like the haphazard of public election, which is *not* the same thing as public *voice*, still less as public *opinion*. What is sauce for the goose is not sauce for the gander. The Celt wants, *and should have*, a totally different form of Government from the Saxon, and if there was any object in keeping up the Celt, then our Govt. should provide a branch legislature suited to his (damnable) idiosyncrasies.

I am utterly sick of the political nostrums prevalent on both sides the Atlantic, and the everlasting peevishness that springs out of our and the others' supposing that the evils of our respective countries are due to the form of Government that we severally enjoy—endure, I mean. Go to—I am cynical.

Have you read Darwin's last book, and what do you say to Pangenesis ? I have gone deeply into the *whole Philosophy of the Subject*—there then—

Apart from the heavy scientific labours of this period the last five years of the Assistant Directorship at Kew were a time of pressure growing more and more intense. Not only did the expansion and reorganisation of the Gardens increase Hooker's own share of administrative work, but the gradual failing of Sir William's power of application and prompt decision threw yet more upon his shoulders. As he tells Darwin (May 26, 1865) :

My dear old Father piles duty on duty, and will neither give in nor give up. I do admire his gallantry, and I do not want to see him give up, but things *do* get into dreadful confusion, and I shall have a heavy day of reckoning.

In addition the death of his trusty Herbarium Clerk was a serious loss. Meantime the departure of the Curator of Pleasure Grounds gave an opportunity which he wished to employ—

to reorganize the whole establishment which is worked to death, and I dread a breakdown of our new Curator, who, what with Garden-duties and accounts, works 16 hours a day : as for myself, who have never done less, this is all very well, but persons not accustomed to it cannot stand it—as matters stand, neither he nor I could leave Kew a week.

Indeed Garden reforms had begun a couple of months earlier :

We have been robbed much by our own people [he tells Darwin on April 7]; and I discharged two foremen, dismissed half a dozen gardeners and labourers, and clapped one fellow in jail for six months. All this is not very agreeable work, but we have really a first rate Curator now (John Smith the second) and I am anxious to put everything straight for him to go on without troubling me.

The late gardeners' neglect during the winter had let many plants perish. In June Darwin is told : ' I hope to have a Botanical Garden worth looking at in a couple of years.'

Ever full of hospitality as he was and delighting to ask his friends to come and be shown the wonders of which he was justly proud, Hooker found that the uninvited ' torrent of visitors,' scientific or otherwise, to the Gardens cut up his time terribly. He often breaks out despairingly to Darwin—e.g. (May 28, 1862) :

I see an everlasting round of visitors whom I (for the most part) wish at Jericho. I broke three solemn engagements to-day.

And (September 20, 1862) :

I am frightfully busy and inundated with d—d visitors. There goes the bell—just as I wrote.

It was at least a relief that the Gardens continued to be closed to the public in the forenoon.

The months brought no relief. In the summer of 1863 it was not only that ' we are overwhelmed and almost knocked

up by visitors and visiting,' but London society, which made
worse inroads upon his time than the extra work involved by
his father's absence.

I cannot see my way to a mean course between dining out
everywhere and nowhere, without a system of prevarication
that would be intolerable, and now that my Father never
goes out, I have double duty that way.

'How opposite our troubles are about society,' rejoins
Darwin; 'you too much, I absolutely none.'

This state of affairs continued till Sir William Hooker's
death in 1865, and his son's succession to the post of Director
at Kew. For this he had long been marked out both as the
foremost botanist in his country and as Assistant Director
since 1855. Nor was there anyone even to stand second to
him. Along with his father he was bound up with the making
and development of Kew. That it had risen to be for botany
pure and applied what Greenwich is for astronomy, the science
that directs the art of navigation, was due to the untiring
energy, the personal devotion, the material private contributions
of father and son in specimens and books. With the appointment of the new Director came the necessary adjustment of
public and private property in the Herbarium, which was,
so to say, the scientific palladium of Kew. This, it will be
remembered, began with Sir William's own collection which
he had brought from Glasgow, and to which he had been constantly adding. In conjunction with the Library it was the
basis of all the scientific work which was reflected over the home
country and the colonies and attracted the botanists of
all other countries. At first it was maintained and housed
entirely at Sir William's expense, but in the first decade it
outgrew all the accommodation within his means. Government consented to provide better accommodation—on terms :
granted a Curator in return for public rights of access. But
it was not taken over bodily nor entirely maintained. Additions still came from Sir William. The gift of Bentham's
fine library and herbarium (of the flowering plants) helped to
fix the national character of the whole collection, and it became

more than ever necessary to put an end to the fundamental anomalies of its ownership.

Sir William's own wish was that the nation should purchase his herbarium—the one valuable piece of property he could bequeath to his son : and he left a memorandum to this effect—though unsigned, for he had procrastinated too long over the matter. Thus, the year after his death the State bought the herbarium, some 1000 volumes from his library, and a matchless collection of botanical drawings, maps, MSS., portraits of botanists, and letters from botanical correspondents, to the number of about 27,000, for the sum of £7,000.

A year later, the Gay Herbarium at Paris came into the market. Hooker purchased it for £400 and presented it to Kew. As it contained a number of specimens which were lacking in the Kew Herbarium, he prided himself on the result. Writing to Berkeley on November 20, 1870, apropos of a new botanical correspondent, with the quality of whose contribution he was much taken, he adds :

> Pray, however, undeceive her about Kew's poverty of European plants, which is rather a cut after my purchase of Gay's Herbarium and presentation of it to Kew ! and which for completeness and perfection beats the Paris ' European ' Herbarium—otherwise the finest in Europe. Having the Gayan I should not feel justified in buying the Pittonian.

The principal change was that the new Director had no Assistant Director. In fact there was no one qualified to take the special post, and the lieutenancy was divided. One subordinate became official assistant in the Gardens : a second in scientific matters, though here Hooker demanded yet another assistant.

The general effect upon Kew of the new appointment is described in a letter to Darwin (November 1865) :

> I am up in heaps with work, and find I shall have a desperate fight to get scientific assistance. I will not give in however. I am prepared to improve the Gardens enormously and will do so, but if the scientific character of the establishment is to go down one iota, I shall intimate that I only hold

the post with a view to retirement when able. My *elevation* brings me no increase of income and a higher scale of living, as I now feel it my duty to give up examinerships &c. that yielded upwards of £300. But I have no fear of not carrying my point, which is a properly educated assistant to be under Oliver. The Curator is in future to be my assistant in *Garden duties*, Oliver, with increased salary, in *scientific matters* ; an excellent arrangement, as there is no one able to be my assistant in both, nor are the functions compatible in any but one who like myself has grown with the growth of the establishment and been educated to it. In the conversation I had with the Board they ' let the cat out of the bag ' in informing me that they abolished the Assistant Directorship because they knew of no one fitted for it !—not only an unintentional compliment to me, but an admission by implication that neither could they find another person fit to be Director ! I took no notice, but have it in hand as ' one for his nob ' if needs be.

You see ' my Dander is up,' as the Yankees say, but pray say nothing about this ; fighting battles before bystanders is only a shade better than in the dark, and one gains nothing by appearing to be in opposition.

CHAPTER XXX

1860-1865: PERSONAL

SINCE his intervention at the Oxford meeting of the British Association in 1860, Hooker was not directly concerned in several bitter controversies which took place during this period, either in attack or defence, though he followed them closely. There was the battle of the brains, human and simian, where Huxley, supported by other anatomists, fulfilled his pledge made at the Oxford meeting to demonstrate the baselessness of Owen's assertions, finally summing up the case in his little book ' Man's Place in Nature ' (1863). There was Owen's attack on Lyell and Lyell's conversion to Darwinism, under cover of a review of ' The Antiquity of Man ' (1863—see C.D. iii. 7 *seq.* ; M.L. i. 238-9). There was the *Athenæum* review (March 28, 1863) of Dr. Carpenter's ' Introduction to the Study of the Foraminifera,' celebrated as having provoked Darwin for the first and, save once only, the last time in his life, to reply on a scientific question in a popular journal. Carpenter had referred to living and extinct Foraminifera as having a common ancestry ; the reviewer took the opportunity of denouncing his Darwinian tendencies and Darwinism itself, propounding instead a wonderful theory of spontaneous generation (Heterogeny). ' Who would ever have thought of the old stupid *Athenæum* taking to Oken-like transcendental philosophy written in Owenian style ! ' [1] exclaimed Darwin to

---

[1] Lorenz Oken (1779-1851), Professor of Natural Science at Jena, Munich, and Zurich successively, set out to deduce all knowledge from certain *a priori* principles, especially of parallelism between the universal and the particular. Thus, as there are five senses in the perfect animal, so there are five main classes of animals each representing the special development of one of the senses—and in the individual the head is in essence the repetition of the trunk.

Though experiment afterwards gave science something not utterly unlike

Hooker. But Darwin's letter saying a word in his own defence, while attacking the 'monstrous article' on Heterogeny (the author of which was Owen himself), only brought forth another skilful appeal to popular prejudice (see C.D. iii. 17–23, and M.L. i. 242).

The whole thing was utterly repugnant to Hooker, who wrote (May 1863):

I cannot abide this lugging of science before the public in *Times* and *Athenæum*, and implore you, my dear fellow, not to do so again. Owen's answer to you is triumphant in the eyes of the public (whom you wish to enlighten) as Manchester's over Natal. The only party that gains by these discussions is the proprietor of the paper ; the only one that loses every way is the maintainer of truth. Science will be much more respected if it keeps its discussions within its own circle.

Similarly, when in 1864 Professor Kölliker [1] wrote a review of the 'Origin,' entirely misconceiving several of Darwin's main positions, and Darwin was strongly inclined to reply, Hooker wrote (September 5):

I did not mean that it was beneath your dignity or really below the dignity of your subject to answer Kölliker, but what I think is, that when such subjects are dragged into periodicals for discussion the public are apt to form a low opinion of them and their disputants. The subject is a

certain of his striking homologies, as in cell development and the relations of heat and light, this kind of transcendentalism was a matter of vague suggestion, not of solid science. With some limitations, Oken's ideas were taken up by Richard Owen in his theory of the archetype and the doctrine that the skull is a virtual repetition of certain vertebræ. But in his method of claiming to be the discoverer of the true theory Owen placed himself in a very equivocal position.

[1] Rudolph Albert von Kölliker (1817–1901 ?), anatomist and embryologist. He studied natural sciences at Zurich, Bonn, and Berlin, and was appointed Professor of Physiology and Comparative Anatomy at Zurich in 1845, and in 1847 took the chair of Anatomy at Würzburg. Among his principal works is his *Handbuch der Gewebelehre des Menschen*, *Die Siphonophora oder Schwimmpolypen von Messina*, the *Challenger* Report on Pennatulida, and *Entwickelungsgeschichte des Menschen u. d. höheren Thiere*. In association with Von Siebold he started the *Zeitschrift für Wissenschaftliche Zoologie*, and in 1858 published with E. Pelikan *Physiologischtoxikologische Untersuchung über die Wirkung des alkoholischen Extractes der Tanghinia venenifera*.

great one, there are acknowledged organs for its discussion,
accessible to all taking a true interest and capable of appreciating the men and their arguments, and to fling these
down to be scrambled for in a weekly periodical is somehow
derogatory. I dare say I do not explain my meaning, nor
should I convince you if I did. Of one thing I can assure you,
that it is never worth *your* while, whose working moments are
worth so much to us, to waste one thought on the discussion.
After all you could only impress outsiders—who would forget
and turn like the wind to the next writer, and it is the dignity
of the subject more than of the proceeding which I am
considering.

In the event Darwin did not reply. He was more than
satisfied with the answer made by Huxley in the *Natural History Review* the following month, entitled ' Criticisms on the
Origin of Species ' (see ' Collected Essays,' ii., Darwiniana,
p. 80).

In a similar strain he adds in his letter of May 1863 how

Falconer has his hands full and goes to Paris to-morrow to
confront Quatrefages, Bouchet and the chemists and anatomists, who to a man say that F. is wrong that both Flints
and jaw [human remains found in a cave with the remains of
extinct animals] are ancient, and *perfide Albion* at its old tricks
of traduction. I met F. last night ; he is beating up for
allies to take over with him. I tell him he should go alone—
it is his only chance of getting fair play. The more go, the
more opposition, the more misunderstanding, the more all
that is bad.[1]

Where, however, the ground of contention was no more
than a reclamation for priority or recognition of material
used, much as he disliked the practice, he exerted himself
privately to bring about a reconciliation. Such were two
public *réclames* made against the veteran Lyell. One was by
Falconer, who complained loudly that his and Prestwich's
researches had not met with proper recognition in ' The
Antiquity of Man ' (1863). Of his idiosyncrasy in suddenly

---

[1] The whole misunderstanding is told in C.D. iii. 14, 19, 21, and M.L. i.
229–41; cp. ii. 377.

discovering and magnifying a grievance, Hooker amusingly
remarks to Darwin, March 29, 1863 :

> Falconer is one of the two classes of Scotchmen that
> Crawfurd distinguishes as ' Scotsmen ' and ' d——d Scots-
> men.' There are two most curiously antagonistic sides to
> his character:

Or as he puts it elsewhere : ' Falconer is a Scotchman, who
when once wrong seems never to get right again,' yet ' one of
the most honourable men I know, except when out of temper.'
The other was by Sir John Lubbock (Lord Avebury) in
1865. Certainly the material in dispute had first been worked
over in English by Lubbock, but it was Danish research contained in a Danish memoir. However, these were uneasy
times, when confidence had been lowered by the methods of
one leader of opinion and those whom he inspired. The
suppressed irritation of a quiet man flared out with unhappy
results. It was too bad, Hooker agreed with Darwin, to treat
an old hero in science thus ; on the other hand, he was not
satisfied with the older man's subsequent *amende*. ' It is not
handsome at all, and from an old Prince of Science to a young
aspirant is not liberal, I think.' In impartial eyes, if the
acerbity of the attack was unwarranted, the explanation was
ungracious. Tenacity was at fault on either side, and as
Huxley pithily put it, the one had failed to set the affair
straight with half a dozen words of frank explanation as he
might have done; the other, ' like all quiet and mild men
who do get a grievance, became about twice as " wud " as
Berserks like you and me.' Hooker, with a sly hit at his
friend's favourite assertion that a ' compiler ' was a greater
man than an ' observer,' wrote to Darwin (June 2, 1865) :

> This comes of *your* divine art of Compilation ! Both, as it
> appears to me, were making capital compilations, and from
> precisely the same sources and to illustrate the same subject.

In both cases, as has been said, Hooker's intimacy with
the parties concerned enabled him to pour oil on the troubled
waters.

It was in connexion with references to Lyell's ingrained caution and similar hesitation elsewhere to speak definitely on the descent of man or religious difficulties, burning questions of the day, that Hooker had occasion to write to Darwin (October 6, 1865) :

It is all very well for Wallace to wonder at scientific men being afraid of saying what they think—he has all 'the freedom of motion in vacuo' in one sense. Had he as many kind and good relations as I have, who would be grieved and pained to hear me say what I think, and had he children who would be placed in predicaments most detrimental to children's minds by such avowals on my part, he would not wonder so much.

Nevertheless if not called upon at the immediate juncture to proclaim his ultimate convictions *urbi et orbi*, Hooker freely gave his support to liberalising movements in the Church. His concern was how to give such support most efficaciously without importing new controversial elements into the affair. This careful temper appears in two letters to Lubbock, apropos of a projected memorial of men of science in favour of the authors of ' Essays and Reviews,' who were being vehemently attacked by unprogressive orthodoxy.

Royal Gardens, Kew: February 29, 1861.

MY DEAR LUBBOCK,—I would sign your memorial with pleasure if I could satisfy myself that it would do good to the cause it so handsomely advocates, but I am far from convinced of this ; and on the contrary I fear that it may do harm.

You see that as matters at present stand, all that have signed may be considered as belonging more or less intimately to one school or party—for the most part they are personally attached by twos or threes : they represent the young progressionists in Science, their opinions are of no weight in religious matters, and the appearance of a large body of such names, unaccompanied by an equally large body of those men of older standing and opposite tendencies (who have nevertheless the confidence of the public), would in my opinion tend to create a fission in the ' body politic ' of scientific men. Now in matters of science I am for no sort of compromise between progression and non-progression, which is retrogression ; but I should be sorry to see anything done that would countenance a belief amongst the outsiders that our scientific differences influenced our religious views—and this would be a very legitimate inference if your memorial was signed wholly or chiefly by men of one way of thinking, in such matters as ' Origin of Species,' ' Age of Man,' &c., &c.

I confess however to have an almost morbid aversion for clique or sectarianism, the spirit of which is around us everywhere and may be evoked at any moment. In the present excited state of the public mind, I think that *our* rushing into the conflict would do more harm than good : we should be listened to more calmly a few months hence, when the futile attempts of the narrow minded shall have demonstrably failed ; and then I shall gladly sign a memorial addressed to the Essayists, thanking them for what they have done and requesting a Second Series of Essays.

Royal Gardens, Kew : March 4, 1861.

MY DEAR LUBBOCK,—I am sorry you cannot be at Linnean on Thursday, for I should have liked to meet you and talk over this affair of the Essays and Reviews ; also because I wanted you to be at meeting in evening.

I should really be glad to join in any effectual method of carrying out your object ; but I think we should be well assured before we start that our plan will be really successful. I assure you I by no means supposed that the names you sent me were either all you had, or all you were likely to get ; they were enough, and more than enough, I thought, to prevent a large body of Naturalists, &c., from signing at all, and I still think that a memorial that embodied the views of a moiety only of a class, and that moiety itself a sub-class, would be prejudicial both to the cause and the interests of science *at this particular juncture.* If taken by the Essayists for more than it was worth it might urge some of them on to some premature step, as leaving the Church, a course which I am not prepared to say I wish to see any of them follow ; if for less than it was worth, its object would be by so much defeated.

I thoroughly sympathise with the Essayists, and their Essays to a very great extent. I would extend to both even

greater countenance than your memorial professes to do ;
but I cannot help thinking that the Essayists are placed in
an extremely critical position as public professors of the
Faith of the Church of England and holders of its benefits :
and as I should wish to do *more* than give my name if called
upon to do so, I feel extremely anxious as to the turn matters
may take any day. What I should suggest would be to give
them *privately* our names in the terms of your memorial,
and offer to rally round them *publicly* when the time comes
for their acting, if they care to have us.

My opinion of the whole thing is that the Essayists cannot
stop where they are ; the public who are now excited by
them, whether to admiration or to determination, have a right
to expect that they will proceed ; they have thrown down
the gauntlet and it *is* taken up ; they must either retreat,
or leave the Church, or justify their position in the Church
by expediency or by honest intentions, and for my part
I am inclined on various grounds to uphold them in the
latter course if they adopt it. Can you not communicate with
them, through A. P. Stanley or otherwise ? If so, and you
can ascertain that such a memorial as you propose, *without*
names such as Owen, Bell, Herschel, Rosse and a host of
others which I fancy you won't get, would be acceptable to
them, I will still sign with pleasure.

Whatever you do, do not suppose I am lukewarm or
snub your memorial.

When a similar attack was made on Bishop Colenso,[1]
he wrote, ' I shall subscribe to the Colenso defence fund on

[1] John William Colenso (1814–83), well known for his school-books on arithmetic and algebra, had become Bishop of the new see of Natal in 1853.   His
critical faculties, already awakened on some theological points, were further
stirred in the course of his translation of the Bible into Zulu, by the plain questions of his converts. His views on the historical authenticity of parts of the
Pentateuch (the first three vols. appeared in 1862–3) led to sentence of deposition (Dec. 23) and excommunication by Dr. Gray, Bishop of Capetown, proceedings quashed on appeal by the judicial committee of the Privy Council. In 1866
he was again upheld by the Rolls Court when the trustees of the Colonial
Bishoprics Fund refused to pay him his episcopal income. His original work
on the Pentateuch was concluded in 1879, having been interrupted by his reply
to the *Speaker's Commentary*, designed to answer him. His *New Bible Commentary Literally Examined* appeared in six parts, 1871–4. ' The result,'
says the *D.N.B.*, ' was not a triumph for the " bishops and other clergy " who
had undertaken to cross lances with him.'   His latter years were taken up with
efforts to obtain justice for certain Zulu chiefs who had been summarily treated.
In one case he was successful ; in the others the alarm of a Zulu invasion, which
ended in the war of 1879, stood in his way.

principle'; but tells Darwin he withholds his name 'as my
poor mother would take it so to heart,' as well as to avoid the
practical unwisdom of seeming to make a party cry of it.
His attitude towards the man and his cause appears from
letters to Brian Hodgson and to Darwin.

To B. H. Hodgson

December 6, 1862.

Of Bishop Colenso and his writings I cannot say much.
I have heard his book discussed repeatedly but have not
read it, and sometimes by clergymen, and by these always
with a total want of candour, but candour in a clergyman
when discussing theological questions is a thing almost
unknown. One will not read the book; another has and
can see nothing in it; a third sees plenty in it and says all
educated clergymen know this, but rightly hide it from the
laity *lest* it should do mischief; as if truth could do mischief!
The most candid clerical disputant I met with would allow
the freest and fullest discussion, *but only in Latin!*

The Press is, I regret to say, not one whit more truthful.
One paper fills its columns with a few mistakes of the author;
another condemns 'cobweb theories' (a curious name for
plain facts); a third considers Arithmetic and common
sense not applicable to the case; a fourth wonders what all
the fuss is about, and says it is all true but of no consequence
and so on. The grave fact that our youth when educated
for clergy are systematically kept in ignorance of there
being two opinions on these subjects, and left till after they
have sworn to an *uncompromising belief*—before they can find
out what they have sworn to—is ignored by all. No doubt
Colenso will be followed by a host of men, good, bad and
indifferent, whose eyes once opened their tongues will be
let loose. The worst of it is that the present condition of
things prevents the rising talent and candid thinkers from
entering the Church at all, and we shall be bepastored with
fools, knaves or imbeciles.

To B. H. Hodgson

April 19, 1863.

Of the biblical question I have heard nothing. I am
not an admirer of McCaul or the Bishop of Manchester, and
as you know I distrust all theologians; there seems to me

a total want of candour and of charity amongst them *in all public matters*, their minds are those of women—a very good type in woman, a very bad one in man. I have glanced at Stanley's sermons and can detect an undercurrent of Colensoism in them very obviously. I had thought that all *educated* clergymen had long ago abandoned the verbal, literal inspiration of the Bible, e.g. the worship of the letter, the Genesis creation, the Flood,[1] tower of Babel, &c., &c., &c., plus much of the so-called Mosaic narrative; but this is either not so, or the educated ones hold their tongues— perhaps the latter is the case, for after all it is curious to observe how few deans, archdeacons and other dignitaries, professors, &c., come forward to condemn Colenso—it is the Bishops and noisy theologians who usurp the press and pulpits and fill them with denunciations.

I think I told you that I stayed a couple of days with Colenso in the country, and was pleased with his calmness, dignity and charity towards his opponents. He is a tall, grave, very striking man, with a quiet determination of mouth, and candid, broad forehead and open eyes, that together produced an impression of power and dignity: He has, however, calculated without his host, and for this he has his education to thank, rather than his judgment or faults. He might in my opinion have said ten times as much as he has *in different language* and he would have created no sensation at all. I think Stanley implies in many of his writings as much at least as Colenso insists upon, but puts a fine spiritual varnish over it all.

To C. Darwin

February 16, 1864.

I am not quite sure about Colenso himself—he ought to go further. My hope is, that after the trial he will go out just to assert his position, and then retire. His holding his Bishopric in Natal can only breed intolerable confusion and do his cause mischief; and as to his going out to convert Zulus, why, he has Christians here to convert, and the Zulus are not worth a thought. He might come back with great

---

[1] As he writes to Darwin, October 25, 1862: ' What a nice simple book Parrott's *Ararat* is : it is refreshing to read his simple faith in the Ark being still under the snow ! '

glory and set up in England as a tutor, abandoning his title
and mitre. I have seen a good deal of him, and consider him
sanguine and unsafe.

His attitude towards the ceremonies of the Church is illustrated by a letter to Huxley, who had asked him to be godfather
to his son, at his wife's desire, though to himself it was an
unmeaning form only to be turned into ' a reality by making it
a bond with one's friends.' ' If,' he adds, ' you have any objections to say ' all this I steadfastly believe,' even by deputy, I
know you will have no hesitation in saying so.'

Kew : January 4, 1861.

MY DEAR HUXLEY,—I will volontiers ' renounce the
Devil and all his works ' for your child, *in spirit*, and chasser
his majesty *in person* from his cradle and bed whenever and
wherever I am called upon to do so. Nay more—I will do
it ' by bell and by book,' for he shall have a coral when his
blessed teeth be coming and a book when he can read it.
Also as the christening is to be done, it is a duty to see it
done properly; ' devoutly, orderly and reverently,' and as
I won't trust these parsons, I will go see it myself. In the
abstract I hate and despise the spiritual element of the
ceremony, but in practice I do not care so much about it as
conscientiously to plead any honest wish to shirk it. I have
a greater objection to say ' all this I steadfastly believe '
by deputy, than in person. I have ∝ conflicting opinions as
to the expediency &c. of doing things by halves, but only one
as to the propriety of being hung for a sheep in preference to
a lamb, and as I have had hitherto, and yet shall have, to
go to Church with other people's bairns, I should be ashamed
to decline to do so with yours. I assure you truthfully that
the pleasure of being in any recognised relationship to your
child will sweeten any pill of doctrine that may be offered, even
if I could not manage to ' sham Abraham ' at the responses,
an unworthy and cowardly resort I affect on such occasions.

Under his critical distrust, however, of theologians and
sacerdotalism generally, he was deeply responsive to the
deep things of the spirit which move humanity in life and
in death. Characteristic in their different ways are letters

touching the death of his father-in-law Henslow in May 1861 ;
of his little daughter Minnie (September 28, 1863) ; of Falconer
and of Sir William Hooker in 1865.

We realise the beauty of Henslow's character from the
words of the friend and close intimate whose intimacy had
only served to increase his admiration and affection. It was
a prolonged deathbed. Bronchitis and congestion of the
lungs aggravated long-standing heart disease, and all through
April and May he was in a hopeless condition. Hooker spent
a long time at Hitcham tending him, for happily his father
was well and active and could spare him from Kew. He writes
to Huxley on April 3 :

> He has bidden farewell to his friends, parishioners and
> little botanical school children, one by one, addressing a few
> words of encouragement and advice to each with a calmness
> and affectionate interest that is quite overpowering.
>
> I am utterly overwhelmed ; to be loved as he was for the
> good he had done I would lay down my science and almost
> turn parson. To me personally the loss will be immeasurable
> —he took interest in everything I did and I loved him—I am
> wrong to think how much.
>
> His loss to this neighbourhood will be incalculable ; there
> is none to take his place, morally, socially or religiously.
> Between his paroxysms he talks ·of all his friends as calmly
> as possible, discourses on Essays and Reviews and all the
> great religious questions with the most perfect openness
> and fairness, and for thorough *appreciation* of the opinions
> of those with whom he differs, his charity is unbounded.
> You know how my associations are sunk in this place and
> can guess how I take tearing them up by the roots—bitterly.

And again on the 11th :

> His brain scarcely indicates a change in its workings. He
> goes on dictating letters when he can, of advice, encourage-
> ment and warnings to all who he thinks may be bettered
> by them. I have written some very touching ones. The
> kindness and wisdom with which he does all this is very
> admirable, not only in counselling individuals to pursue some
> innocent substitute for their besetting sin, but recommending

them to mutual friends, of integrity, resources and inflexible purpose, who will encourage and quiet them if they will take his advice and use his instructions.

To Anderson in Calcutta he also opens his heart :

[April 22, 1861.]

It is a grievous break-up in many ways, and I for one had little idea of the enormous extent and power of Henslow's influence, socially, morally and religiously, till called here to his dying bed and to witness the extent of sympathy his illness creates and the huge blank his death will cause. It is like gouging a piece out of the face of the country. His death-bed is wonderful and makes one wish to have led his life and almost reconciles one to [his] having been a parson ! Well, my dear Anderson, we shall never be like him, a man who never turned back on friend or foe, and never spoke or thought ill of another, a man who with strong enough religious convictions of his own, had the biggest charity for every heresy so long as it was conscientiously entertained.

And finally on May 23 :

Henslow has left a blank in my existence never to be replaced. Quite apart from considerations matrimonial, H. had more influence over my life and conduct than any other man, so good, so calm, so wise, so far above all taint of pride, prejudice or passion, so magnanimous in short was he in all situations of life. More than all this, I miss his knowledge of loads of matters bearing on Botany which I never knew or took up but through him, and of loads of kindred subjects in which I have keenly interested myself, ever since I knew him. He was one of those friends formed late in life to be a lamp unto our path whom we never go ahead of as we do with the instructors of our youth. I know what death and losses are, but this is the first of which ' the funeral over ' is no relief. His loss hangs like a dead weight upon me: I feel as if a bit of each faculty was gone for ever, for he sharpened every faculty I had, and created some too. You knew enough of him to understand all this.

His little daughter, who had died almost suddenly on September 28, was six years old.

Kew, October 1, 1863.

DEAR OLD DARWIN,—I have just buried my darling
little girl and read your kind note. I tried hard to make no
difference between her and the other children, but she was
my very own, the flower of my flock in every one's eyes, the
companion of my walks, the first of my children who has
shown any· love for music and flowers, and the sweetest
tempered, affectionate little thing that ever I knew. It
will be long before I cease to hear her voice in my ears, or
feel her little hand stealing into mine ; by the fireside and in
the Garden, wherever I go she is there.

The funeral service had no more effect on me than on
her : the association with her personally snapped as the
ceremonial left my door, and oddly enough, I felt nothing
at seeing the little white coffin go into the vault, my mind
was wandering amongst sweeter memories elsewhere.

And now I can calmly think of what sorrows I am spared.
Hers was no contagious disease, threatening the whole
family for weeks afterwards ; she suffered comparatively
little ; and above all do I rejoice that she was yet so young
and happy, that death did not enter her little head during
her illness, and I was spared the agony of seeing my darling
pass through the 'valley of the shadow of death.' Then too,
strangely enough,'I never knew she was dying till 3 minutes
before the breath left her body. For 3 hours I was blind
to every one of those symptoms of rapidly approaching
dissolution, that every nurse knows and every novelist describes, and I have seen myself so often. The doctor came
in just 3 minutes before she died and told me to my horror
she was dying. I knew the extreme danger, but assumed
she had many hours to live. The retrospect of that last
night is thus in some respects comforting, in others hideous,
and I can still feel the cold shudder that every misinterpreted
symptom still sent through me, during that long night of
agony and suspense.

A month later, October 23 :

I am very well, but it will be long before I get over this
craving for my child, or the bitterness of that last night.
To nurse grief I hold is a deadly sin, but I shall never cease
to wish my child back in my arms, as long as I live.

Three years afterwards, writing to Darwin, whose sister, Mrs. Langton, was hopelessly ill, he is pursued by the same memory.

I have been so haunted by death and his darts this 6 or 8 years, that I can hardly bear to look at my children asleep in bed. I used to think a child asleep not only the loveliest thing in creation, but the most gratifying in every respect —leaving nothing to be desired except that it would not grow older. All is changed now.

The death of Falconer in January 1865 took away an old and warm-hearted friend of both Darwin and Hooker.

Poor old Falconer! how my mind runs back to those happiest of all my days, that I used to spend at Down 20 years ago—when I left your house with my heart in my mouth like a school-boy.

What a mountainous mass of admirable and accurate information dies with our dear old friend. I shall miss him greatly, not only personally, but as a scientific man of unflinching and uncompromising integrity, and of great weight in Murchisonian and other counsels, where ballast is sadly needed. The inconceivability of our being born for nothing better than such a petty existence as ours is, gives me some hope of meeting in a better world. What does it all mean? . . . When we think what millions upon millions of lives and intellects it has taken to work up to a knowledge of gravity and natural selection, we really do seem a contemptible creature intellectually, and when we feel the death of friends more keenly the older we grow we do strike me as being corporeally most miserable, for we have no pleasures to compensate fully for our griefs and pains : these alone are unalloyed.

Three years later Falconer's ' Palæontological Memoirs ' appeared—and Hooker wrote to Darwin :

Feb. 1, 1868.

What a fine work Dr. Murchison [1] has made of dear old Falconer's Memoirs; it strikes me that it will be most

---

[1] Charles Murchison (1830–79), F.R.S. 1866, a cousin of Sir Roderick, was a distinguished physician, who served in India 1853–5, and was Professor of Chemistry at Calcutta. Returning to London he won a high reputation both as a practitioner and as a lecturer at several of the great hospitals. He made many contributions to medical science, and was a considerable geologist.

useful. I sigh when I think how poor my reprinted Memoirs
would appear beside them, if any injudicious post-mortem
friend were to issue them. There is something grand in
the blunt force of Falconer's writings, and when he mounts
the Pegasus of Theory, he reminds me of the picture of
Sintram (ask Henrietta)—with him the very thought of a
Speculation is sin, and a *very serious thing*—it is the original
sin, besetting sin, of the scientific man—but when he speculated himself, as on the perfection of the post-Tertiary record,
how lame and impotent he was. He sinned and suffered
in short.

Of the contrast between the death of the old and of the
young he writes to Charles Darwin, September 26, 1865.

How strange is the difference between the loss of an aged
parent and child : my father has been my companion as well
as parent for 25 years, our intimacy has never been broken;
our aims have been one as much as those of father and son
ever could by possibility be ; but I have to *reflect* on his loss
before I realise it and swell with grief. How different in
my child's case ! I cannot see that it is altogether natural,
though it is so in the main. Is my grief for him more selfish
than that for my child ? I cannot feel it to be so. I do suppose we have a pure nature, independent of conditions (and
of Darwinism applied !), but what it is we can only hope to
know if we realize a future state.

I am gratified by your expressions about my father ;
he was one of the most truly liberal and modest men I ever
knew ; he had not an atom of self in him, always thought
nothing of himself, and never took any self-seeking steps
to raise himself in the estimation of the Government or
of scientific men. With one-tenth of the exertion that
Murchison displayed, he would have had honors and titles
showered on him, and I hate the Royal Soc. for never recog-
nizing the obligations science is under to him. He never
received any honor, distinction or reward from the Crown
or Government for all his public services, because he never
would put himself into the way of them. I thought the boast
of the R.S. was that they sought out such as had similar
claims upon science. I know I am not agreed with, but I
will not give in.

When Darwin was very ill the following February, he was allowed to see no one : and Hooker, who had spent the week-end near by at the Lubbocks', writes feelingly :

> I yearned to go over and see Mrs. Darwin, but it would have been too great a punishment to both of us (you and me). I cannot tell which I crave for most, another little girl, or for you to get well.

And as the anniversary of his loss came round, he wrote (September 16, 1864 : the British Association was meeting at Bath) :

> I go to Bath to-morrow for two or three days. I am glad to do so, though I go with a very heavy heart ; on principle I think we should not keep anniversaries of great sorrows, but as the day draws nearer I feel all the misery of last year crawling over me, and my lost child's face and voice accompany me everywhere by day and by night ; so that I now dread an attack of what were more the horrors of delirium tremens than the chastened sorrows of a sensible man. I am sure however that there is no fear of that now ; time, as you told me it would, has done its inevitable work. What queer mortals we are ! Poor Grove's far more dreadful blow reconciles me to my loss, in a real though irrational manner.[1] I have felt for him exceedingly. It is too bad of me to write on such selfish subjects to you, and I am sure Mrs. Darwin must be angry with me for doing so—but your affection for your children has been a great example to me, and there is no other living soul with whom I can talk of the subject; it would make my wife ill if I went on so to her. She is wonderfully different from me, the loss simply made her very ill almost dangerously so. I am of tougher, coarser material and, like Rawdon Crawley, have greater capacity for feelings, which when once roused run riot.

Here may conveniently be added later expressions on these and similar subjects. The first is a letter to Darwin, January 7, 1873 :

---

[1] Probably Sir George Grove (1820–1900), writer on music and first director of the Royal College of Music, whose daughter died about this time.

Greg's [1] 'Enigmas' is one of the most eloquent books I ever read, and it quite fascinated me by its manner, not by its matter, which is singularly weak and inconclusive. I wrote to him combating some of his positions, and met him soon after and had a delightful conversation. As to the poor man's faith, he frankly admitted to me that, as I put it, all scientific evidence is in favor of extinction upon death, and that any reasoning to the contrary was 'ingenious wriggling.' I quite agreed with him, however, that this was not conclusive and that there was no inexcusable presumption in the conclusion, that there was a future state. It is a book that cannot but be disappointing : remember all it pretends to do is, not to crush hope, but to foster the presumption of hope being tenable—barely tenable perhaps !

We have just returned from a visit to Cardwell's, near Godalming ; both he and his wife are singularly pleasing persons at home. He is almost a religious man, or I should say a devout one perhaps. We had some long talks about faith and prayer ; he was very frank, admitting to the full how much more difficult it was for a scientific man to believe than for any other ; that the Miracles were open questions, of evidence entirely ; and that prayer in the common sense was wrong ; he much regretted such occasional outbursts as Huxley's, but blamed the clergy more. He was singularly earnest, candid and calm, even on such matters as Darwinism ! which he only a little believes—much disliking some of the results (Monkeydom &c.), but could see even to this no opposition to any religion worth holding.

The other two citations are from letters to Huxley. One— Huxley had sent him proofs of his chapter ' On the Reception of the " Origin of Species," ' which was to appear in the ' Life and Letters of Charles Darwin.' On October 21, 1886, he

---

[1] William Rathbone Greg (1809–1881) began life as a cotton spinner, following in his father's footsteps. His literary activities and his removal from Bury to the Lakes on account of his wife's health hindered him in his business, and he gave up his mill in 1850. In 1856 he was appointed a Commissioner on the Board of Customs, and from 1864–77 was Comptroller of the Stationery Office. He was distinguished as a thoughtful and prolific essayist on religious, political, and economic subjects, and equally ardent in his philanthropy and his disinterested love of truth, which balanced generous enthusiasms with an unflinching view of the difficulty and complexity of modern problems. His first book, *The Creed of Christendom*, appeared in 1851 ; the *Enigmas of Life*, mentioned here, in 1872.

replies, agreeing with the suggestion that a paragraph or two should be added with ' the two chief objections made formerly and now to Darwin—the one that it is introducing " chance " as a factor in nature, and the other that it is atheistic.'

You must deal with the ' Chance ' objection, and that involves the atheistic ; but this you can do better than any one, briefly and effectively.

The haziness of ordinary people's minds in regard to both Theism and Atheism, and the idea that either can be supported or negatived by reasoning—e.g. from little fishes is wonderful.

As you say, Theism and Atheism are just where they were in the days of Job and his comforters.

The other is apropos of the first volume of the ' Collected Essays ' which Huxley had just sent to him. (October 8, 1893.)

The ' Inequality of Man ' is thoroughly well dealt with, and leaves nothing to be desired. There is much that merits consideration (would that it could be action) in the conception of a National Church at p. 284. Something is wanted in the present day, that would systematically foster, in the young especially, a spirit of reverence for the higher aims and aspirations of the best men towards the attainment of knowledge, truth and pure living. My old friend W. R. Greg used to discuss this with me, and would have had me proceed on these lines !

A thousand thanks for the coming volume.

As Sir William Hooker advanced in years, the possibilities that would open out at his death inevitably presented themselves both to himself and to his son. To the latter the thought was odious. If he should be compelled to shoulder the burden of continuing his beloved father's work alone, it would of course have to be done ; but he would gladly have renounced an official position for quiet research. Administrative work with its official shackles and its shadow of official honours made little appeal to him : much less the open lionising of science, or its exploitation as a stepping-stone to knighthoods and the like. Thus when in 1863 the Indian Government

talked of commissioning him to do the Flora Indica, he writes
to Darwin (October ?) :

> Pay would tempt me, but only because it would hold
> out a prospect of early retirement from the struggle of
> scientific work for one's livelihood, and shaking the dust
> off my feet at the Govt. and Kew Gardens—but for God's
> sake let this go no further. I regard succession to my father
> with horror. Not that a better scientific place exists in the
> world, except my own. I am beginning too to hate the οἱ
> πολλοι of science. Huxley, Lubbock and half a dozen others
> are enough for me of the workers, outside my own imme-
> diate pale, which includes only yourself, Bentham, Oliver and
> Thomson. As to Murchisonian science and all that sort
> of thing, like K.C.B 's, it makes me sick to read his science at
> the Newcastle Meeting.

Still, when it came, Sir William's death, the dividing point
between the two eras in his son's life, came as a sudden blow.
To the last he had been wonderfully active. Though now past
eighty, on the Monday he had escorted Queen Emma of the
Sandwich Islands and her party over the Gardens—' I never
saw him more lively and active.' Next day he was out and
about both morning and afternoon, first walking over to see
the subtropical plants in Battersea Park, then taking friends
over Kew. On the Wednesday he developed what we to-day
should call a septic throat with utter prostration, which was
epidemic in Kew, and on Saturday the 12th died very quietly
and almost without pain. ' He never realised his danger,
nd altogether his illness and end were unspeakably peaceful
and happy for himself and those around him.'

For the first two days his son and his faithful servant nursed
him. The other members of the family were away at Yarmouth,
owing to the domestic exigencies of house-painting. Lady
Hooker returned on the Thursday, but Mrs. Hooker and the
children were forbidden to come for the next fortnight, owing
to the epidemic. But at this critical moment Joseph Hooker
himself, to his intense grief, was himself stricken down. On
the Wednesday night he had slept on the floor of a dressing-room
by which he was airing his father's room. As he slept, the wind

got up, and he awoke with pain and stiffness all over, and though he held out on the Thursday, was down with his old enemy, rheumatic fever, next day. For three weeks he lay in great pain, distracted by his inability to render help when it was so much needed: Happily his friends Berkeley and Thomson, who were at Kew, took over the examinations for Assistant Surgeons which he had in hand.

Then Dr. Campbell, his old Darjiling friend, backed by a London doctor, carried him off to his own house in Notting Hill, whence, as he got better, he was sent to complete his cure in the bracing air of Buxton, being forbidden to return to work until October 20, a leave subsequently extended to the end of the month.

The enforced leisure of convalescence afforded much opportunity for miscellaneous reading. From time to time the letters which passed between Darwin and Hooker contain references to novels, for Darwin, as we know, constantly had novels read to him when unable to work, and Hooker, from his wife's and his own reading, would offer suggestions or criticisms. Thus in 1863 Hooker recommends ' The Admiral's Daughter ' by the author of ' Emilia Wyndham,' which on re-reading he had found as deeply interesting as on his first reading twenty-five years before ; but this was barred as ending too sadly. Next year ' Quits ' is more successful ; on a return recommendation, Hooker at Bath cannot get ' Beppo,' but borrows ' Romola,' ' which is ponderous.' In April 1865, having received from Darwin the serial numbers of Wilkie Collins' novel, Hooker replies, ' I have nearly finished " Can you Forgive Her ? " and have made up my mind that I cannot at all do so, and don't care whether she minds it or no.'

Now the unexpected scope of holiday reading appears from two letters to Darwin. Indeed he was so much tickled by the idea of having been reduced to reading ' Clarissa Harlowe, that he repeated the announcement to Huxley, with a ' Figurez vous, mon cher Huxley.'

September 26, 1865.

Out of the utter idleness of my mind I write to you, you dear blessed ultima thule of my fatuous correspondence, to whom I can write in my folly, as well as in my sorrow and perplexity. Don't you see I am better ? We have read Uncle Silas, isn't it creepy ? and crawly too. One should have a brandy bottle and sal volatile to get through it in safety alone. How splendidly the interest is kept up. Then I took the ' Mill on the Floss ' and am ravished with it ; what a clever person the authoress is, I like it even better than ' Adam Bede.' How evidently the authoress belongs to the class of life of her heroines, with whom first love is an animal passion with nothing to elevate it. How splendid are her analyses of the mixed motives of human action in the young, but not in the old, and yet how vividly she represents the acts and conversation of the old. Then I took a dose of Jamieson's paper on the Glacial period of Scotland,[1] and wrote him a long letter praising it. Still I am sure there was a time when the contour of submerged Scotland was ploughed by icebergs moving in definite direction (S.W. to N.E., or rather vice versa !). Given a submerged Great Britain a hundred miles or so off Victoria Land, and the Bergs would plough it in a direction S.W. to N.E.—Bergs some of them 10 miles long and 700 feet below water ! I can fancy no other explanation of the parallelism of the great Scotch valleys but this, and as there are *not* more things in Heaven and Earth than are dreamt of in &c., it follows as a matter of course.

October 6, 1865.

And now for a confession—I have read ' Clarissa Harlowe'! I feel that this is self damnatory and can only plead my illness and the tedium of a watering-place. As however ' frank confession is good for the soul,' I will tell you the first 5 volumes are simply illegible, so dull, so poor, so attenuated, that had I stopped there I should have considered the former popularity of the book as one of those things which ' no fellow can be expected to understand,' as Uncle Sam has it ; the 6th and 7th (*horresco referens*) opened my eyes however ; though to me they had no merit or interest what-

---

[1] In the *Quarterly Journal*, Geological Society, vol. xix., p. 235, 1863. This paper brought forward further evidence as to the existence of glacial barriers damming the mouth of Glen Roy, &c., and so forming lakes, the margins of which are still marked by the famous ' Parallel Roads.' Jamieson's work converted Darwin from his earlier theory of raised sea-beaches, which was the only explanation possible in the then state of knowledge (see M.L. ii. 172).

ever as a tale, I could quite understand the deep interest
they must have had in an artificial and vicious age, when
alone such compositions could be put by mothers into the
hands of virtuous daughters with injunctions to *study them*,
and the immense good they may have done. In an age when
men of fashion had no honor, and when the prejudices of
education or absence of it, and want of public journals kept
women in the dark as to the means men employed, and when
maudlin sensational writing *did* act on the brain in a way it
does not now, it is obvious to me that Richardson's works
must have frightened hosts of young women into caution
at any rate, and stimulated a few to good works. Be this as
it may, there is no doubt I suppose that his works were
perused by thousands as standard literature for young ladies
in 1750–1770, and that the change of manners was so rapid,
that in 1780 I find by the life of Reynolds (I am ashamed of
owning that I have been reading a solid book) both Richardson's and Fielding's works were considered as too coarse
for young ladies.

The regret at politics clashing with science finds repeated
expression.

I gnash my teeth when I think of Lubbock going
into Parliament [he exclaims to Darwin, April 19, 1865].
The awful waste of time, of energy, of brain, of life and
all that makes life worth living—always except a man
goes in for Politics, Finance or Self-aggrandizement—for
such the up-hill drag through mire of all kinds, dinners,
Committees, Deputations, Lady P.'s receptions, Levees, &c.
&c.—all this and more, may be worth a man's undergoing
who has a clear calling that way, and a prospect of some 25
years of political superiority or supremacy at the end of it.

And (on the 7th):

I grudge so good a man from Science, and have a presenti-
ment that it will inaugurate a very trying life for him. I
believe I am no end of way happier in avoiding every avenue
to ambitious ends in my small walk of life, and so long as
one's mind is fully occupied, there is nothing to regret in a
life of mere drudgery.

Miscellanea from the correspondence of these years may appropriately close this section.

Of a slashing writer :

He goes like a desert whirlwind over the ground, scorching, blasting and suffocating all opposing objects, and leaving nothing but dry bones on the ground. The vegetation he withers was one of vile weeds to be sure, but vile *weeds* are green, and all is *black* after him.

A photograph goes to Darwin on March 17, 1862, with the criticism :

As regards my Photograph, I believe I have very little expression. I have often remarked that I am not recognized except by those who know me tolerably well, that I have often to introduce myself, added to which all my photographs and portraits make me look either silly or stupid or affected. Artists find nothing salient, nothing to idealize upon. Poor Richmond, who generally knocks off his chalk heads in two sittings, gave me eight I think, and grumbled all the time, and has turned me out a very lackadaisical young gentleman.

In return, Darwin sends his photograph in June 1864 saying, ' Funnily enough the boys declared it was like Moses ' :

Glorified friend—Your photograph tells me where Herbert got his Moses for the fresco in the House of Lords—horns and halo and all. Well done William.

Darwin had reported that all the doctors seemed to think him a case of suppressed gout.

What the devil is this ' suppressed gout ' upon which doctors fasten every ill they cannot name ? If it is *suppressed* how do they know it is gout ? If it is apparent, why the devil do they call it suppressed ? I hate the use of cant terms to cloak ignorance. (January 1865.)

In lighter vein he writes to Darwin on April 29 (?), 1864 :

Frank Palgrave told me a good story last night : He met a Frenchman who talked largely of art, and asked him if he knew Ary Scheffer. ' Oui,' he answered, with enthusiasm, 'je pose quelquefois pour M. Scheffer comme Jésus Christ, et *quelquefois aussi* pour le diable!' If you don't laugh I will hate you.

The same light touch enlivens the description of a burglary at Kew, when 'a nice young man who introduced himself to the maids' made off with the contents of the plate-basket, so that 'I have had tears, groans, hysterics, Police inspectors and all the other evidences of civilisation in the house.' But strange to say, the 'nice young man' overlooked a 'lovely teapot,' Darwin's gift, and various solid but unattractive articles.

I am disgusted at their not taking the candlesticks, which are of no use to me a bit, and at their assuming your teapot to be plated! or they surely would have taken it—so 'there is no pleasing some people' you will say. (May 5, 1862.)

An epigrammatic piece of characterisation is that of J. E. Gray, the anatomist and zoological keeper at the British Museum. Gray had a loose-tongued habit, if any one came under his criticism, of heaping reckless abuse upon him, quite without foundation and often self-contradictory. On one of these occasions Hooker tells Darwin how he took him to task (May 13, 1863).

I pitched into him hot and strong, and made him eat all his assertions. I think I made him heartily ashamed of himself. I never heard such a slanderer in my whole life. I suppose it is because he so overdoes it that he makes so few real enemies thereby.

And on the 24th he expands his judgment.

Dr. Gray is really not *malignant* . . . he has all the attributes of malignancy except malignance—there then!—or rather, he talks like a malignant man without feeling in the least malignant. I never knew Gray to do an action that sprang from an unkind motive or feeling. He abounds in all the active attributes of unkindness and malignancy without being either in heart.

Another character study is in reply to one of Darwin's appeals for help to a clever young man who had submitted some original observations to him.

I am afraid A.B. is a man who cannot be helped . . . he is one of those men whom love of knowledge makes to forget that man is not born for self alone, or rather, that the only way of serving self effectually is to do it by proxy, and make yourself a useful self-supporting member of society. The man frankly says ' l am fit for nothing but what " won't pay," this is the world's fault, not mine.' A love of science, however pure, may be practically as selfish a love as any vice. A.B. should have been born to £1000 a year and no ties domestic, social or territorial—in short should not be called upon to take his part in the ' struggle for life.' I have known many such—most amongst artists—next most amongst scientific young men. No one such ever succeeded, even in science, and depend upon it after 10 years A.B. would be as used up as a man of science as he now is as a man of mental energy.

Tyndall, Faraday, Huxley, Graham, Lindley, &c. all began by establishing themselves as useful self-supporting members of Society, and, that accomplished, they gradually shook off the disagreeable work as they took on science. A.B. has not established himself as a useful member of society, knows it, owns it, and blames the world for it. Now, my dear Darwin, you may depend on it, that such men are no more able to cut a figure in science than in life—useful drudges for a time they may be and are, gradually the feeling grows that their drudgery is other men's fame and bread, and they become pestilent fellows. My dear old friend, my heart sinks sometimes, and I could cry like a child, when appeals for charity come to me from cases to which I must apply your theory in all its force, and come to the conclusion that in giving I am hastening the fall.

As regards letters which reveal the personal affection and happy intimacy with Darwin, one of the very best is unfortunately missing. It is the letter written to Darwin when the proposal to award him the Copley Medal in 1863 failed. Only Darwin's reply is given in M.L. ii. 338. ' Your

Hastings note, my dear old fellow, was a Copley Medal to me and more than a Copley Medal.' But in 1864, at Falconer's proposal, the Copley was awarded to him, and Hooker writes in ironical delight on November 23 (?) :

> I have not got over the shock of your getting the Copley. I had so made up my mind that you were too far ahead of your day to be appreciated, that I was [flabbergasted ?]. I thought it took [word illegible] like *me* and Huxley and Lubbock to see so far ahead as you are of the ruck of candidates whom the Council bring forward for (Copley) medals. However it is best as it is ! ! ! and I am resigned to the feeling that if they could not appreciate you, they could appreciate (or fear) the opinions of those who brought you forward. I am curious to see the President's address.[1]

General Sabine, the President of the Royal Society, was notoriously anti-Darwinian and willing to deliver a left-handed blow at the medallist. The sequel, which is referred to in C.D. iii. 28, and fully told in M.L. ii. 255, including the quotation from the 'Life of T. H. Huxley,' is sufficiently described in the following to Darwin, December 2, 1864 :

> Have you heard of the small breeze at R.S. apropos of your award ? Busk told me thus : Sabine said, in his address, that in awarding you the Copley 'all consideration of your " Origin " was expressly excluded.' After the address, Huxley gets up and asks how this is, and being assured it is so, he insists on the Minutes of the Council being produced and read, in which of course there was no such exclusion or indeed any allusion to the ' Origin.' Busk and Sabine afterwards were discussing the point, Sabine saying that no allusion = express exclusion, and shuffling as usual, when up comes Falconer, and to Busk's horror compliments Sabine's address unreservedly. Busk, thinking that F. had overheard the discussion, said nothing at the time, but

---

[1] The same spirit of happy banter occurs in a note of 1865, when Darwin had been, as it were, reading the *Origin* for the first time, as he was collecting material for a second French edition, and laughingly declared ' Upon my life, my dear fellow, it is a very good book, but oh my goodness, it is tough reading.' Thereupon Hooker retorted : ' I am egregiously delighted with your calm judgment on the *Origin*. Do you know I have re-read some of my papers with the same result, and NEVER WAS WRONG ONCE IN MY OPINION.'

calls Falconer to account afterwards, upon which F. is grievously put out at finding out what he has done and forthwith
goes and writes a letter to Sabine on the subject.   May the
Lord have mercy on S. is all I can say ; for F. will have
none.   This is the story as I believe Busk to have told it
to me yesterday ; but as it has thus passed through two
hands I do not doubt it is damaged in the process, so pray
take it for no more than it is worth.

Moreover, having been asked to supply a statement as to
Darwin's botanical discoveries, Hooker, on reading Sabine's
complete address, which he found ' very good on the whole,'
expressed himself to his friend as

indignant and disgusted at the mutilation and emasculation
of what I wrote—especially about *Lythrum* and *Linum*,[1]
which he has made nonsense of, and the use your observations will be in interpreting no end of phenomena not
yet guessed at.   (January 1, 1865.)

He has certainly not praised you too much as to your
Botany, but I do suppose that your merits as a Geologist
and Zoologist are AUDACIOUSLY EXAGGERATED—there then !

A year or so later, Sir Charles Lyell was desirous that the
same recognition for his great scientific labours should be
given to Hooker.   The latter, however, was by no means of
this mind, and frankly tells Darwin :

After his funny and not-at-all-agreeable-to-me fashion of
telling me all about it, of course I must not tell him so, but
it is God's truth, that not only shall I never think I deserve
it if I get it, but that if I did deserve it, it would be far too
dear at the cost of an after-dinner speech.   These are things,
however, which must take their courses.

Darwin's rejoinder was emphatic :

As for thinking that you do not deserve the Copley Medal,[2]
that I declare is mere insanity.

---

[1] I.e. the two and threefold relations between the pistil and stamens of
certain plants, ensuring cross-fertilisation.
[2] He received it in 1887.

It was during this period that Hooker took up the hobby of collecting Wedgwood ware, which became a subject of much cheery banter between him and his friend.

By the way—now don't despise me—I am collecting Wedgwoods, simply and solely because they are pretty and I love them. I have not even a Grayan excuse, they afford me pleasure—*voilà tout.*

Darwin, who declared that he drew the line at collecting stamps, was much amused, but confessed his family to be ' degenerate descendants of old Josiah W., for we have not a bit of pretty ware in the house ' (see C.D. iii. 4), and to Hooker's enthusiasm retorts : ' You cannot imagine what pleasure your plants give me (far more than your Wedgwood ware can give you).'

To Charles Darwin

January 6, 1863.

I am quite aware of your insensibility to Wedgwood ware. Were it otherwise I do not think I could have *gone into* the foible, for I should have bored you out of your life to beg, buy, borrow and steal for me (do not tell Henrietta). As it is, I do not go further than little Medallions and such matters—such gorgeous things as you had on slates are not for the like of me, and as to the chimney-pots on your chimney-piece in the dining-room, they are not worth carriage.

And next day enjoyment bubbles over :

It is rather jolly this writing about matters non-scientific —let's give up science when you have done the three vols., and take to gossip. I quite agree with you, that a holiday is an unendurable bore, but depend on it, that is because we have no vices to indulge in, and if you will only join me in some good vice, such as talking about and writing about what will do no good to our neighbours and some harm to ourselves—we shall get on capitally, and scratch away.

At a time when he declared he found life too great a worry to allow him to bring mind or time to bear on botanical experiments, he exclaims (March 5, 1863) :

I do assure you that without joking, Wedgwoods are an unspeakable relief to me. I look over them every Sunday morning, and poke into all the little second-hand shops I pass in London, seeking medallions. The prices of vases are quite incredible : I saw a lovely butter-boat and was quite determined to go up to 30s. for it, at the dirtiest little pigstye of a subterranean hole in the wall of a shop you ever were in, the price was £25. All this amuses me vastly and is an enjoyable contrast to grim science. No lady enjoys *bonnets* more heartily.

So he tells Hodgson :

I have gone mad after Wedgwood ware, and especially the medallions—things of another world. If you come across any good specimens of old Wedgwood, pray beg, buy, borrow and steal for me.

And to his uncle the Rev. J. Gunn, a sympathiser in such things, he writes (January 29, 1863) :

When are you coming up ? I have some absolutely stifling Wedgwoods to shew you : a plaque 18 in. long, with Achilles dragging Hector round the walls of Troy, of Flaxman's grandest time and manner—it will make your hair curl to look at it ; an oval medallion of Goldsmith, 18 in. diam. ; Mitten and Erasmus in white on pink clay, and the Prince and Princess of Wales on pea-green clay ; besides about forty other portraits of sorts.

Darwin fed the hobby with mingled grain and chaff.

I had a whole box full of small Wedgwood medallions [he tells his friend in April] ; but, drat the children, everything in this house gets lost and wasted ; I can only find about a dozen little things as big as shillings, and I presume worth nothing ; but you shall look at them when here and take them if worth pocketing.

He got his sister to send Hooker one of her black and brown vases, but—

You sent us a gratuitous insult about the 'chimney-pots' in dining-room, for you *shan't* have them ; nor are they Wedgwood ware.

From Darwin Hooker borrowed a medallion of his grand-father Erasmus, and had a cast carefully made by Woolner the sculptor for the Kew Museum. Through Darwin also he made acquaintance with the Wedgwoods of Etruria and visited them there, where he ' dabbled among the moulds ' to his heart's content, and chose several fine plaques which the Wedgwoods kindly reproduced for him.

Jesting allusions constantly recur on either side, especially to the value of the hobby as a standard of intellectual activity. Hooker sums up the scientific worth of ' Juventus Mundi ' by declaring that ' Wedgwood is a science to it.' Mr. Gladstone, it may be remembered, was also a collector of Wedgwood ware. So too as a guide to history. Speaking of what a picturesque Joan of Arc Miss Susan Horner would make, he remarks :

N.B. My ideas of J.A. are wholly derived from Etty's and Millais' pictures. I do not know even in whose reign she lived, if in any, and as I have no Wedgwood medallion of her, I have no means of knowing.

But by this time (May 13, 1866) the hobby had perforce to go slowly :

My pursuit of that blue art is over, and the crockery shops know me no more. I have never time to go to London now, and hope never to have again.

Still though the hope of filling up certain gaps at a sale after the death of Mrs. Langton, Darwin's sister, failed because all the medallions were bought in, he continued to buy when occasion offered.

CHAPTER XXXI

KEW, ST. PETERSBURG, AND MAROCCO

HOOKER returned to Kew in the autumn of 1865 'really extremely well, though still a little stiff in the joints.' 'I am taking to gardening,' he tells Darwin, and the share of outdoor occupation certainly made for health in his strenuous life. 'I am very busy,' he adds (September 28, 1866), 'out of doors six hours a day and delighting in my occupation. I can make even Kew 50 per cent. better than it is.' In June 1867, 'I am turning into a landscape gardener, getting up cheerfully at 6 and before it, and sleeping like a ploughboy in consequence, or rather in spite of it.' And by February 1868, 'I am getting very proud of the Gardens, in which I really have worked tremendously hard for now two years.'

But this portion of outdoor life never quenched the deep-seated desire for travel in the wilds. Being bidden by the Admiralty in 1866 to look out for two 'high class' naturalists for voyages to Corea and the Straits of Magellan, he exclaims: 'I wish I could go!' And to one of these, Dr. Cunningham, who had sailed on the latter expedition, he repeats:

> I know no life so enjoyable as camping out, and I never met a man worth his salt that did not keenly relish it, under whatever hardships, discomforts and dangers. If I have an ardent wish (which alas is not even tempered by a hope) it is to camp out again for a month or two in a savage country —the worst of it is, it is confoundedly bad for collecting, preserving and stowing away specimens.

Happily fate still reserved two more expeditions for him.

The actual supervision of the Gardens was the least part of the official work at Kew, though fuller organisation proceeded apace and he could tell Darwin (November 19, 1867) :

In the Garden I am very busy laying out grounds and planting all over, and doing a vast deal for better or for worse. Also I have induced the Board to put the whole heating apparatus (which has been messed and jobbed till Curator and Foremen are driven wild) into my hands instead of the Surveyor of Works, and I have elaborated a plan for rearranging the whole in 25 houses and 3 Museums, and have put out all for estimates from 3 tradesmen. I shall effect an enormous saving, and have all properly heated too. Also I am planning one new range of houses to supersede 7 old ones, and which will not only save 6 fires, but save Smith and myself a deal of labor.

And though illness deprived him of his Curator for some time :

The whole garden system is in such good order that I can conduct the out of door duties in his absence with pleasure. I can trust all my 7 foremen and Oliver reigns.

The correspondence was vast, and constantly increasing, alike with foreign and colonial establishments, and with contributors of specimens and inquirers seeking identifications of plants or seeds.   The more successful the Gardens, the greater the army of special visitors who broke in on working hours. Kew, as he exclaimed in 1884, had become the house of call for all nations. Again and again, as in 1869, 1875 and 1878, the threatened loss of working time made him deprecate fresh proposals by would-be popularisers of Kew, who knew nothing of its inner workings, to open the Gardens at 10 A.M.

As he told Asa Gray, when he had to report on the matter officially (July 25, 1869) :

I feel it is inevitable *and right* ; but it will require a complete reorganisation and great increase of outlay, and *dish me*. They cannot make it up to me in any way. I do not want more pay—no wish for more—I am one of those who live from hand to mouth, with always a small balance on the wrong side of my bank book, and the more I get somehow the larger that balance gets !

Then, too, no one can help me much—no one can write this letter to you ! and having grown with the growth of this Establishment, I know too much and can do too much : ' Knowledge is power '—till it becomes overpowering. I shall certainly go in for an aid to Smith (the Curator) ; he will else break down. I am of tougher metal and coarser fibre.

But six years later he proclaims to Huxley with grateful astonishment the merits of a Government so rarely anything but grudging towards science : ' My Lords snubbed a deputation in favour of opening the Garden in the forenoon on the ground of its being injurious to Botanical Science ! '

His official position as Director also demanded some sacrifices to Society in the way of dining out in London ; but delightful as such meetings with friends might be, at the Lyells' or the Spottiswoodes', for instance, he had to confess, as he breaks off from writing to Darwin in order to get on with ' Genera Plantarum,' that ' these London dinners are the ruin of science.'

A fixed income and a family of six helped to tie him down. Moreover :

There is no fun in a holiday when you know that work is piling up mountains high at home meanwhile. So I shall carry on, with stunsails alow and aloft, till the end of the chapter. (To Sir W. Macleay, September 4, 1868.)

So to Darwin also he lets himself go in 1869 apropos of the accumulations of correspondence awaiting him on his return from St. Petersburg and the general pressure of official demands upon his time. June 24 :

How I wish I could join you in Wales, but it is impossible. I have a pile of letters that appal myself, and I am not easily frightened—plus a large unopened box of documents and pamphlets accumulated during my absence. I too sometimes wish myself in a tomb, though I hold that the balance of life is always on the side of enjoyment, and that the bitterness of the bitterest loss is an insufficient measure of the enjoyment we had in the object lost.

And August 13 :

I suppose I must read the N.B. [The *North British Review*, where Prof. Fleeming Jenkin's [1] review of Darwinism touched on Hooker also], but I never read now, and am getting very tired of the struggle, not for life, thank God, but of life, and am getting overweighted with duties for the Colonial and Foreign Office which want endless supplies of seeds and forest trees, &c., that I alone can procure, and I only through personal correspondence with people, who would snap their fingers at official requests. The D. of A. [Duke of Argyll, now Secretary for India] has further *requested* me to superintend the publication of a Flora Sylvatica of India, that will give me a lot of trouble. I think he is paying me off for my kick at Nat. Theology Address [Presidential Address at the meeting of the British Association in 1868: see below, p. 118].

While the 'Genera Plantarum' continued its laborious course, and the no less laborious 'Flora of British India' advanced for the first stage of publication in 1872,[2] Hooker was still busy with other botanical work. Some consisted of important works left unfinished by the death of their authors, but which no one else was prepared to complete. Thus he writes to Darwin (November 19, 1867):

As for me, I have been, and am, *sic vos non vobissing* rather too much even for my liking—and I really do like that sort of dilettanteing for my neighbours. I have just concluded Boott's Carices, and am at the distribution of the copies (as much bother as anything). I am printing Harvey's 'Genera of Cape Plants,' and revising the English edition of De Candolle's 'Laws of Botanical Nomenclature,' which will be a good thick pamphlet.

---

[1] Henry Charles Fleeming Jenkin (1833–85), Professor of Engineering at Edinburgh, in 1868 criticised Natural Selection on mathematical grounds. It was, he urged, an infinitesimal chance that an individual with a particular variation should meet with a similarly varying mate and so propagate the variation. At that time neither the frequency and extent of variation nor the actuarial 'expectation' of its reproduction had been investigated.

[2] Seven vols. 8vo. Vol. i., 1872–5. The seventh volume was not completed till 1897.

During these years when several larger books were on hand
the Appendix reveals a smaller number of technical papers
than in preceding years. This period contains six contributions to botanical journals, of which that on *Nepenthes* pre-
figures the 1874 paper on Carnivorous Plants. He contributed
furthermore Rosaceae to Martius' 'Flora Braziliensis,' and the
descriptions of four orders to Oliver's ' Flora of Tropical Africa.'
For the ' Admiralty Manual of Scientific Enquiry ' he revised
his father's article on botany. The year 1869 was mainly
devoted to a work of especial interest, alike as a continuation
of his father's work and a pursuit of his own inclinations.

To Charles Darwin

(End of 1868.)

Now lift up your hands and eyes, when I tell you that I
am doing a British Flora ! My Father's British Flora is just
out of print, and Arnott, his coadjutor, is dead, and both
Balfour and other Scotch Professors have been at me to
write another Flora that shall be better adapted to students'
purposes—more scientific, with references to such observations as yours, with attention to various points that in
structure and morphology [are] not usually noted, and
with rather more complete and uniform Genera descriptions
than the former editions. Bentham's is far the best Flora,
but he slurs over very distinct subspecies &c., his English
names are an abomination to the Professors, and his phrase-
ology is not scientific enough for a class book, that should
impress terms that express definite morphological combi-
nations and structures of flower, fruit, &c. I have long
wished to write a book of this sort, and shall have famous
help from Oliver in all scientific points, and Baker [1] as
to critical species, &c. I should like too to write a good
brief introduction to the principles of plant-classification,
with a map or two of orders such as we have often spoken of.
It is an awful task and you may wish me well through it ;
but by my wife acting as amanuensis the descriptive part
goes on very smoothly. It will, if well done, be the class
book for Edinburgh, Glasgow, Dublin and U. College, London,
and perhaps other schools, and hence have a good sale, a

[1] See p. 242, note.

matter of importance to me now as the children grow up and my income is yearly more inelastic.

This was published in 1870 under the title of ' The Students' Flora of the British Islands ' ; it reached a second edition in 1878, and a third in 1884.

Its aim was to ' supply students and field botanists with a fuller account of the plants of the British Isles than the manuals hitherto in use aim at giving.' Nor is this all that English students and lovers of our native plants owe to him. In 1887, after Bentham's death, he edited the fifth edition of Bentham's ' Handbook of the British Flora.' To quote Professor Bower, ' Both of these still hold the field, though they require to be brought up to date in point of classification and nomenclature.'

In the spring of 1867 Hooker went officially to Paris as Juror in the botanical section of the Exposition. Similarly in 1869 he

was threatened with being sent to St. Petersburg by Govt. to represent British Botanists and Horticulturists (God help them) at the approaching Congress which the Emperor has taken up. I hate the sort of thing, but shall have to go.

He goes on to tell Darwin (March 11) how he is ' mugging up French as hard as he can ' with the help of a French Baron from London two hours daily, besides French novels with his wife and French conversation with Miss Symonds, who was staying at Kew. He was also getting three months ahead with his current duties in hopes of extending his travels from St. Petersburg for a couple of months to the South-East.

In the end, however, the Treasury refused to send him, and he went, accompanied by his wife, independently, and not as a delegate. His tour, which did not take him to any new botanical regions beyond Moscow, lasted six weeks, from May 7 to June 23, going by Berlin and returning by Stockholm.

As a traveller, his delight in flowers and scenery remained vivid as ever, and as of old, he pointed his descriptions of strange places by references to familiar scenes. To his mother he tells of the wistarias in the beautiful gardens at Brussels, trained like pyramidal standard trees and covered with gorgeous masses of bloom ; the Ardennes country, so different from the rest of the route to Berlin, is like Derbyshire, with rocky wooded glens, brawling streams and so forth, while the entrance to Stockholm is through miles of rocky wooded islets and long bays like the Kyles of Bute, clothed with luxuriant forests, and the rocks carpeted with mosses and wild flowers, especially lilies of the valley, anemones and yellow tulips. As for the city :

If you can imagine 5 or 6 St. Peter's Ports of Guernsey on as many rocky headlands fingering in and out in all directions, some into the sea on one side, others on the other side into the fresh water lake, you have some idea of Stockholm.

To his mother also he makes a point of mentioning an interview with the Princess of Wales and Princess Alice at Sans Souci, and their kindly recollections of his father.

On the forty-four hours of train from Berlin an unexpected fellow-traveller made himself very friendly ; this was General Todleben, the defender of Sebastopol, who had been fêted in England some five years before, ' a grand old fellow,' full of wounds and honours—' lame of both legs—an English bullet in *one*, and a French in the other, which shattered the bones ; he has a huge hole in the neck, caused by a bayonet thrust, and a wound *through* the bridge of the nose, from a Turkish poniard.'

At St. Petersburg, thanks to the vast distances and the imperfect arrangements of the secretary to the Congress, it was difficult to find several friends to whom they had introductions, but they met with a warm welcome from Hooker's second cousin, Dr. de Wahl, and General Manderstjerna, who had married another second cousin of his, a de Rosen, and who, as A.D.C. to the Emperor, ' was immensely useful to us—nothing indeed can exceed the kindness of these Scandinavians and Scythians.'

The inefficient secretary, who had taken too much upon his own shoulders, to the disgust and effacement of the well-known Russian botanists, had made no sort of preparations

to receive us, or to introduce us, or in any other way to put us *en rapport* with the Russians. We tumbled into the city and continued for the best part of a week unknown and unrecognised, unable to speak a word of the language, and utterly helpless. We fortunately had many kind personal friends, and I was able through them to do much of this duty. (To Asa Gray, June 25, 1869.)

To Darwin he writes both of science and society, June 6, 1869.

At the Academy I was much interested with old Brandt, the Zoological Director, who declares that all the Bos'es longifrons and Co. are one species, that there is but one fossil elephant, and that the Dinotherium is simpliciter a Mammoth! He believes in you with a vengeance, and I hope I do not misinterpret his ideas. I saw the Rhinoceros tichorinus with skin and hair on head and feet found in East Russia. I was not aware, or had forgotten, that this animal had been found with the soft parts preserved. The Mammoth is certainly a magnificent thing. The skin is preserved in huge masses. But the Syremi Stelleri (I cannot spell the name) of which they have a complete skeleton, is even more interesting ; the texture of all the bones is like the hardest ivory, and the proportion of bone to size of animal I should think exceeds that of any other animal : this and its curious organization rendered it to my eyes the most curious thing I ever saw. The Birds and beasts at the Academy are most admirably stuffed and set up, and the series of varieties of Rodents &c. is most instructive in the variability point of view.

With St. P. I was a good deal disappointed ; it is huge, tasteless and void of all national architecture, except the Churches which are sublime, and the choral services celestial ; beside these our emasculated Anglican service, with its halting imagery and puling intonation, is contemptible ;— if we are to have music and gesticulation and incense and gold and jewels, give them me hot and strong, and the Russo-Greek Church is the place for my money. The altar screens and chapels are literally ablaze with jewels, and every jewel given is a full and perfect sacrifice for some real stunning crime, sin or misdemeanour committed by this most immoral people.[1]

[1] 'As regards the people, their devotion is emotional wholly ; they understand not a word, but go to worship with a blind faith and feeling of the deepest humiliation : it is *Adoration* in fact, pure and simple, not worship in any intellectual sense. We combine (or endeavour to combine) both, and not always harmoniously.' (To his Mother, May 23, 1869.)

The Palaces are gorgeous, but one gets quite sick of French decoration, and endless cabinets of diamonds and rubies. The streets are enormous and horribly paved, distances are tremendous, living very expensive, and the place terribly unhealthy. I never saw such sickly children, and I am assured that the mortality exceeds the births by 6000 per annum; immigration of French and Germans keeps up the population.

The Exposition was very fair, but the arrangements extremely bad. The Emperor was most polite; received a lot of us at his palace of Tsarskoe Selo, showed us himself over the private apartments that were of historic interest, gave us two dejeuners, and at the end decorated a dozen or so of the savants and expositors. As I declined a decoration, he has sent me a pair of beautiful jasper vases from a private mine he has in Tomsk which he reserves for such purposes. I was sorely puzzled what to do about the decoration, not wishing to be rude on one hand, and on the other anxious to avoid it, lest my motives in coming, after the refusal of Lowe to send me, should be misunderstood. So I said that as one could not wear foreign decorations at our Court, I would decline, adding that to Englishmen of science they were not of the same value as to foreigners. His functionaries were most civil about it, and he consequently sent to me the vases and to my two compatriots, Murray and Hogg, each a malachite table.[1]

With Moscow we were enchanted, and could have spent weeks there with pleasure; it is as eminently national as St. P. is the contrary.

To avoid the weariness of the train journey to Berlin, a return was made by Stockholm and Upsala, Copenhagen and Hamburg, then by Hanover, Utrecht, Amsterdam, the Hague and Leyden to Rotterdam, inspecting the Botanical Gardens and their Museums throughout.

[1] 'Medals were distributed by the score, and some thirty or forty decorations distributed. They offered me a high one on my arrival, independent of the Congress, and I declined it on various grounds. The *Gardeners' Chronicle* has stated that it was owing to Dr. Hooker's advice that decorations were not given to Delegates from England. This is utterly untrue. I was never consulted about them, and the decoration offered to me was before the meeting of the Congress and independent of it.' (To Asa Gray, June 25.)

I got very tired of it [he tells Darwin, June 24], though
it was excessively interesting, but the constant packing and
moving got odious. Such lots of people asked for you.
Even at the Hague I found a young Frenchman busy making
notes on the pictures, so I pointed out the Dodo to him, and
he immediately asked me whether it was alluded to in Darwin's
last book on Animals and Plants, which he had read.

At Upsala he received ' a regular ovation and Latin
speech from old Fries, a noble old septuagenarian,' which he
had to answer in English. At Stockholm and still more at
Copenhagen he is struck by the Ethnographical Museums,
illustrating the lives and arts of native races from the Stone
Age to modern civilisation. ' We have nothing in England at
all to compare with it.' At Herrenhausen, near Hanover, the
Palms, which he made this special pilgrimage to see, appeared
' the finest in Europe, far surpassing Kew in number and good
cultivation, and a few in height too.'

Taken all round, this trip was no less interesting than
agreeable, especially in the making of new scientific acquaint-
ances or the renewing of old ones, such as with Professor
Miquel and his family. Nevertheless, railways and hotels
proved most wearisome, and he confides to Asa Gray :

> It will take a great deal to get me to travel again in
> civilised countries. I do long to get into the jungles and
> live in tents or have *my own* cabin at sea.

Minor excursions of this time combine active holiday-
making with the companionship of equally energetic friends.
Thus in April 1867 he spent a fortnight in Brittany with
Huxley and Lubbock, exploring the monuments of the ancient
big-stone builders ; he had another ' perfect April fortnight '
with Huxley in the Snowdon country the next year. In 1870
he tells Darwin of

> the jolly tour I took with Huxley (April 14–24) to the Eifel,
> with my boy Charlie, to whom H. has taken a great fancy.
> We dabbled a little in the Geology, which is most curious,
> took long walks, ate very heartily, and came back quite as
> well as we went.

In 1871, at the age of fifty-three, he accomplished another of his important botanical travels. This was to the little known country of Marocco, and included the first ascent of the Great Atlas.

Marocco, indeed, was the China of the West, jealously guarded from foreign eyes lest the discovery of mineral treasures should bring in the hated Christians. Its botany was even more scantily known than its geography ; the Alpine regions of the Great Atlas, untrodden by European foot, probably held the key to important problems of botanical distribution. As in Sikkim, science was spiced with adventure, and here too Hooker's Himalayan experiences enabled him to deal success-fully with suspicious natives, blending firmness with reason, and never suffering the dignity of a party under Imperial authority to be slighted.

The trip had been planned for some time with George Maw, whose business was pottery and his pleasure gardening and botany, ' the best friend the Garden ever had in many ways:' Hooker knew him for an excellent companion, as well as ' a capital plant-hunter and grower, and fair Geologist.' As plans took shape, John Ball [1] asked to join the party, ' so old a friend and so good a man, that we shall take him with pleasure.'

The general plan is outlined in the following letter.

To Charles Darwin
March 19, 1871.

I am off for Marocco on the 1st, and shall be glad of any commands from you. I go partly to try and bake out my

[1] John Ball (1818–89), man of science, politician, and Alpine traveller. At Cambridge he came under the influence of Henslow, and on his subsequent travels through Europe, did much botanical work, notably a paper on the botany of Sicily, while also studying Glaciers. He was in Parliament 1852–8, and as Under-Secretary for the Colonies after 1855 was instrumental in seeking out the best route for Trans-Canadian railway communication, and in securing Government support for Sir William Hooker's efforts to publish floras of all our colonies on a definite system, which he himself drew up.

He was the first president of the Alpine Club (1857), and in his famous *Alpine Guide* (1863–8) united the scientific and practical points of view. In *Marocco and the Great Atlas* (1878) he completed the story of this expedition, which Hooker had been compelled to lay aside, and his *Spicilegium Florae Maroccanae* (1878) was a classical memorial of their joint researches. In 1882 he also made a five months' voyage in S. America, described in his *Notes of a Naturalist in South America* (1887).

rheumatism, partly in faint hopes of connecting the Atlantic Flora [i.e. that surviving in the Canaries and Madeira] with the African, and (perhaps most of all) to taste the delights of savagery again. Lord Granville [1] has applied to the Sultan for permission and escort for self and Maw to visit the highest points S. of the city of Marocco—but this permit is not yet arrived, and probably will not be granted. We take P. and O. to Gibraltar, thence cross to Tangier and botanize there as far as we can go with safety under the aegis of Sir J. D. Hay [the Minister]. Our future movements will depend on circumstances ; if there is a chance of the Greater Atlas we shall take the steamer to Mogador, and thence head Eastwards. We shall not be gone many weeks, and as the success of the whole project is dubious, I do not care to have much talked about it. I expect Alpine Maroccan Botany to be the most novel and interesting of any W. of Central Asia in the Old World. Of course we take tents, saddles, and such like, soups, tea, old watches, musical boxes, &c., no end of paper for drying plants, and so forth.

I am busy clearing off arrears and prospective work, and have not read your book yet [2]—very much because every one asks me and worries me about it, and it is safest to say I have not even looked into it. I shall take it with me.

The fortnight's botanising in the north was over beaten ground, but served to determine the relations between the floras on either side of the Straits of Gibraltar, and to emphasise the antiquity of the severance between them.

Hooker and Ball rode as far as Ceuta, and crossed to Algeciras, meaning to take the daily boat on to Gibraltar, and so cross again to Tangier. But they were held up, all the steamers having been taken off the line owing to a great bull fight at Cadiz. However, he writes to Professor Oliver on April 12 :

The day at Algeciras was very instructive, as enabling us to compare Spain and the opposite coast at the same season ; the general character of the vegetation was the same, but the civilisation of this, the least civilised country in Europe,

[1] Lord Granville (1815–91), the second earl, was at this time (1870–4) Secretary for Foreign Affairs.
[2] *The Descent of Man.*

is so far ahead of the barbarism of the Moor, that there might
be hundreds of miles between them. . . .

I say that Ball finds this and that, because he beats me
hollow in botanising, and is making a splendid Herbarium.
I find my eyesight quite fails me as a collector ; indeed I have
been remarking for two years now, that I cannot read the
garden labels with my spectacles even, except I stoop down.
Mr. Maw has a marvellous eye also, especially for
bulbs ; and the aggregate knowledge of Ball and Maw, as
to European plants, is simply astounding. Ball knows the
smallest flowering scrap of hundreds of obscure things
(*Medicago, Carex,* and such like), and Maw recognises the
bulbs by leaf, however long the tall grass they grow amongst.

On April 20 they left Tangier for Mogador (April 26–29)
and, reaching the city of Marocco on the evening of May 3,
left it again on the 8th.

Sir John Drummond Hay,[1] our representative in Tangier,
had obtained the proper permit from the Sultan. At Marocco
it was necessary to interview El Graoui, governor of the mountain district they wished to explore, in order to make detailed
plans of travel. Incidentally Hooker was able to play off the
goodwill of El Graoui and the Viceroy, the Sultan's son, against
the discourtesy of the fanatical governor of the town, and to
get the better of him at the first encounter.

Writing to his mother on May 6, he tells of his success so
far ' in botanising and getting about in this barbarous country '
and the delays of the local officials, while reassuring her alarms.
It was only on May 5 that his main object was secured.

Yesterday I went to El Graoui, the Governor of the
Atlas Provinces, whom the Sultan had given orders to
facilitate my travels and objects in every way, and this he
will do now, sending a guard of soldiers and providing me
with food every day for myself and all my party. El
Graoui is an ignorant man, almost a Negro, with a pleasant
face and, like all these people, extremely courteous in his
manner.

---

[1] Sir John Hay Drummond-Hay (1816–93) was first assistant, and then
successor (1845) to his father as Consul-General of Marocco, finally becoming
Minister Resident 1860–72 and Plenipotentiary 1872–86.

The Sultan's power is absolute, where acknowledged,
which is not over more than one-third of his dominions ;
where it is, there is absolute safety of life and property ;
where it is not, he will not allow any one to go under his
orders.

The Mountain people we shall visit for two or three
weeks (till May 29) before returning to Mogador (June 3–7)
and home (June 21) are said to be a very fine race, and as
I have lots of presents for them in knives, scissors, handker-
chiefs, watches, musical-boxes, opera-glasses, &c., I expect
to be well treated and received, over and above the food
and respect which the Sultan's orders ensure. . . .

I am now anxious about getting home,[1] but the chance
of exploring so new and hitherto unvisited and inaccessible
a region as the Greater Atlas must not be thrown away, or
I should be disgraced everlastingly. Nothing short of the
strongest representations on Sir J. Hay's part and the
assurance that I had no political or commercial object and
would not explore the mineral riches of the mountains, together with the assurance that a refusal would be unfriendly
to the English Sultana, whose *Hakim* and Gardener I was !
compelled the Sultan to yield the point, and then Sir J.
H[ay] did not think all secure till he insisted on my being the
bearer of an autograph letter of the Sultan's to the Chief
at Mogador ordering him to put me on the journey to Marocco
and hand me over to El Graoui, to whom and the Viceroy
he (the Sultan) had sent orders to treat me properly and send
me to the Atlas with liberty to pursue my investigations.

With these restrictions, they were unable to examine the
rocks openly, or to secure geological specimens ; while as to
botany, the only acceptable pretext was to give out that they
were commissioned to collect the plants of the country,
especially those useful in medicine. As improved by the
interpreter and camp talk, the belief among their followers
undoubtedly was that the Sultana of England had heard that
there was somewhere in Marocco a plant that would make

[1] Writing to Mrs. Hooker on the 19th, he repeats : ' We are all perfectly
well, but I am most anxious to get home, if only to relieve Smith [his curator].
This is the last expedition of the kind I shall ever undertake. At my age
one has had too much experience and sees too well how much he leaves undone
to enjoy such feats as of yore.'

her live for ever, and that she had sent her own *hakim* to find
it for her.

When, in the course of our journey, it was seen that our
botanical pursuits entailed rather severe labour, the commentary was : ' The Sultana of England is a severe woman,
and she has threatened to give them stick (the bastinado)
if they do not find the herb she wants ! '

Though the chief captain of the escort was a surly, extor-
tionate fellow, apt at contriving local opposition so as to escape
the discomforts of the mountains, Hooker reduced him to order
by threatening an appeal to the Viceroy, and the substitution
of another officer who would take his place and his perquisites.
Thus the Atlas summits were reached from two directions,
and an excellent botanical collection made, in spite of frequently
unfavourable conditions.   The amount of moisture in the air
made drying a great difficulty, and the labour of dealing with
the rich harvest of the hills was increased by the exigencies
of mountain travel.

Those who have had experience in this line [their book
records (' Marocco and the Great Atlas,' p. 273)] know that the
labour of a botanical collector is not light, and in truth it
would be almost intolerable if it were not for its compensating
pleasures.   It often happened that the solitary candle was
in use throughout the entire night, Ball working till two
o'clock or later, when Hooker would rise, more or less refreshed,
and keep up work till daylight.

Save for one day in the low country, early in the journey,
Hooker kept an unbroken record of good health during his
Marocco trip.   Exertion and exposure only increased his
physical fitness, and up the mountain passes his stride was
indefatigable.

So on July 4 he writes to Darwin :

Well, I am back, as usual, like a bad shilling ! after a
very pleasant cruise.   I must get up a readable account of it
in a small volume, and shall publish the Bot. Geog. in Linn.
Soc., I hope with Ball.   The results are mainly negative,
the Atlas being the dying out of the European Flora.

And on the 6th:

> I really believe that the Moraines are the only points in my journey worth much ; except the negative results of no Alpines on the Atlas !

Darwin, who published his ' Expression of the Emotions in Man and Animals ' in 1872, had asked him to make observations on this point among the Moors and Berbers. The same letter adds :

> I tried for expression, but the people are too civilised, and so taciturn and unpleasant with Christians, that their features were too constrained to make anything out of.

Before long, however, the book of Moorish travel was brought to a standstill. He writes to Ball on September 17, 1871 :

> In primis I wanted to tell you, that I see no prospects of my publishing my book on Marocco *within any reasonable time,* and I therefore hope you may publish whenever and wherever you choose at home and abroad. I have a little narrative on the stocks. I had begun and made progress with it, but have been worried out of my life with Ayrton, —Gen. Plant.—and Flora Indica, and in my mother's state of health I cannot count on finishing it for some months.
>
> The Marocco plants I have all ticketed and thrown *roughly* into species.
>
> I should still like to join you in a work on Marocco Botany in any shape you please, and take any of the drudgery.

Later, the additional duties as President of the Royal Society still more emphatically blocked the way, and the book was ultimately brought out by Ball in 1878, using Hooker's journal and fragment of narrative as well as his own and Maw's journal. Hooker also contributed three botanical appendices, on some economic plants of Marocco and comparisons of the Marocco Flora with those of the Canaries and of Tropical Africa. Ball's general description of Marocco Botany has already been mentioned.

The following letter on social economics also arises out of the Marocco experiences.

To T. H. Huxley

August 31, 1871.

My visit to the Moors has led me to think a good deal on the real source of the wealth of a people, and I am disposed to attribute it to the development of mere artificial wants in the main and ' au fond.' Do you know any good book on the subject—French more likely than English ? I have read Adam Smith twice, years ago, and though much admiring did not think that he went to the bottom of the thing, and do not care to read him again now. Marocco is retro- grading, though food is abundant and cheap,—because the state of the people and their laws are hostile to the display of any wealth but that of food, slaves and women. You see there neither fine arms, jewels, horses or furniture—and from highest to lowest, the food is materially the same and the table services of the coarsest and commonest description from the Sultan to the Slave. Grain, butter and honey are hoarded and rotted by the Chiefs, money is buried by every one. The population is stationary or dwindling—the natural increase being checked by wars, climatic famines, the locust and cholera. I doubt if there has been any material change in the country since the Moors were driven back from Spain ; the successes of the Riff Pirates and the Sallee Rovers cannot have contributed materially to the wealth of the country, except through boat building (for they then had fleets, and now have none whatever). Give security to life and property within a ten mile radius of any Port and wealth would flow in and be utilized, not to supply nature's wants, but artificial wants, and most of them imported. Of the many hundred articles I call necessaries of life, very few contribute to my health, sustenance, or daily labor, nine- tenths are to make me more comfortable, luxurious or happy. Corollary.—The more dense the population, the easier it is to find something to do : so the means of obtain- ing a livelihood increase with the population which has to get a livelihood. So it is all bosh to say that it is every year more difficult to find places for your sons. Q.E.D. by J. D. H.

It is interesting to note that the Marocco expedition cost about £110. Though Kew was to benefit by his collections, he was careful to pay everything, even the wages of the Garden man whom he took out. In cases of this kind he preferred to be in a position where no question could possibly be raised. The home carriage of the plants was all the expense that fell on the Gardens.

CHAPTER XXXII

DARWINIAN INTERESTS

THE special interest of 1866 was the discussion of Insular Floras. As one of the crucial points of the great question of Distribution, it had been a frequent subject of discussion in the correspondence with Darwin from the very first. Hooker now chose this as the subject of an address before the British Association at Nottingham. In the course of the summer, while the lecture was being prepared, the correspondence was very full, and is largely quoted in M.L. i. 479 *seq.*

Two hypotheses were in the field to account for the problem involved—one, the more obvious and sweeping, that of continental extensions ; the other, that of migration or accidental transport. Darwin was a migrationist ; Forbes and others pushed the extension theory to excess. In the then state of knowledge, before the soundings taken on the *Challenger* expedition, which put unlimited extensionists out of court, Hooker found either possible, but neither proved. The difficulties were not all met by the arguments adduced, and in discussing the subject he found himself without the stimulus of a thesis to defend, or a side to take.

' I think I know Origin by heart in relation to the subject,' he tells Darwin ; and it was reading the ' Origin ' that had suggested questions as to ice-transport for European plants, betokened by boulders in the Azores, and the European character of the Madeiran birds. But while he deliberately raised all the difficulties that would have to be overcome by Darwin's arguments, he added :

You must not suppose me to be a champion of Continental Connection, because I am not agreeable to trans-oceanic migration. I have no fixed opinion on the subject, and am much in the state regarding this point that the Vestiges left me in regarding species. What we want is, not new facts, but new ideas analogous to yours of Natural Selection in its application to origin. Either hypothesis appears to me well to cover the facts of Oceanic Floras, but there are grave objections to both, Botanical to yours, Geological to Forbes'. I intend to discuss the point with as little prejudice as I can —in fact to d——n both hypotheses, or, if you like, to d——n Forbes's and double d——n yours! for I suppose that is how you will take my *fair play*. I own that it is *most disgusting* to have no side, and I cannot tell you how it dispirits me with the whole thing. I shall make up for it by blessing Nat. Selection and Variation—and they shall be blest— as necessary to either hypothesis, and therefore proving them to be twice as right as if they only fitted one! (July 31, 1866.)

However, on August 6 he adds, ' You need not fear my not doing justice to your objections to the Continental Hypothesis ! ' And on the 7th :

You must not let me worry you. I am an obstinate pig, but you must not be miserable at my looking at the same thing in a different light from you. I must get to the bottom of this question, and that is all I can do. Some cleverer fellow one day will knock the bottom out of it, and see his way to explain what to a Botanist, without a theory to support, must be very great difficulties. True enough, all may be explained as you reason it will be, I quite grant this ; but meanwhile all is not so explained, and I cannot accept a hypothesis that leaves so many facts unaccounted for. . . .
I do want to sum up impartially, leaving verdict to Jury. I cannot do this without putting all difficulties most clearly—how do you know how you would fare with me if you were a continentalist ! Then too, we must recollect that I have to meet a host who are all on the continental side, in fact pretty nearly *all* the thinkers, Forbes, Hartung,[1] Heer,

---

[1] G. Hartung, joint author with Dr. K. von Fritsch of *Tenerife Geologisch und topographisch dargestellt*, published 1867.

Unger,[1] Wollaston, Lowe, (Wallace I suppose), and now Andrew Murray.[2] I do not regard all these, I snap my fingers at all but you ; in my inmost soul I conscientiously say I incline to your theory—but I cannot accept it as an established truth or unexceptionable hypothesis.

And finally, on August 9 :

If my letters did not gêner you, it is impossible that you should suppose that yours were of no use to me ! I would throw up the whole thing were it not for correspondence with you, which is the only bit of silver in the affair. I do feel it disgusting to have to make a point of a speciality, in which one cannot see one's way a bit further than I could before I began. To be sure I have a very much clearer notion of the pros and cons on both sides (though these were rather forgotten facts than re-discoveries). I see the sides of the well further down and more distinctly, but the bottom is as obscure as ever.

After all, the lecture proved a great success despite the cry : ' I am worked and worried to death with this Lecture, and curse myself as a soft headed and hearted imbecile to have accepted it.' ' It cost me much midnight oil and more phosphorus of the brain,' he tells Sir W. Macleay, ' and yet the deuce take it these luminous principles cast very little light on the subject. I delivered myself to about 2000 persons in the Theatre, and gave them a pounding about Darwinism till they jumped from their seats.'

It was, as he promised, a judicial survey of the facts which clamoured for explanation and the rival theories that would explain them. Thus, though in Madeira, for example, the predominant Flora is European, specialising with a certain number of varieties and distinct species, there exists in the heart of the island a set of non-European plants—' Atlantic types '—recur-

[1] Franz Unger (1800–70), an Austrian botanist and palæontologist ; Professor of Botany at Vienna from 1850. He published in 1866 *Die Insel Cypern einst und jetzt.* He was noted for his researches in the anatomy and physiology of plants and in fossil botany.

[2] Andrew Murray (1812–78), naturalist ; abandoned law and took up natural science. F.R.S. Edinburgh 1857 ; President of the Edinburgh Botanical Society 1858 ; Secretary of the Royal Horticultural Society 1860 ; F.L.S. 1861 ; its scientific Director 1877. Wrote on botany and entomology.

ring in the Canaries and Azores, with unique genera, species and varieties occurring on the outlying islets. This is the wreck of an ancient flora, now surviving in Asia and America, which is only found as fossil in Europe, having succumbed to the Northern and Eastern floras that now hold that continent.

The Canaries, so much nearer to Africa, have not more than a sprinkling of African plants ; nor have the Azores, so much nearer America, more American plants than the others. The tropical Cape de Verde Islands, while showing some affinities with the Canaries and Madeira, have a mainly Saharan flora.

Going further afield, the indigenous flora of St. Helena in the isolation of the S. Atlantic, is mainly S. African. Linked with this are the scanty plants of Ascension, though emphasising the effects of isolation. In the S. Indian Ocean, the Kerguelen Land flora is clearly Fuegian, though the island lies nearer to S. Africa and New Zealand than to S. America, and its most notable plant, the *Pringlea* or Kerguelen Land Cabbage, has no ally in the Southern hemisphere.

Thus with all their peculiarities—the result, on Darwinian principles, of isolation in survivals and modifications—no island flora is an independent one. What was the link that made immigration possible, whether ancient or recent ? This question Hooker called ' the *bête noire* of botanists.' Now geology did not favour extensions to the volcanic islands of the ocean : the absence of land mammals and batrachians, and sundry great gaps in the flora, also told against continental extension. The difficulties of ocean transport in relation to prevailing winds and currents, the vitality of seeds in sea-water or in the crops of birds, or in mud sticking to their feet, the chances of land insects reaching Oceanic islands [1] had been matter of long discussion with Darwin.

If he could pronounce for neither theory, still his balance of opinion appears from his words :

[1] He writes to Dr. Cunningham (see p. 80), May 18, 1867 : ' Your observations on the abundance of terrestrial insects seen so far out at sea alive are very curious. Pray collect all such evidence carefully and collate it, it bears so strongly on Darwin's theory of populating Oceanic Islands from Continents.'

The great objection to the continental extension is, that it may be said to account for everything, but to explain nothing ; it proves too much ; whilst the hypothesis of trans-oceanic migration, though it leaves a multitude of facts unexplained, offers a rational solution of many of the most puzzling phenomena that oceanic islands present : phenomena which, under the hypothesis of intermediate continents, are barren facts, literally of no scientific interest— are curiosities of science, no doubt, but are not scientific curiosities.

He wound up with an amusing apologue upon the reception of new ideas.

You have all read of uncivilised races of mankind that regard every month's moon as a new creation of their gods, who, they say, eat the old moons, not for their sustenance, but for their glory, and to prove to mortals that they can make new ones ; and they regard your denial that their gods do monthly make a new moon as equivalent to denying that they could do so if they would.

It is not so long since it was held by most scientific men (and is so by some few still) that species of plants and animals were, like the savages' moons, created in as many spots as we meet them in, and in as great numbers as they were found at the times and places of their discovery. To deny that species were thus created was, in the opinion of many persons, equivalent to denying that they could have been so created.

And I have twice been present at the annual gathering of tribes in such a state of advancement as this, but after they had come into contact with the missionaries of the most enlightened nations of mankind. These missionaries attempted to teach them, amongst other matters, the true theory of the moon's motions, and at the first of the gatherings the subject was discussed by them. The presiding Sachem shook his head and his spear. The priests first attacked the new doctrine, and with fury, their temples were ornamented with symbols of the old creed, and their religious chants and rites were worded and arranged in accordance with it. The medicine men, however, being divided among themselves (as medicine men are apt to be in all countries), some of them sided with the missionaries—many from spite to the priests, but a few, I could see, from conviction—and putting my trust in the latter, I never doubted what the upshot would be.

Upwards of six years elapsed before I was again present at a similar gathering of these tribes ; and I then found the presiding Sachem treating the missionaries' theory of the moon's motions as an accepted fact, and the people applauding the new creed!

Do you ask what tribes these were, and where their annual gatherings took place, and when ? I will tell you. The first was in 1860, when the Derivative doctrine of species was first brought before the bar of a scientific assembly, and that the British Association at Oxford ; and I need not tell those who heard *our* presiding Sachem's [1] address last Wednesday evening, that the last was at Nottingham.

Hooker's qualms about lecturing happily came to nothing.

Nottingham : Tuesday, August 28, 1866.

DEAR OLD DARWIN,—The whole thing went off last night in very good style. The audience were well fed and conformable, they followed the whole lecture with admirable good nature, and were sent into fits by the conclusion. I made myself well and easily heard without unreasonable effort, and have all the more reason to bless my stars that I have not earlier given way to popular lecturing, for which I am already besought! I never was so glad to get a thing out of hand and mind, and now I must in the course of the winter cast it into scientific form for publication.

I am awfully busy as you may suppose, and only just beginning to enjoy the fun.

Huxley is getting on splendidly in Section D. He returned thanks for my Lecture in the most skilful, graceful and perfect way. I never heard anything so hearty and thoroughly good—no coarse flattery or fulsome praise—but an earnest, thoughtful and, I believe, truthful eulogy of what he thought good and happy in the treatment of the subject, with a really affectionate tribute to myself.

Ever your affectionate,
JOS. D. HOOKER.

[1] Sir W. R. Grove.

Darwin replied (August 30) :

I have seldom been more pleased in my life than at hearing how successfully your lecture went off. Mrs. H. Wedgwood sent us an account, saying that you read capitally and were listened to with profound attention and great applause. She says when your final allegory began ' For a moment or two we were all mystified, and then came such bursts of applause from the audience. It was thoroughly enjoyed amid roars of laughter and noise, making a most brilliant conclusion.' I am rejoiced that you will publish your lecture, and felt sure that sooner or later it would come to this ; indeed it would have been a sin if you had not done so. I am especially rejoiced, as you give the arguments for occasional transport with such perfect fairness ; these will now receive a fair share of attention, as coming from you, a professed Botanist.

Hooker's response includes a description of the President's address, and the aim it had in view.

Kew : Tuesday, September 4, 1866.

Dear old Darwin,—I am very proud of your letter. I thought I might have exaggerated the effect I produced on my audience, and did not like to think too much of it. I do now pray to be another ' single speech Robinson ' ! I wish you could have heard Huxley's éloge, it pleased me so immensely, and was so much better than all the applause. I had set my head, heart and mind on gaining yours, Grove's, Huxley's, Tyndall's and the Lubbocks' (especially Lady L's !) good opinion, I cared little for other people's. I have not seen Tyndall since, nor heard how he liked it. He came up to me in the forenoon, evidently most anxious for my success, and questioned me about it. When I told him that it was a written discourse and that I intended to *read* it, his countenance fell and I saw he was cut. He turned away first, but came back and with great delicacy and loving-kindness gave me some hints, to learn passages by heart, &c. (I had done this copiously already), and to put myself *en rapport* with the audience, &c., &c. I saw in short that he prognosticated a dead failure, and I spared no pains that afternoon in preparing myself to succeed in his eyes. I hope I did.

Huxley made a capital President of Section D. and was
very conciliatory, prudent and amusing too. I really heard
few papers, and none of any consequence. Wallace's was no
doubt the best in our line.

As to Grove's address, I can quite understand your
disappointment at the Species part of it—I only wonder he
did it so well, for when I have talked the subject with him,
he has shown so little appreciation of its *difficulties*, that I
was rather pleased than otherwise that he thought it needful
to discuss it. I knew too that he had left it all to me—
indeed he, on accepting the Presidentship, *retained* me as
champion of the cause. I wished him at the Devil, but felt
flattered at the selection, puzzled as I was then, and am now,
to make out why he should have thought me worthy of so
responsible a post on so critical an occasion. I had always
a notion that he looked on me as a very weak vessel, and my
branches of Botany as mild child's play. Then too he had
no hints or instructions for me. I was ' to back him up '
and ' to carry Darwinism through the ranks of the enemy '
after he had sounded the charge ; and whether or no his
' Continuity ' Address was well received. In short I was
a stinkpot, which he was to pitch into the enemies' decks,
whether sinking or swimming himself.

I am so glad you are succeeding with *Acropera.* I should
not like you to be beat by any Orchid.

The lecture was published in instalments in the *Gardeners'
Chronicle* of January 5, 12, 19, and 26, and afterwards reprinted
in pamphlet form. But the lecture was only a stage in the
discussion of this crucial question. Further criticisms were
sent by Darwin (see M.L. i. 492–4, ii. 1–4), who was the only
critic able to detect an anomaly on page 9 of the reprint.
Hooker replied :

I see you ' smell a rat ' in the matter of insular plants that
are related to those of [a] distant continent being common.
Yes, my beloved friend, let me make a clean breast of it—
I only found it out after the Lecture was in print! and by Jingo
it has played the very devil with me ever since. I have
been waiting ever since to ' think it out ' and write to you
about it coherently. I thought it best to *squeeze* it in, any

how or where, rather than leave so curious a fact unnoticed. I am glad that you are the only one who has twigged it and its importance.

Here may be added two extracts from Hooker's letter of February 4, apropos of the reprint.

The only thing I do not like, and could not conscientiously consult you about, was the passage about a wise Providence ordering &c., &c., or something of that sort (I forget the words, it matters little).[1] It is bosh and unscientific, but I could not resist the opportunity of turning the tables of Providence over those who think and argue the contrary of its intentions, and showing those who will have a Providence in the affair, that yours is the God one, theirs the Devil's. I always felt that if I had to print the Lecture, I should wish these passages cut out, but that this would be dishonest, so it e'en went forth in G.C., and now will further.

What I mean about Providence is this :—
I think and believe that all reasoning upon the subject is utterly futile, that there is no such thing in a scientific sense—but that whereas those who deal in it hold that the theory of fixed types is the only one consonant with a belief in a Providence, I hold that they are wrong and that the theory of continuity and variation is the only one consonant with the belief.

Bentham is doing Umbelliferae for Gen. Plant., and finds that the two remarkable umbelliferous genera of Madeira, *Monizia* and *Melanoselinum,* are only species of *Thapsia,* a Mediterranean genus of most remarkable and exceptional habit. Now this is one of those cases of Genera *confined* to the Island, being then created out of a Continental form ; the genus I suspect not having ever existed on the Continent. It appears to me that it will always be difficult to say whether a genus *that has continental allies,* is an Insular development,

---

[1] ' By a wise ordinance it is ruled, that amongst living beings like shall never produce its exact like ; that as no two circumstances in time or place are absolutely synchronous, or equal, or similar, so shall no two beings be born alike ; that a variety in the environing conditions in which the progeny of a living being may be placed, shall be met by variety in the progeny itself. A wise ordinance it is, that ensures the succession of being, not by multiplying absolutely identical forms, but by varying these, so that the right form may fill its right place in Nature's ever varying economy' (p. 12).

or an old, now extinct Continental genus ; the utter want of
fixed system upon which genera are and *must always* be
formed, will always throw insuperable obstacles in the way
of this inquiry—it is easy enough with regard to the Laurels,
and other things having no continental affinities.

Many more botanical relations required careful analysis
before definite conclusions as to origin could be reached, and
Insular Floras were a study of much concern during the following years, capped with the wish that it were possible to write
a new Essay on the subject. He dealt with it again, however,
in the 1881 address on Geographical Distribution.

At the same time he was always ready to meet his friend's
challenge with some excellent scientific fooling. To test the
hypothesis that bright seeds attract birds which thus help in
their wide dispersal, he recommended Darwin to pass some
through a fowl. Darwin thereupon experimented with seeds of

the Mimoseous tree, of which the pods open and wind spirally
outwards and display a lining like yellow silk studded with
these crimson seeds, and look gorgeous.

But he was disappointed.

I gave two seeds to a confounded old cock, but his gizzard
ground them up. . . . Please Mr. Deputy Wriggler explain
to me why these seeds and pods hang long and look gorgeous,
if Birds only grind up the seeds, for I do not suppose they
can be covered with any pulp.

Hooker then replied (December 14, 1866) :

The scarlet seed is that of *Adenanthera pavonina*, a native
of India. I am well acquainted with its self and with its
habits from the year $-\infty$ [minus infinity]. At that rather
(geologically) early period it was a low bush, and the seeds
were all black (an allied species has seeds half black and half
red, which proves this statement). Gallinaceous birds were,
after its creation, introduced into the part of the Globe where
I first saw it, and these sought the seeds with avidity : so
that finally only those vars. of climbing habit survived
and thus got out of the way of the gallinaceous birds (which
are not perchers) ; its chances of dissemination being thus

diminished, the tendency to scarlet next developed itself
in excess, being determined by the perchers (whose gizzard
would not grind the seeds) and which were attracted by the
color, and soon led to the extinction of all but the full scarlet
forms.

Nonsense apart, I should suppose that it is to imitate a
scarlet insect and thus attract insectivorous birds, or frugi-
vorous perchers, of weak digestions, that the color is acquired.
The plant is a very common Indian one, and it would be
easy to ascertain how far it is a prey to birds.

Early in 1867 he was urged to accept nomination at the
next meeting of the British Association for the Presidency of
1868. This honour he at first declined ; as he wrote to Darwin
(February 4, 1867) :

The fact is that I have an insuperable aversion to high
places ; the acceptance would have been bad dreams in
anticipation for 18 months, and a downright surgical operation at the end of it ! I believe I inherit this from my father,
who never would put himself forward, or be put forward,
and I am sure it *paid in the end*. I was also actuated by the
fact that I can see no way to a good ' Address.' I played
out my trump card at Nottingham, knowing that if I were
called upon to be President (which I had already good reason
to expect) and accepted, I was throwing away my last chance
of success. Lastly it would stand terribly in the way of my
work—both Genera Plantarum and Insular Botany. Here-
above is a pretty dose of egotism even from one friend to
another.

Darwin's approval was a relief, and he begged for support
in his resolution, if the subject cropped up, against his friends
of the *x* Club, whose joint attack he had much difficulty in
beating off, though ' with a heavy heart, for I would fain have
obliged them.' They dwelt on the scientific need of it, especially
after the choice for 1867 of social prestige instead of scientific
distinction, in the person of the Duke of Buccleuch.

But though he doubted whether the post, with all its dis-
tractions from research, was one for the *most* scientific men of
the day to aspire to, he had to yield to the insistence of all

the botanists he respected, and on March 14, ' in a state of
deep dejection,' bids Darwin pity him. ' However, in for a
penny, in for a pound, and if I am in good health and keep
a so at the time, I will do my very best.'

The matter that most interested him at this time outside
his own work, was Darwin's ' Variation of Animals and Plants
under Domestication ' (published January 30, 1868) with the
speculation ' which will be called a mad dream,' said its author,
of Pangenesis. Several letters bear on this.

To Charles Darwin

March 20, 1867.

I am dying to understand Pangenesis, that haunts me at
night. Huxley told me that he had referred you to something
of the kind in Bonnet. I cannot conceive a Pangenesis
without a correlative Panexodus (the Great God Pan is not
dead yet, that's clear). What I mean is this, that if every
previous attribute (infinitely subdivided) of all its ancestors
exists in an organism, any of these may come out (turn up)
in its progeny—but I suspect I am talking nonsense to you.
I was so long blind to the force of the derivative hypothesis,
that I always feel too inclined to take your views *au coup de*
(I forget what ; I am coaching up French, hard, for Paris
Exposition).

Darwin answered that Pangenesis by no means implied
that every previous attribute of all the ancestors exists in an
organism, ' but I fear my dear Pang. will appear bosh to all
you Sceptics.'

Until the middle of November, Darwin was very busy with
proofs of the book, and Hooker, knowing this, abstained from
writing ; but after the book appeared, he wrote at some
length.

To Charles Darwin

[This replies to Darwin's letter of the 23rd, C.D. iii. 77.]

February 26, 1868.

I am extremely obliged for your candid record of opinions
on Pangenesis. I was talking it over with Huxley, who made

a very clever remark, so deuced clever that I cannot quite
recollect it, and still less write it down—to the effect that the
cell might not contain germs or gemmules, but a potentiality
in shape of a homogeneous mass, whose exact future condition, or the exact future of whose elements, depended
on an impulse consummated at moment of evolution. I
suppose he meant, just as a crystallizable compound, that
presents various isomorphic forms, depends on some unknown
influence for the crystalline form it ultimately does take—
but this is only my guess at his meaning, I will try and get it
more clearly. I fear you will laugh at my density, but I
cannot see that in Pangenesis you are doing aught but
formulating what I have always supposed to a fundamental
idea in all development doctrines—viz. the transference to
the progeny of any or every quality (property) the parent
possessed ; or at least the potentiality of reproducing these
qualities—and it was the inconceivability of grasping this
idea that was always a great barrier to my accepting the
development doctrine. You transmit this potentiality in
a cell—you diffuse it from that cell throughout the whole
living organism, and you regard a spermatic cell as neither
more nor less charged than others with this potentiality.
Of this point I am not quite sure, I must read up every point
again of your argument. This was always with me an
essential condition of the Development Doctrine, and I do
not see what you gain by putting it in an imagery of germs
and gemmules analogous to a chemist's atoms. A chemist's
atoms are useful imagery, for they convey definite ideas of
proportions and have an exact meaning as *relative* values.
If Biology enabled us so to convey definite ideas through
your gemmules, they would have their use—but inasmuch
as organisms are not given to unite in definite proportions,
I do not see what you gain.

Be all this as it may, I regard your Pangenesis chapter as
the most wonderful in the book, and intensely interesting—
it is so full of thought, of genuine *mind* ; and you do so love
it yourself. I should not care a farthing were I you what
people thought of it. Not one Naturalist in a hundred can
follow it I am sure. Spencer, Huxley, and Lubbock (if he
has time) may. I have not yet mastered it. The ' throwing
off gemmules ' is hard to hold in head, as a real vital process

—if you say that each cell ' diffuses an influence,' that *is*
intelligible ! ! !

I wish I could help you anent sexuality — the male
element affecting the mother plant or animal is your strong
point, nothing I suppose can explain that, but what is or is
akin to Pangenesis.

Next morning. After re-reading all this vaporous letter,
I shall try to answer your last page in a concrete manner
(to adopt the current literary slang). I can neither answer
nor explain nor account for any of the facts you put to me,
except on the supposition that every mother cell thrown off
by the parent and destined to reproduce the kind, must
contain within itself and diffuse throughout every cell to
which it gives rise any or all the properties of the parent.

I have put this in another form on a separate piece of
paper—how does it tally with Pangenesis ? Please *postulate*
Pangenesis as I have my crudity.

To this Darwin replied on the 28th (see C.D. iii. 81); and
Hooker wrote again on March 3 :

Tuesday.

Your letter has delighted me, and I want to answer it at
length, which I shall do from Norwich where I go for two
days on Friday. I now quite understand your Pangenesis.

I wrote all the first part of my letter by fits and starts,
and no doubt made a precious muddle. It is all true what
you say that the satisfaction which Pan. may give will depend on mental constitution, or as I call it, Mental Parallax.

I never *arrived* at any such *conclusion*, nor did I ever in
any way shape my thoughts or reason towards it, because it
was simply self-evident. What I have always instinctively
held and thought and never could help seeing is, that in all
cases of descent ' *all the properties of the parents are transmitted in the one cell* ' (a wonder of wonders it always was
and is to me) *and were diffused to every part of the future
offspring*—my examples being the reproductive power of
single cells of most lower orders of plants, of *Bryum andro-
gynum* and many Ferns, and of *Malax paludosa* on the one
hand, and the fertilized cell of all organisms on the other.

I do not see how any one who ever thought of the matter
of descent could escape this conclusion—that the properties

are not only transmitted by the one cell, but diffused thereafter throughout the future individual. It is so hard to
conceive this, or rather to grasp this, for individuals, that
when you come to extend it to species, genera, orders, classes,
&c., it may very well form a stumblingblock to the accepta-
tion of the ' Development of Species Doctrine '—as it did
with me.

So far I have instinctively held your doctrine but never
as a postulated or formulated theory or hypothesis—it was
merely as part of the doctrine of descent, the most ordinary
phenomena of descent being simply inconceivable to me
without it. Much less did I ever ask myself whether the
most obscure facts of reproduction were explicable on any
other hypothesis.

So far we are agreed ; when you come to your atoms and
germs and gemmules and so forth we do not part company,
but move off a little—I do not see my way. Tyndall believes
he feels atoms, as firmly as St. Paul believed he saw Christ.[1]
I do not say that atoms do not exist, but I rather suppose
that they may be like minutes of time or inches of space
or any other *purely arbitrary quantities*. Your doctrine of
atoms thrown off in no way furthers my perceptions or
advances my ideas.

I have again read Part I. of Pan. and with literally renewed delight. I do think Pan. as fine a thing as you ever
writ, the idea of germs and atoms notwithstanding. As to
[my] laying claim to having by any logical process or reasoning
arrived at such a doctrine, in any scientific sense, i.e. by
testing it as you have done, do not for a moment entertain
it. I always held, *as part and parcel of the development
doctrine*, that the potentiality of the parent was not only
transmitted by a cell, but indefinitely diffused therefrom,
and hence, as I told you from the first, I could not see what
there was new in your theory, except the idea of atoms, &c.,
which I could not grasp.

To Asa Gray

March 12, 1868.

Have you read Darwin's last book, and what do you
say to Pangenesis ? I have gone deeply into the *whole
philosophy of the Subject*—there then.

[1] Cp. p. 359.

I must say that the Pangenesis chapters are in themselves admirable—so careful and so good ; but what he gains by clothing what appears to be a simple, necessary and inevitable belief with all who accept the derivative hypothesis, in a garb of atoms, germs and gemmules I do not see. When I accepted the derivative hypothesis, I accepted the fact, that each individual must contribute by a cell to its progeny more or less of any or all the properties of all its forefathers ; and that such properties, or the potentiality to reproduce them, must be diffused from that cell more or less throughout the mass of the plant. E.G. a single cell of tip of leaf of *Malaxis paludosa* will reproduce a whole *Malaxis paludosa*, with any or all the properties of its parent and grandparents so diffused through its mass from that parent cell, that each of the cells of its leaf will do ditto. This always appeared to me a fundamental doctrine in the history of propagation of individuals from parent to offspring. If you accept this for the propagation of individuals, and reduce the origin of species to the same category as the propagation of individuals comes under, you must accept it for these too.

A better instance than *Malaxis* is *Begonia phyllomaniaca*, and a better still any cellular Alga that propagates by any constituent cell. This power of packing into a cell the potentiality of an indefinite number of the indefinite properties of its ancestors, is as much beyond our comprehension as atoms, or ethics, or time, or space, or gravity, or God. And as any definite conceptions of God are to be had only and solely by anthropomorphising him or his attributes, so are our only ideas of the potentiality to propagate all qualities by a cell, only to be formulated by calling the contents of that cell atoms, gemmules, and so forth. My upshot is that it is not necessary to formulate or postulate such subjects at all, and better not to do so.

To Charles Darwin

May 20, 1868.

You greatly underrate the interest of your [book]; it is capital reading, putting aside all question of its matter, which will, if foreigners deign to read it at all, do you more credit in their eyes than all your other works put together. (I have not read a quarter of it yet.) Bentham has, and now I think unreservedly, acknowledged himself a convert to Darwinism! this, I quite expected, would be the case with many : a few will still hold back and flaunt the ' rag of protection ' till your next part appears, holding that cultivation is no argument, when, the said rag being worn back to the rope and no longer visible, they will gracefully haul it down.

. . . I have finished the Reign of Law [by the Duke of Argyll] with utter disgust and uncontrollable indignation [for] his suppressed sneers at you. . . . I like a man to sneer at me out of malice and envy, but cannot stand a man's sneering at me from atop of a high horse. The preliminary reasoning on the principles of flight appears to me radically unsound. The idea of God being compelled to dab on rudimentary organs *to keep up appearances*! as it were, is very droll. He writes extremely well and expresses himself with admirable facility—in fact he has a fatal facility for handling things he does not fully understand, and which he has not the time, and probably not the power to grasp the principles of.

I am used up and have nothing more to say. I feel my barrenness of scientific matter to communicate creeping over me every day now, and the tide of scientific literature is already up to my knees. The time was when I had now and then something to communicate that you cared to know— that is all changed now, and I feel very low at times about it. I begin to despair of doing anything, even at Insular Floras again, wherein I see that I could still do much. Perhaps when this Norwich meeting is over I shall feel more at ease. I would give 100 guineas that it were over, even with a failure, a fiasco, or worse. The address is *nowhere* yet, and I look on its prospects with a loathing that cannot be uttered. To-morrow I go to see Fergusson to *encourage* him about his prospective Lecture at the Meeting! God pity us both— the blind leading the blind. I shall have to play the hypocrite with a vengeance.

These letters reveal how greatly his mind was taken up with the progress of Darwinism, while he was still casting about for a good subject for the Norwich address. He had already written to Darwin on April 7 :

I get more and more unhappy about the Address as the time draws on. Nothing on earth would induce me to do a thing so *damned indelicate* as to force such a position on an unwilling soul. Science might go to the Devil before I would do so by an enemy even. You see I am working up myself to the starting point.

I have often thought of a History of great steps in Botany, but it would take a deal of reading, and I have no time for any, and then when we came down to later years I should offend everybody. And after all, should a President's Address be a ' scientific thesis ' ? I think not. Who ever consulted such addresses, or regarded such as *authorities* ?

Finally, as the pressure of administrative work at Kew forbade any more recondite study, he fell back on the Darwinian interests that engaged him as his main theme, with others that were specially topical.

To Charles Darwin

July 12, 1868.

If I cannot get to Down before you go to the Isle of Wight, do you think that I might see you there for a day in August ? I shudder at the thoughts of bringing you my Address and at the same time cannot bear the cowardice of not doing so.

I have utterly broken down in every attempt to compose a solemn scientific harangue, or a philosophical résumé of the progress of Botany, or a dilatation on the correlation of Botany with other sciences. I cannot possibly give the three clear weeks of continuous application that such subjects demand, and I am going to say so. I have sketched out a sort of see-saw discourse on several subjects that are germane to the Association and the Norwich Meeting par excellence : some of them are practical (as Museums), others theoretical, as the influence of your labors on Botany—and Pangenesis (God help it)—others touch ' Tom Tiddler's Ground,' as the early history of mankind apropos of religious teaching and the International Prehistoric Congress, which part I feel convinced you will advise me to burn if I read it to you, which is hence doubtful, as I shan't burn it, but will read it if I burn for it. I do not intend to show any part of the Address to my wife, from the conviction that she would burn

it all, nor shall I worry myself by telling anybody else anything about it. I have written very little of it as yet, and I
will not go touting about for matter or illustrations.

The work was completed under a great strain. His youngest
child nearly died.

It did indeed make the Address repulsive [he writes on
July 29], but on the other hand it druv' me to it and made me
work. You know the horrid way a man who has his work
at home, loafs about the house when a child is ill.

I have just concluded the rough sketch of what I shall
say (if not hissed down), for by George I would hiss anybody
who would eruct such stuff as I have written *under any other
circumstances* than a Presidential martyrdom.

Apart from a description of the ideal organisation of a local
museum, such as that at Norwich, with its educational possibilities before the still unattained epoch when teachers should
be trained in science, the main theme, as has been indicated,
was the progress of the ' Origin ' and an estimate of Darwin's
contributions to botany, alike in observation and in fertile
theory. For this task none was so well qualified as Hooker
himself, and none could take greater pleasure in it. In the
' Fertilisation of Orchids,' in the almost more wonderful discoveries of the twofold and threefold mechanisms to ensure
cross-fertilisation in the primrose, the flax and the loosestrife,
and in the ' Habits and Movements of Climbing Plants,' he
found a wealth of observation which made the greatest of
living botanists ' feel that his botanical knowledge of these
homely plants had been but little deeper than Peter Bell's,'
while at the same time it opened up entirely new fields of
research and discovered new and important principles that
apply to the whole vegetable kingdom.

Then, turning to Darwin's ' Animals and Plants under
Domestication,' so eagerly awaited as one of the *pièces justificatives* of the ' Origin,' he exclaimed :

It is hard to say whether this book is most remarkable
for the number and value of the new facts it discloses, or
for its array of small forgotten or overlooked observations,

neglected by some naturalists and discarded by others, which, under his mind and eye, prove to be of first-rate scientific importance.

One of Darwin's characteristic faculties, as Sir James Paget put it, was this power of utilising the waste materials of other men's laboratories. As to the theory of Pangenesis, Hooker frankly admitted that the hypothetical ' gemmules ' invoked as the mechanism of inheritance, were ' not proven,' but like other assumed mechanisms which escape the senses, could serve as an orderly basis for reasoned investigation till a more plausible hypothesis be brought forward. Meantime, whatever the scientific value of the ' gemmules,' the statement of the theory was ' the clearest and most systematic résumé of the many wonderful phenomena of reproduction and inheritance that has yet appeared.'

To the critics of Natural Selection, whether on metaphysical or physical grounds, he made firm reply. Those who reject it on metaphysical grounds with customary appeal to the *odium theologicum*, are, so far, outside the pale of scientific criticism.

Having myself [he said] been a student of Moral Philosophy in a Northern University, I entered on my scientific career full of hopes that Metaphysics would prove a useful mentor, if not a guide in Science. I soon however found that it availed me nothing, and I long ago arrived at the conclusion, so well put by Agassiz, when he says, ' We trust that the time is not distant when it will be universally understood that the battle of the evidences will have to be fought on the field of Physical Science, and not on that of Metaphysical.

On the score of geology, there were still some, a dwindling minority, who relied for criticism on the assumed perfection of the Geological Record. This gave occasion for the well-known tribute to Sir Charles Lyell, who after upholding the fixity of species for forty years, was led by the researches of his old pupil to abandon it in the tenth edition of the

'Principles.' 'I know of no brighter example of heroism of its kind,' he exclaims, and adds in telling phrase :

Well may he be proud of a super-structure, raised on the foundations of an insecure doctrine, when he finds that he can underpin it and substitute a new foundation ; and after all is finished, survey his edifice, not only more secure, but more harmonious in its proportions than it was before ; for assuredly the biological chapters of the tenth edition of the ' Principles ' are more in harmony with the doctrine of slow changes in the history of our planet, than were their counter-parts in the former editions.

To the astronomer critics he pointed out the limits of mathe-matical infallibility : as was said on another occasion, mathe-matics is a mill which grinds out results very accurately, but the value of the results depends on the material put into the mill. Did the physicists, calculating (on somewhat uncertain data) the rate of the earth's cooling, assert that evolution claimed an impossibly long period, the biologists replied that they took their time from geology, and if the geological clock needed speeding up, they would automatically follow suit.

Finally he turned to the new science of Pre-historic Archæ-ology, which was holding its first International Congress at Norwich. It was a science which led men where hitherto they had not ventured to tread—where science clashes with the old accepted Scripture chronology, where separation can hardly be made between its physical and its spiritual aspect. Yet as Truth (in Disraeli's words) is the sovereign passion of mankind, religion and science must speak peace to one another.

But [he added] if they would thus work in harmony, both parties must beware how they fence with that most dangerous of all two-edged weapons, Natural Theology ; a science, falsely so called, when, not content with trustfully accepting truths hostile to any presumptuous standard it may set up, it seeks to weigh the infinite in the balance of the finite, and shifts its ground to meet the requirements of every new fact that science establishes, and every old error that science exposes. Thus pursued, Natural Theology is to the scientific man a delusion, and to the religious man a snare, leading too often to disordered intellects and to atheism.

Thus only, with mutual recognition that, as Herbert Spencer had put it, the ultimate power of the universe is inscrutable, can religion and science proceed at peace on their common but disparate search into the whence and whither of man's existence, that passionate aspiration of the stanzas from Francis Palgrave's poem, ' The Reign of Law,' with which the Address concluded.

He wrote at once to Darwin :

It is all well over, though I broke down in what I least expected—voice—the place was atrocious to speak in, and the desk so badly placed that I could with difficulty read—so about the middle I got husky, but recovered towards the end and am said to have done the agony bits and the poetry very well. I modified two or three things, left out the allusion to Gray's being superseded, and something else.

All is going off well. Huxley spoke nicely after it of our sea-faring life, and Tyndall warmly of *you* and me being types of ' unconscious merit ' ! ! ! !

To Charles Darwin

August 30, 1868.

A thousand thanks for your letter—a regular sunbeam it was. What a pother the papers kick up about my mild theology ! An Aberdeen one calls me an Atheist and all that is bad : to me, who do not intend to answer their abuse, misquotations, garbled extracts and blunders, it is all really very good fun. There were gentle disapproving allusions at Kew church to-day I am told ! I am beginning to feel quite a great man !

Tyndall most assuredly did couple our *names* most prominently, unequivocally and unmistakeably, as the *two* modestest men in science ! ! !

. . . The Cathedral service was glorious, the Anthem was chosen for *me*, ' What though I know each herb and flower,' and brought tears to my eyes, and Dr. Magee's discourse was the grandest ever heard by Tyndall, Berkeley, Spottiswoode, Hirst and myself.

. . . I forgot to tell you that I read all over about you to Thomson, who thought I had 'drawn it very mild.' Bentham and Oliver do not think that I said a word too much.

The astronomers do not quite like my allusions to them. I had a long talk with Adams,[1] who is a most charming fellow. He will not agree with me, but won't give me any definite answer. He does not allow that Astronomy is in fault in the matter of the sun's distance—no more it is in one sense ; but astronomers are, and the science of Astronomy is simply the exponent of astronomers' knowledge.

For the toil of the concluding day, with more than twelve hours of continuous committees and councils and lectures and social functions, punctuated with speechmaking at each, he paid with a sleepless night and consequent fatigue. The redeeming point was the evident enjoyment of his wife, who was able to make her gracious presence felt everywhere. She

did enjoy it all most thoroughly, and proved herself ' as strong as a woman.' I am sure that without her the whole thing would have been to me simply intolerable.

As he told Macleay (September 4), ' Without her I really should have been miserable, I was so disappointed at her not being present at Oxford and Cambridge when I was doctored. I feel I do want somebody who can help me to take so much more than my deserts.'

To Charles Darwin

January 18, 1869.

I have got tremendously pitched into for quoting (Spencer) in my address, as I expected ; and for declaring the power above us to be inscrutable. My last flagellation is from Pritchard the Astronomer, who blames me for not being complimentary enough to the Almighty. I have answered him that I think the concluding three verses of Palgrave's poem is enough for the occasion.

[1] John Couch Adams (1819–92), the Cambridge astronomer and co-discoverer of Neptune ; President of the Royal Astronomical Society 1851–3 and 1874–6 ; Copley Medallist 1848.

To Charles Darwin

(Undated.)

Babington is 'very much surprised at Dr. Hooker's advocacy of Darwinian views at Norwich, and observes that it has greatly disappointed many of Dr. Hooker's friends and well-wishers.' I feel like the Parrot which was in the habit of saying in a tone of great contempt after the family prayers were over, ' My God,' or like the Turk in Hogarth's picture, calmly smoking his pipe as he gazes in through the window of a Church where the congregation are in a state of religious excitement.

Other questions arose directly from Darwin's wide-ranging work. Such were the cause of variation, the transmission of acquired characters, the Descent of Man (Darwin published his book in February 1871), and the introduction of life to our globe by meteors.

The following are criticisms on passages in Darwin's fifth edition of the 'Origin,' published in May 1869 (see especially p. 151), on which he asked Hooker's assistance (December 5, 1868). Nägeli in his ' Entstehung und Begriff der Naturhistorischen Arten ' objected that Darwin's ' useful adaptations ' are exclusively of a physiological kind—i.e. showing the formation or transformation of an organ to a special function. He knew of no morphological modification in plants which could be explained on utilitarian principles. (See M.L. ii. 375, where the editors point out that this is a truism, since Natural Selection is assumed to work upon structures which have a function, while on the other hand a difficulty arises from the various meanings given to the word ' morphological.')

To Charles Darwin

January 15, 1869.

I do not quite like the starting by shirking the question of what is a ' morphological character '—you *imply* that it is a term of indefinite meaning. You talk of what ' *he* calls M. characters ' and of what ' *I* presume likewise to be M. characters.' I think that non-scientific readers will at once

say, ' How little these men know of what they write so much
about, when their fundamental terms have no definite
meaning.' All characters, i.e. all departures from a given
structure, are and must be morphological. All originate in
the fact that every individual varies from its parents ; and
this from being subject to ' the direct and definite action
of the conditions of life '—(an admirable definition ; Weis-
mann's is not intelligible to me, if sense at all).

P. 3 at A. This is very mildly put ; would it not better
meet Nägeli's objection, which seems to point to *histological characters* (and to which and symmetry he probably
confines his use of term ' Morphology '), to add ' nor do we
know the uses of all the special tissues of any one organ.'

P. 4 at B. Furthermore, though these arrangements of
leaves are reducible to mathematical laws and might hence
be presupposed to be the most constant of all the laws of
vegetable growth, and to be absolute and irrefragable, they
prove not to be so—shewing that even here is variation
which no one could call progressive ! capable of transmission
and ready for the action of selection.

To Charles Darwin

January 18, 1869.

I do not see either how you can avoid using the term
' morphological,' but can you not use it without leaving the
reader to suppose that it has no definite sense : a very slight
modification of what you say when alluding to Nägeli's
limitation of it would effect this, I think.

I should not have implied that variations in leaf divergence
were transmitted, but that they might be inherited (like
any other variation)—but that if such a variation occurs,
there is no reason why it should not be transmitted, and if
transmitted why N.S. should not determine its prevalence
and subsequent constancy in a specific mark. If you have
kept my letter, please look and let me know if I have implied
more than this. I should extremely like to graft a Chestnut
branch if such a variation from the normal leaf divergence
occurred, and sow the seed [which] a similar branch produced.

I know no case of ovules differing in position in the
different flowers of one plant, except perhaps in monsters.

I think Henslow gave me a Primrose in which the ovules
were basal (as normally they should be) in most flowers,
and they were parietal in others. It was otherwise
monstrous.

I was much struck with your conclusion that the near
approach to uniformity in an organ throughout a group
implied its functional inutility—it is no doubt true. I had
a sort of gleam of this truth when considering the fact you
once pointed out to me, that the calli of *Oncidium*, though
essential to the plant for physiological purpose, are still so
very variable. It then suggested the converse which you
have so well evolved. But what an apparent contradiction
it involves—or paradox at least—that classification and
system is founded on the least useful modifications, and
this explains a very common observation, that Physiology,
i.e. the operations of active plant life, does not much help
the systematist. And yet there is something uncomfortable
in the idea that system is based on modifications the active
exigency of which is no longer in play. It seems frightfully
paradoxical to say that the quinary arrangement of Dicotyledons is a matter of no moment to the Dicotyledon as such :
and yet that this is true is proved by the fact that such
Dicots. as are ternary or quaternary are as good Dicots.
as their quinary brethren. It is a tremendous upset to
Owen's *doctrines*, or rather his *writings*, for these in no
way rise to the dignity of doctrines. The ' law of necessary
correlation ' is—nowheres.

Monday (January 1869).

Just one last thought anent Genetic characters of no
value to the plant : is not the fact, that characters of primary
value in system are so often of no use, an argument in favour
of your conclusion, that such characters as are of no use, if
not in any way detrimental, are not necessarily eliminated
but may be retained *ad infinitum* ?

On the other hand, is it not an argument against the
theory of characters acquired by the *individual* being hereditary—thus, if hereditary modifications that never come
into play do not die out, is it likely that non-hereditary
modifications brought into play by the *individual* (for its
own special use) should be transmitted ?

The following answers Darwin's letter of August 7 (M.L. i. 314, where it is partly quoted in a footnote) apropos of Hallett having found some varieties of wheat which could not be improved in certain desirable qualities as quickly as at first.

August 13, 1869.

I did not mean to imply that Hallett affirmed that all variation stopped, far from it, he maintains the contrary, but, if I understand him aright, he soon arrives at a point beyond which any further accumulation in the direction sought is so small and so slow that practically a fixity of type (not absolute fixity however) is the result. Also that coincident with this point is that the plant is also very slow to vary in other directions than that it was bred to accumulate. This, I supposed, correlation would account for, viz. that while you are *knowingly* accumulating in one direction, correlation obliges you *unknowingly* to be accumulating in others.

To Charles Darwin

July 17, 1869.

I have had a queer Strasburg Mathematician here with me this morning about Phyllotaxy, &c., and we have had a long chat, during which he has expounded certain queer aspects of scientific theories—e.g. that the original primordial cell, from which all organized creatures were developed, was that of Man, inasmuch as it has attained its highest development in Man. I told him that Pangenesis would demand this, for the original cell must have contained the original gemmules which enter into the composition of every cell of Man.

To Charles Darwin

July 18, 1870.

I had a long talk with the D. of Argyll last night, with whom I dined, about origin of man, and found him a ' cleft stick ' about Wallace, believing him to be right in the fact about man, but allowing that he must be wrong in his argument ! (he had not read that paper of Wallace's). What a clever little beggar it is ! But I cannot follow his views about man, or quite see what he would have us believe. His chief quarrel with the ' Origin ' is that you do not state

that the order of evolution is preordained, though he believes
that you would admit this. I told him that I did not think
this was any business of yours—that you did not pretend
to go into the origin of life, only into its phenomena. I
could not, before his wife and children especially, go into this
matter, and avow my own (and I suppose yours) belief that
all speculations on preordination are utterly idle in the
absence of better materials than theologies and cosmogonies
supply us with—that in fact the whole subject is beyond
the range of our conceptions.

March 26, 1871.

The success of your book ['The Descent of Man'] delights
me to hear of—5500 copies! it is tremendous. I hear that
ladies think it delightful reading, but that it does not do to
talk about it, which no doubt promotes the sale—the only
way to get it being to order it on the sly! I dined out three
days last week, and at every table heard evolution talked of
as an accepted fact, and the descent of man with calmness.
I take it to read in P. and O. in intervals of sea-sickness.[1]

A man called yesterday who had been up to my most
distant passes in the Himalaya—the first man to do it
since 1848!—a Mr. Elwes,[2] formerly I believe a Guardsman,
who has taken enthusiastically to Ornithology—one of the
Blanfords accompanied him. I must be vain enough to
tell you that he found my book a 'miracle of accuracy,' and
that he could find nothing I had not taken note of. I dare
say that Blanford[3] will tell a different story! 'Sufficient
for the day is the Kudos thereof.'

I fear for Huxley, who (his wife tells me) is running a
fearful rig of work. . . .

What I most dislike is, this unsettlement for any
future scientific or self-sustaining work: his love of exer-
cising his marvellous intellectual power over men is leading
him on—and on—and on—God knows to where—here he is

---

[1] I.e. on the forthcoming voyage to Marocco.
[2] See i. 271.
[3] Henry Francis Blanford (1834–93) studied as a geologist and from 1855–
62 was on the Geological Survey of India. From 1862–72 he held a professorship at the Presidency College, Calcutta. Then, having devoted himself to
meteorology, he was appointed meteorological reporter first to Bengal, and
later to the Government of India till he retired in 1888. He became F.R.S.
1880, President of the Asiatic Society of Bengal 1884–5.

now, at Owen's College, Manchester, on Friday, and lecturing
again to working men at Liverpool yesterday, and to be
back in London to-night!

The following deals with Sir William Thomson (Lord
Kelvin's) address at the Edinburgh meeting of the British
Association in 1871. In it he had suggested that life had
been brought to the earth by ' seed-bearing aerolites.' Huxley,
who was present, welcomed the implicit acceptance of evolution
by such a theory, however improbable in itself, and whatever
the criticisms passed on Darwin's views of the working of
evolution.

To Charles Darwin

August 5, 1871.

I have been reading W. Thomson's [1] address, and am
anxious to hear your opinion of it. What a belly-full it is,
and how Scotchy! It seems to be very able indeed, and what
a good notion it gives of the gigantic achievements of mathematicians and physicists—it really makes one giddy to
read of them. I do not think Huxley will thank him for his
reference to him as a positive unbeliever in spontaneous
generation—these mathematicians do not seem to me to
distinguish between un-belief and a-belief—I know no other
name for the state of mind that is traduced under the term
scepticism. I had no idea before that pure mathematics
had achieved such wonders in practical science, and I wonder
how far Thomson's statements will be contested. The total
absence of any allusion to Tyndall's labors, even when comets
are his theme, seems strange to me.

The notion of introducing life by Meteors [2] is astounding
and very unphilosophical, as being dragged in head and

---

[1] William Thomson (1824–1907), afterwards Lord Kelvin, was the son of
James Thomson, who became Professor of Mathematics at Glasgow. Inherit-
ing his father's powers in an intensified degree, he entered the college at
the early age of eleven, and thus, though Hooker's junior by seven years,
managed to be in the same class with him.

[2] Huxley wrote to Hooker (August 11): ' What do you think of Thomson's
" creation by cockshy "—God Almighty sitting like an idle boy at the seaside
and shying aerolites (with germs), mostly missing, but sometimes hitting a
planet!' Following which Hooker adds to Darwin on the 15th: ' Huxley calls
Thomson's the " Cock-shy theory "—God makes a cockshy of the world. I
hear that he baited T. awfully in section D.'

shoulders apropos of the speculations of the 'Origin' of
life from or amongst existing matter—seeing that Meteorites
are after all composed of the same matter as the Globe is.
Does he suppose that God's breathing upon Meteors or their
progenitors is more philosophical than breathing on the
face of the earth ? I thought too that Meteors arrived on the
earth in a state of incandescence,—the condition under
which T. assumes that the world itself could not have
sustained life. For my part I would as soon believe in the
Phœnix as in the Meteoric import of life. After all the
worst objections are to be found in the distribution of life,
and the total want of evidence of renewal by importation
such as meteoric visitations would suggest the constant
recurrence of. The quotation of Herschel's very early
objection to Nat. Selection is surely not fair, if indeed
correct, and again highly unphilosophical—what real objection is it to Nat. Selection that it should be too Laputan ?
Surely Columbus and the egg might have occurred to him,
and to call this (Herschel's objection) ' a most valuable and
instructive criticism ' ! I wish he, or any one else, could tell
me the logical significance of the phrase ' *the* argument from
design.' I understand design well enough,·but ' *the* argument '
from it is just what the arguer pleases to argue. He means
I suppose ' a certain conclusion from design,' assuming always
that his idea of design is God's idea too. Again, how the
Deuce can ' proofs of intelligent design ' (in Nature) show us
' through nature the influence of a free will ' ?

What will Huxley say to the phrase ' metaphysical *or*
scientific ' ? If Metaphysics are anything, they are in his
opinion as good science as aught else scientific. Are the
Commentators on Paley a bit worse than Paley himself ?

I am pleased with his praise of old Sabine, because I
think there has been too much disposition to overlook his
really great scientific merits, and his indomitable perse-
verance—just as I think Humboldt is underrated now-a-
days. Well, these were our Gods my friend, and I still
worship at their shrines a little.

I am hammering away at a narrative of my Marocco
trip (see p. 95), and find it harder work than ever ; I suspect
that systematic and descriptive writing hurts head and hand
for other writing, though you preserve freshness of style
with any amount of purely scientific writing.

The letters that follow are concerned with the attack made on Darwin by Mr. St. George Mivart [1] openly in his ' Genesis of Species ' and anonymously, but from internal evidence indubitably, in the *Quarterly Review*. The reply made by Huxley in the *Contemporary Review* for November 1871 (see ' Collected Essays ' ii. 120) under the title of ' Mr. Darwin's Critics,' was one of the most deadly in the history of controversy. Mivart, *inter alia*, had attempted to show that evolution, properly garnished with limitations as to man acceptable to the priesthood, had been accepted in advance by the Fathers of the Roman Church. Turning up the authorities quoted, Huxley found the precise opposite stated, and with delicious irony was able to pose as the defender of Catholic orthodoxy against a heterodox son of the Church, while combating his philosophy and psychology. At the same time he was full of cold anger against the man who was writing privately to express his friendship for Darwin, yet, as the anonymous Quarterly Reviewer, treated Mr. Darwin in a manner ' alike unjust and unbecoming,' sneering at his candour and the mutually generous relations between him and Wallace over the enunciation of Natural Selection.

Writing to Mrs. Darwin on September 16, apropos of her daughter's marriage to Mr. Litchfield, Hooker also refers to the impending reply.

I had not seen the marriage in the paper—I hope all passed off with the least possible ' putting about.' I am accused of once having uttered the horrid sentiment, that I would rather go to two burials than one marriage, any day.

I heard from Mr. Huxley yesterday—threatening to ' pin out ' Mr. Mivart, for his insolent attack on Mr. Darwin,

[1] St. George Jackson Mivart (1827–1900), F.L.S. 1862, Sec. 1874–80, F.R.S. 1869, biologist and brilliant anatomist, who, having embraced Roman Catholicism, formally opposed Darwinism, while supporting evolution by the side wind of derivative creation. But though he employed his great knowledge and polemical adroitness in the service of his spiritual advisers, his liberalising philosophy finally led to his excommunication. Mivart's biographer in the *D.N.B.* speaks of the criticisms mentioned in the text as ' an assertion of the right of private judgment which led to an estrangement from both Darwin and Huxley.' This is not the fact. True that they resented, and Mivart privately apologised for, the personalities of his *Quarterly* article ; the breach took place three years later owing to a repetition of the offence in a peculiarly hurtful form.

and adding that he was reading up Suarez and the Jesuit
Fathers and found that Mivart either misquoted or misunderstood him, and he (H.) proposed to vindicate the
Catholic Fathers! What an irony his life is becoming. I
call him a ' Polemician.'

To T. H. Huxley

Kew : September 17, 1871.

DEAR H.—There is an irony in your going in for Suarez
in Scotland—were not his works burnt in public by James I ?
I have just glanced again at Mivart's last chapter ; it is
curious for the illustrations it introduces pro and con his
views, which seem to have been sought with zeal and produced without discretion. The pages on the attributes of
an Almighty God are hopelessly vague and commonplace,
and I never had much respect for the God who *originates—
derivatively.* His ' God inscrutable ' is no better or worse for
me than Spencer's ' God unknowable ' whom he won't have !
Given a God who can be in two places at once—and it
is mighty little odds whether you call him inscrutable or
unknowable in reference either to his disposal of events, or
to our consideration of him or his attributes !

The whole scheme of ' Derivative Creation ' in its religious
aspect always seemed to me a poor makeshift—a sweet to
the physic of evolution ; and I should indeed be astonished
if the Jesuit Fathers' conceptions of creation squared with
this. All they contended for, I assume, was that God made
beasts and birds, &c. out of solids, and not out of vacuum.

I see that as far as possible Mivart gives Providence a wide
berth—well for him. If I understand him aright, he believes
in an original creation of Soul in every man (not a derivative
one)—it is a pity that he had not expounded that idea ; he
could scarce have escaped the pitfall of Heredity in reference
to the attributes of the Soul, i.e. of all we know of what we
call Soul—which I take it is simply a mixed idea.

I shall be most curious to read your paper.

To Charles Darwin

Kew : Monday (November ? 1871).

DEAR DARWIN,—I return Huxley's article [' Mr. Darwin's
Critics ' : *Contemporary Review*, Nov. 17] which I have read

with all the admiration I can express. What a wonderful Essayist he is, and incomparable critic and defender of the faithful. Well, I think you are avenged of your enemy—but are not the happier for that—though you must be for the spirit and body which the avenger has given to the subject, and above all for the grand use he has made of your own arguments for confuting your enemy. What you must feel, and always feel, is, that peculiar and quite unreasonable bitter sorrowing which a man excites who praises you to your face and abuses you behind your back. Why should this excite anything but contempt at worst, or pity at best? And yet there is no man with generous emotions but feels more sad and sorry over such treatment than either angry or vindictive.

The Psychological passages seem to me to be wonderfully clear and good—how tight he clothes a difficult idea in language. I was particularly struck with the paragraphs on Neurosis and Psychosis—consciousness and its physical basis—but really it is difficult to single out either passages or subjects, all is so good and there is so much power and acumen in the treatment of every branch of his subject—you may call it an Essay, a critique, an exposition, a discussion, an enquiry, or what else you wish—you may read for one and all of these aims.

The exposition of Mivart's presumptuous ignorance in citing the Catholic Fathers is delicious—that's the last pitfall the poor devil expected to be snared into. The *tumbling over* Wallace is, however, if not an equal feat, a far, far greater service to Science.[1]

The appeal to conscience in the matter of the clergy and the 6 days is very powerful, and must make many a

---

[1] Wallace in his *Contributions to the Theory of Natural Selection*, 1870, p. 359, urged that Natural Selection accounted for the evolution of man's bodily frame from the simian stock, but that from this point on some extraneous power had inspired him with his mentality, and with a future purpose in view had provided the mere savage with a brain disproportionate to his requirements, whether compared with civilised man or with the brutes. Thereafter the struggle for existence among men had operated mainly through their mental abilities, with the consequence that the human body retained comparative fixity of type.

Against the argument adduced Huxley quoted Wallace's own words in *Instinct in Men and Animals*, describing the vast calls upon the intelligence even in a savage's life, and pointed out that by parity of reasoning wolves likewise had brains too large for their requirements, and must therefore have been supernaturally bred up to prepare them to become dogs.

poor devil wince in the pulpit. And all the quiet contempt with which he treats the Squires and Parsons is extraordinarily humorous in its manner.

Well, the article has been a god-send to me, for I am very low, and cannot get my spirits up, about my poor Mother's state. I have just returned from Torquay. I am also in the most detestable position that a scientific man, or an officer, or a gentleman can be with my Lord and Master, Ayrton, whom I have officially denounced to the First Lord of the Treasury for his conduct to me and to Kew; and I need not say that our lives are not the happiest after such an explosion! How it will all end God knows. I began the battle with heart and spirit—and gloried in it—but my Mother's condition has poisoned the whole, and I left my sister very ill, even for her—so I am in a state of utter disquiet, not caring a farthing what the Treasury or Ayrton do. What a poor lot we men are—a woman would be twice as rational as I am, under twice the hard lines. God bless you, dear old friend.

The reference in the preceding paragraph is to a long-drawn and perfectly gratuitous quarrel between Hooker and his official superior who was appointed to the Board of Works in Mr. Gladstone's Government of 1870. The story of this is told in a subsequent chapter.

CHAPTER XXXIII

THE PRESIDENCY OF THE ROYAL SOCIETY

FROM 1873 to 1878 Hooker was President of the Royal Society. Pre-eminent as a philosophic botanist, successful in administration, trusted for his cool judgment and knowledge of men, he was clearly marked out for the most honourable and most responsible service which Science can claim of her representatives, and on the resignation of Sir George Airy, he was nominated by a three-fourths majority of the Council.

The duty was one which he could not refuse, but which he undertook with great reluctance. It meant the restriction of his own botanical work; it meant a good deal of speech-making and an inevitable pressure to accept the long over-due knighthood for thirty years of official services which so far he had, on various pretexts, managed to evade. *Nolo episcopari* was his cry when he knew he was to be nominated.

To Charles Darwin

January 12, 1873.

I quite agree as to the awful honor of P.R.S., and its inestimable value to me in my position, and under existing circumstances—but my dear fellow, I don't want to be crowned head of science. I dread it—'Uneasy is the head,' &c.—and then my beloved Gen. Plant. will be grievously impeded. The dream of my later days is to be let alone, where I am and as I am—I want no higher position, no dignities nor honors. I cannot undertake to represent

132

Science officially, and refuse the inevitables that flow from
it, or come with it, and stick to you for the rest of your life.
This may be all very selfish, but so it is. I would fain die
as I now live.

By the way, have you seen the lovely compliment that
R. Strachey pays us [1] at the end of his paper on the Scope
of Scientific Geography, in the last number of Geog. Soc.
Proc.—p. 450—has he not ' pointed his moral and adorned
his *tail* ' with our names ! I was and am astonished indeed.
I hope Owen will see it.

I sent Gladstone a Wedgwood medallion of my Father,
and he writes so nice and characteristic a letter that I must
enclose it for your perusal.

Ever, dear old fellow, Yours,

J. D. HOOKER.

The nomination was made on January 16. Hooker's
acceptance was marked by a new procedure.

Sabine had held the Presidency from 1861 to 1872 ; Airy,
the astronomer, who was already seventy-one years of age,
during 1872–3. Following up some previous discussion of
the matter, Hooker made it a condition that his tenure of
office should not be of indefinite length, but ' only from year
to year,' thus ensuring elasticity to the working of the Society,
without the breach of continuity involved in a fixed short
term.

As President, he did much to consolidate the organisation
of scientific interests which had so long been his great concern,
Since the forties, the Royal Society had been steadily becoming
a more strictly scientific body. By its original constitution
it had found social and financial support in the admission of
distinguished persons with presumably a general interest in
' natural philosophy '—*naturae curiosi* as some learned societies
called them. With the advance of professional science, this
element of support became less valuable. It crowded out
working men of science. In administrative presidential
positions it was more ornamental than useful. Now, under

[1] As being the two most modest men of science.

Hooker's auspices, this 'privileged' non-scientific element was further restricted.[1]

Next to be noted is the completion to date of the catalogue of all the scientific papers published by the Society, the lack of which had been a serious handicap to scientific workers. The cost of preparing the catalogue was borne by the Society ; that of printing by the Government. By 1875 six volumes had already been issued ; two more on the same scale covering the decade 1864–1873, and including 95,000 titles, appeared in 1876.

A subsidiary improvement in the publication of the Transactions made it easier to obtain separate copies of the papers when required.

Most memorable, however, and of widest benefit to the Society at large, was the idea which took shape as the final

[1] In this connexion mention may be made of Hooker's membership of the Philosophical Club. At the time of his election to the Royal Society (April 22, 1847) there was much dissatisfaction owing to the indiscriminate election of men of rank and fashion to the Fellowship, often to the neglect of real workers in science. A strong reforming party, in opposition to the then president, the Marquis of Northampton, and many influential Fellows, had carried a resolution in favour of the present system of election, with a limit of fifteen each year, selected and recommended by the Council from the whole body of candidates. Hooker himself was among the last batch elected under the old rules.

The leading advocates of this and other reforms were Sir Henry De la Bèche, Sir William Grove, Leonard Horner, and Sir Charles Lyell, all intimate friends of Hooker's. On April 12, 1847, the reformers and a number of friends met and decided to found a dining club to be called ' The Philosophical Club '—its objects the discussion of questions affecting the prosperity of the R.S. and the provision of opportunities for the early announcement and discussion of new discoveries. It was resolved that the number of members be forty-seven, to commemorate the date of foundation, and at the third meeting, on June 3, Hooker was one of those co-opted to make up the number. He was very regular in his attendance both before and after his journey to India, and in 1854 became Treasurer.

The Club played a useful part in its informal work for science, though for Hooker its personal attraction was eclipsed by that of the x Club from 1863 onwards. So long as the meetings of the R.S. were held in the evenings, both the Philosophical and its antitype the Royal Society Club (founded in 1743) were well attended ; but when the R.S. meetings were held in the afternoons, there was a great falling off. Proposals were made for fusion of the two clubs, for the old grounds of difference had disappeared. Of the three original members surviving, Sir William Grove and Sir William Bowman somewhat reluctantly consented ; Hooker stood out, saying that there might still be work for the Club to do in resisting abuses. His final regret, in a farewell letter to the Treasurer, Professor Judd, after the last meeting he attended on April 24, 1890, his 43rd anniversary, was the inevitable move with the times which had substituted more elaborate menus for the old time simplicity of the dinners.

act of his Presidency. The high fees payable by the Fellows
pressed hard on men of small means and often prevented them
from coming forward as candidates for election. In half a
dozen cases, indeed, the fees had been remitted by special
resolution.
The inception of the scheme is told in the following letter.

To Charles Darwin

June 9, 1878.

I have long had at heart a scheme of reducing the
monstrous heavy fees (in future) of F.R.S. by establishing a
'publication fund,' which by relieving the income of part
of the expenditure on publication, would eventually set free
the desired amount for reduction of fees to the standard
of other Societies.

To this end I induced my old friend Young of Kelly to
give me £1000, and the Council has entered into my scheme,
accepted the £1000 as the first contribution to the fund
and sanctioned my taking any honest course towards increasing it.

Spottiswoode has gone into the matter for me, and finds
that £10,000 would suffice, and further he thinks that an
effort should be made to raise this sum at once amongst
the Fellows by subscriptions varying from £50 (which is
as much as I can afford) to £1000, out of which a few swells
may be cozened !

I need hardly say that I am ambitious to confer this
boon on the Society and on Science before I leave the Chair.
I am sure of your sympathy, but can well suppose that you
cannot help and shall not be surprised to be told so.

The response was immediate. There was no need to go
tentatively and fund the first subscriptions until the minimum
of £10,000 should be reached. In a few weeks £8000 was
subscribed in large donations, and the remainder of the
minimum soon followed. Mr. T. Phillips Jodrell, who had
given the Royal Society a large sum for research purposes
in the same spirit of liberality with which he had endowed
science at the Universities and built a physiological laboratory
at Kew, consented to transfer £1000 to this fund ; Sir Joseph

Whitworth[1] gave £2000, Sir W. G. Armstrong[2] (afterwards Lord Armstrong) £1000, the Duke of Devonshire, Mr. De la Rue, Messrs. Eyre and Spottiswoode, Dr. Siemens,[3] and the Earl of Derby, £500 each, and Dr. Gladstone [4] £250. The remainder was contributed by thirty-two Fellows of the Society.

Thanks to the fund thus raised, new Fellows were relieved of the entrance fee and paid an annual subscription of only £3. No man henceforth need be kept outside the Society on the score of money.

To Hooker's administrative work at Kew was now added the ordinary administrative work which falls to the P.R.S., not to mention the fact that he was furthermore *ex officio* a Trustee of the British Museum. Council days are described by him as ' great pulls, 1–6 P.M. *continuous*—then dinner, followed by the Meeting at 8½.' The internal affairs of the R.S. covered a wide range of business, on this occasion including a long negotiation with the Treasury as to the tenure of the rooms at Burling-

---

[1] Sir Joseph Whitworth (1803–87), the mechanical genius who deliberately set himself to become a perfect craftsman by entering one great engineering shop after another as a workman, thereafter setting up as a toolmaker in Manchester. His great discovery of how to make a truly plane surface was the basis of a device for ensuring the accuracy to $\frac{1}{10000}$ of an inch of his standard measures and gauges, which revolutionised engineering. His new rifle and cannon, the result of patient experiment at the request of the Board of Ordnance, anticipated modern developments, but were rejected by the officials. After this ' Battle of the guns ' Whitworth made his other great discovery in the forging of steel under hydraulic pressure when fluid. His patient thoroughness in scientific investigation has been compared to Darwin's. He became F.R.S. in 1857. The bulk of his great fortune was finally devoted to educational and charitable purposes.

[2] Sir William George Armstrong (1810–1900; knight 1859, baron 1887) applied his mechanical genius to many inventions, especially in hydraulic machinery and the manufacture of guns, and was the founder and organiser of the great Elswick Works at Newcastle. Elected F.R.S. in 1846, he continued his scientific researches not only in mechanics, but also in electrical science. He was President of the British Association in 1863 when it met at Newcastle.

[3] Sir William Siemens (1823–83), metallurgist, electrician and inventor. Three of his brothers, like himself, turned to the practical applications of science. In 1843 he left Hanover to dispose of an electroplating invention in England, becoming naturalised in 1859. With his name are associated such diverse inventions as a water-meter, the regenerative furnace, insulation for submarine cables, the dynamo, as well as the application of electric lighting. He was President of the British Association in 1882.

[4] John Hall Gladstone, Ph.D., F.R.S. (1827–1902), chemist and physicist, was Professor of Chemistry at the Royal Institution 1874–7 ; President of the Physical Society 1874–6 and of the Chemical Society 1877–9. At this time he was also a member of the London School Board.

ton House, when Hooker succeeded in renewing tenure on the original terms of 1856. President and Council are responsible for the grounds on which the medals are awarded each year, and for scrutiny of the merits of the fifteen candidates whom they recommend for election annually. Now also the whole question of fixing this number at fifteen was redebated. The Society's trust funds for research have to be allotted and recommendations made for the allotment of the Government grant. The President sits on the several committees which had charge of these and other questions taken up by the R.S. either on its own account or as advisers to the Government. These matters are multifarious, for the P.R.S. is, so to say, the Attorney-General of science. At the head of his Council and supported by hard-working secretaries, he keeps in touch with the progress and needs of science on the one hand, and on the other, speaks for science in the official world, giving advice or drawing up the instructions required, when public money is to be given for exploration or research or in pensions to science workers or their dependents. In effect, we see Hooker consulted alike about great things and small ; now deep in complex details as to the work done by our Observatories and Admiralty Hydrographer, when it was proposed to establish a new Meteorological Department, and now asked to recommend a head gardener for the Phœnix Park in Dublin.

Per contra, he was able on occasion to use the dignity of the Royal Society as a lever against official apathy or niggardliness. There was a small house next the Kew Herbarium, which had been empty for half a century, and standing as it did within the demesne, could not be let to anyone unconnected with the establishment. For more than a year Hooker vainly urged the Treasury to have it done up at a trifling cost and assign it, rent free, to J. G. Baker, the Herbarium First Assistant. After two refusals he went in person to one of the under-secretaries, and ' insisted on a reversal of the refusals, telling him that Mr. Baker was an F.R.S., that so was the Chancellor of the Exchequer (Sir Stafford Northcote), and that if it was not done, a representation would be made by some Fellows of the R.S. to the C. of E. and a scandal ensue. After yet another refusal, Hooker had the long-deferred pleasure of seeing the wasteful penny-wise policy reversed to the advantage of one whom he described as ' the most hard-working useful man [whose] services to this establishment have been most self-sacrificing.' (To Darwin, Dec. 14, 1878.)

Among the more important matters which passed through Hooker's hands as President were the arrangements with Government for the expeditions to observe the solar eclipses of 1875 and 1878 and the transit of Venus, also in 1875, which involved the transport of the astronomers and their instruments to stations on the other side of the world, and any miscalculations in which meant the loss alike of public money and scientific results. To the latter expedition, naturalists and geologists were also attached, who made full investigations on the remote Oceanic islands of Rodriguez and Kerguelen's Land.

For the Polar expedition of 1875 under Captain (Sir) George Nares,[1] naturalists were selected, and a scientific manual drawn up, ' The Natural History, Geology, and Physics of the Arctic Regions,' which, with the Scientific Instructions, made a book of over 800 pages.[2] A further suggestion was carried out, that deep sea research should be made on board the store ship of the expedition on its way to Davis' Straits.

Another such matter was the publication of the meteorological and magnetic observations which had been carried on since 1851 in an observatory in Travancore, the first volume appearing in 1875. This specially enlisted Hooker's interest and help, and the valuable results won a Royal medal in 1878 for the observer, Mr. J. Allan Broun.

Of special importance again was the Naturalist's Report

[1] Captain Sir George Strong Nares, R.N., K.C.B., F.R.S. (1831–1915), had already had experience in Arctic travel and in surveying the coast of Australia and Torres Straits, before being appointed to the command of the *Challenger* on her great scientific voyage, 1872, from which he was recalled in 1874 to lead the attempt to reach the North Pole in the *Alert* and the *Discovery*, 1875–6.

[2] ' Have you any botanical suggestions for the Arctic Expedition ? If so, please let me have them *at once*. I recommend special attention to insect action and fertilisation, hybrids, &c., sowing earth from Icebergs. Also to try experiments on germination of seeds exposed to various degrees of cold.' (To Darwin, April 15, 1875.)

from the Voyage of the *Challenger*,[1] and the care and working
out of the valuable collections made. Here, as usual with
scientific expeditions, the Royal Society had furnished the
Government with ' instructions ' for the scientific plan of
campaign, had found the workers and published results. On
the return of the expedition the Natural History Department
of the British Museum, moved by Professor Owen, laid claim
to the collections and the right of describing them, though
the British Museum authorities had never shown the smallest
interest in the expedition, and the Museum itself was a place
where a naturalist could only work for a very limited time
daily, and inaccessible to Sir Wyville Thomson at Edinburgh.
It fell to Hooker to uphold the credit of the Society and incidentally of its President, easily showing that the precedents
invoked were irrelevant and the allegations unfounded, that
the sending of Ross's collections to Kew instead of the British
Museum was diversion of public property. The collections
were Ross's own, to be disposed of as he pleased.

All this additional work was a heavy burden, and at Kew
the period after his wife's death was one of excessive mental
and physical strain. This was accentuated by the refusal of
the Office of Works to forward his application for assistance,
so that he was compelled to appeal to the Treasury direct.
Yet he was able to write indomitably to his old friend on
January 14, 1875 :

To Charles Darwin

I have 15 Committees of the R.S. to attend to. I cannot
tell you what relief they are to me—matters are so ably and
quietly conducted by Stokes, Huxley, and Spottiswoode
[the Secretaries and Treasurer] that to me they are the

---

[1] This voyage of oceanic research lasted from Dec. 1872 to May 1876. The
scientific observers were under (Sir) C. Wyville Thomson. [Sir Charles Wyville
Thomson (1830–82), after holding several professorships in Ireland, was elected
to the chair of Natural History at Edinburgh in 1870. The value of his deep
sea researches with Carpenter and Gwyn Jeffreys on the *Lightning* and the
Porcupine led to the despatch of the *Challenger* expedition (1872-6) with
Thomson as head of the scientific staff. Besides scientific papers, he wrote *The
Depths of the Sea*, 1873, and *The Voyage of the Challenger in the Atlantic*, 1877.
But he did not live to complete the reports on the *Challenger* material, which
were entrusted to (Sir) John Murray.]

same sort of relaxation that Metaphysics are to Huxley. I have no sense of weariness after them. Of course I must expect some rows and difficulties in the Society, and they will come when least expected, you will say, — but meanwhile let me enjoy my illusions.

Much labour and correspondence were also involved in preparing the Presidential Address for the Anniversary meetings, when the medallists and their services to science were announced and the work of the Society for the past year summarised.

Thus he writes to Darwin, October 11, 1874 :

I am busy with my address for R.S., which I am advised to make a purely business one, and confine it to the operations of the Society, its Committees, Funds, labours under Government and private affairs, about which it appears that the Fellows in general are absolutely ignorant. They know nothing of the Donation Fund, Government Grant, &c. Relief Fund, and the dozen or so Committees, many of them Standing Committees, that involve an amount of work on the part of the officers that not only justifies paying the Secretaries but makes it expedient for the Society to do so, and necessary to support themselves.

To summarise his Presidential Addresses : the first, in 1874, reviewed the finances and work of the Society ; the second, in 1875, dealt with various scientific expeditions initiated or directed by the Royal Society ; the third, in 1876, with the relations between the R.S. and Government, the Government Grant Fund, the Vivisection Act, the Loan Collection of Scientific Apparatus, the Meteorological Office, and the return of the *Challenger* ; the fourth, in 1877, with Nares' Polar expedition, the American Flora, and the relation between the Cretaceous and Tertiary fossils ; the fifth and last, in 1878, with the reduction of fees to Fellows, recent discoveries, Palæobotany and modern development of botanical science, notably Darwin's work and the sequel to Burdon Sanderson's discovery of electromotive properties in plants, and the new world of knowledge opened by bacteriology and its bearing on the theory of spontaneous generation.

Of the final address, delivered on November 30, 1878, he writes to Darwin, October 4 :

My Address for Royal is *nowhere.* I have not thought of a word for it, and every time I try, it makes my head and heart ache. One's last Address ought to be good. I have this last half-hour (moved thereto by your letter) maundered over the matter and written to De la Rue for some information relative to Electric discharges apropos of Spottiswoode's researches. Hitherto I have not (like my predecessors) sponged on my Fellows for matter for my Addresses. Now I must, if, as I am advised, I am to give a résumé of some of the advances in Physical and Biological Sciences that have rendered the Society's labours noteworthy during my Presidentship. Would Frank[1] give me some crude data in reference to your and his labours ? and as to what they point to ? I would work them up. Pray do not allude to it to him if you think better not. I should like to give a short analysis of the question of biogenesis—and so forth, but it makes me giddy to think of it. I shall consult the godlike Huxley on this. I must keep off controversial questions.

And again after the address had been delivered (December 14) :

I am immensely gratified with your praise of the Address, which I was most anxious about, and feared would be a failure. I have to thank Frank for the gist of the story about your works, and Dyer gave me great help in vegetable Physiology—the rest cost me a deal of coaching up. I left out the Palæontology because I dosed them with it in last year's Address and I could not grapple with Zoology in the time and space. I felt very sorry to leave the Chair, but the relief is very great.

In 1875 also a successful experiment was made by holding two evening receptions of a less formal character than the annual conversazione, in order to bring the Fellows together socially. Popularity, however, has its drawbacks, and of the more formal gathering in 1874 he tells Asa Gray that it was

[1] (Sir) F. Darwin, who was working with his father, and especially extending research among carnivorous plants.

a tremendous affair, I suppose the fullest known for many years ; twice as many as ever known, but very fatiguing for me. How I did pity the President of the United States !

And in 1877 he gets home from the same function at half-past one in the morning, ' with a crick in my shoulder and " phalangitis " from pump-handling some 500 people.'

In regard to the vivisection question, Hooker, as a botanist, was less actively concerned with the agitation of the seventies than were the biologists. But with entire appreciation of the interests and principles involved, he cordially joined in the protest of science against the sweeping prohibitions which perhaps did more honour to the heart than the head of their proposers, who seemed to make no moral distinction between the wanton infliction of pain and the infliction of pain *per se,* and to justify their attitude by denying what was the cumulative experience of peace and war, the value to suffering mankind of treatment based upon experiments on living animals.[1]

Examples may here be given of some of the difficulties which beset the ' referees ' on whose judgment depends the acceptance or rejection of a paper submitted to a learned society.

In 1866 a paper had been submitted to the Linnean Society dealing with a subject on which Hooker's friend Col. Munro, the authority on Indian grasses, was at work. Munro hoped to have early sight of it to quote in relation to his own research. When Hooker, seeing that it would not be helpful in providing systematic references, simply wrote that he was not sending it, just as he would have written if Munro had never heard of its existence, his friend apparently was seized with alarm lest Hooker should have some hidden meaning. Hooker hastened to explain how he had misled him, or forgotten something which he ought to have remembered.

---

[1] Of Lord Carnarvon's ' Act to amend the law relating to Cruelty to Animals ' which followed the report of the Commission in 1876, a writer in *Nature* (1876, p. 248) remarks : ' The evidence on the strength of which legislation was recommended went beyond the facts ; the report went beyond the evidence, and the bill can hardly be said to have gone beyond the recommendations, but rather to have contradicted them.'

To Col. Munro

April 24, 1866.

A paper comes from L.S. which was expected to be of value and, as such, to be printed, in which case every one would have wished you to see it at once, and before printing, knowing that you would make no unfair use of it. It turns out to be worthless, and we therefore propose not to hamper you with it ; both because it would not help you, and because it avoids all suspicion of our having had a sinister object in sending it to you.

No thought of your cutting out X. ever occurred to B[entham] or self ; we would both of us readily swear by your honour and not only by this but by your generosity, yea, even to your own detriment, and we felt sure that a sight of this crude performance would be a bad service to you, and to science. If you did see it you might have found quarries of gold that you would wish to quote and must have quoted, and would then be open to be accused of not quoting more, or of using more without quotation ; and the paper remaining unpublished another would be able to say what he liked uncontradicted. Believe me, my dear friend, that it was in consideration of your scrupulous honour and generosity that we acted as we did ; and to avoid embarrassing you.

Similarly he writes to Darwin in April 1870 :

I am now in a frightful state of mind. The R.S. have referred to me ——'s [address], and I find it so full of *perfect* trash that I am compelled to recommend its non-publication. It will be a knock-down blow to the poor man. The systematic part is very meagre indeed, the vegetable anatomy miserable and often utterly wrong ; the affinities more often mere guess-work than not, and as to the theories and speculations, they would make your hair stand on end. . . . Altogether the affair has cut me up terribly, and I would rather have burnt my fingers than performed so painful a duty. The curse of Cain will cleave to me. By the way he pooh-poohed my Greenland paper— this has only just come into my head, and does not mend matters—for he will, if he hears of it, put my sinister report down to spite, whereas I would fain have heaped coals of fire on the poor devil's head by a gushing (not crushing) report.

So also, when Huxley as referee got into hot water for rejecting a paper and in the usual course retaining it among the R.S. archives, or as the author said 'suppressing' it, Hooker wrote to him :

December 28, 1874.

You could not have answered T. better. I have long thought that the retention of rejected papers was a course that had its awkward side ; it is so often regarded, however unreasonably, as 'suppression' of the papers, which, added to rejection, piles the horrors. We must be unfettered in our power of rejection, and we must keep the originals as our *pièces justificatives*, and I see no middle course but that of offering copies to be made at the author's expense.

CHAPTER XXXIV

THE PRESIDENCY (*continued*)

IN 1877 Hooker received knighthood in the Order of the Star of India. He had received his C.B. in 1869. There being then no vacant K.C.B., this was offered together with knighthood *en attendant* a K.C.B. after longer services. That the first honour of the kind to be offered him should be the C.B. was quite unexpected. It would have been more appropriate and in many ways more acceptable had he been offered Companionship of the Star of India. His services to Indian science had begun before his official connexion with Kew, and had continued since gratuitously. The Court of Directors snubbed him before he set out, refusing him assistance and official letters of introduction to India and even a passage out. Both the Court and its successor, the India Board, made no move of recognition, though they constantly wrote to him for information and for recommendations in filling up appointments : though he rescued all Falconer's, Griffith's and Helfer's collections from destruction, and himself distributed them all over Europe and America, with catalogues and numbers. It was Hooker who surveyed and mapped the whole province of Sikkim and opened up the resources of Darjiling at the cost of captivity in Sikkim and the consequent loss of all his instruments and part of his notes and collections. Yet the India Board actually sold on Government behalf the presents the Rajah made him after his release, though they owed the annexation of the province and the Government sites of the Tea and Cinchona cultivation to his misfortunes and his energy. On his return to England, although they were spending £30,000 on the explorations of the four Schlagintweit brothers, they would not give a shilling to pay the printing even of the ' Flora Indica,' nor subscribe for copies. And at this very time they had sent an Indian officer—who was actually using Hooker's MSS. which he placed at his disposition—on full pay and on service time to Kew, to publish a Forest Flora of the N.W. Provinces, instructing him to prepare it ' under Dr. Hooker's advice and directions.'

Here were services given freely to India for which, had Hooker cared to do so, he could quite properly have claimed a mark of official recognition. But he had no personal ambition in such a direction : he did not covet the C.S.I. either as a vanity or as a reward of scientific eminence. So far as he could be said to covet such at all, it would be as a recognition of actual services rendered by his science or himself, a recognition that he could accept if offered, but would never ask.

Thus to the offer of 1869 he replied by return of post, accepting the C.B. *for services* and declining knighthood.

The latter [he tells Darwin on November 14] I did on various grounds, partly because it signifies nothing, whilst the C.B. recognises services, which is the only recognition I care for—and because if they wanted to knight me (and I do not *wish* for Knighthood) they might have offered it in an order that indicates special services.

His friends, however, were eager that he should be offered such a recognition. His services in India were at least as noteworthy as those at home. Murchison and Lyell therefore approached the Duke of Argyll at the India Office, recommending Hooker for the Star of India.

The Duke replied more than once that he had set his subordinates to seek official information on which to act. Little were they likely to find, for the East India Company had consistently ignored Hooker, and refused him countenance and assistance !

On hearing what was afoot, Hooker begged that he might

not be put forward ; not being in the Indian service, he doubted his eligibility.

But [he continues to Darwin] I do not think there is the least chance of my getting the offer of it.

The K.C.S.I. is so rare an honour that I might well be proud to have it, for my Indian services ; but I really do not desire Knighthood, and would infinitely rather be plain Dr. Hooker with C.B. to testify to my having done my duty as well as others who have *that certificate.* So if it comes I shall be proud of it ; if not, I shall be as well content. Please say nothing about it. The fact is the Duke might do it with a stroke of the pen, but he don't like my Darwinism and my Address and I am right proud of that ! And now to more congenial subjects.

A week later there is a pointed postscript :

Pray do not C.B. your letters to me—I can't stand it. I own C.B. gratifies me in a *service* point of view, and it is very useful officially in Indian and Colonial correspondence, but scientifically I rather dislike it.

The result did not satisfy Lyell, who a little time afterwards —it appears to have been early in 1870—took Hooker to task for his refusals. The latter unbosoms himself very freely to his closest friend.

To Charles Darwin

Monday.

Lyell had a long talk with me about Knighthood, which he *seems greatly to regret* my not accepting. I could not tell him how gladly I ' not accepted,' nor how much I do *not* regret it, nor could I make him understand my feelings in the matter,—that I have no wish whatever to be Sir J. at all, but that if this must be some day, the least objectionable way is clearly in the Bath, the Bath being a recognition of *public* services &c. I have no *wish* even for that, but I quite feel the force of all my friends wishing it, of my public position suiting it, and of keeping up (what I and my relations do, and my children will take a reasonable pride in) the public recognition won by so many members of my family without Court or aristocratic influence. If I had raised myself, wholly unaided, to my scientific position, as Lyell himself, and my father, and Murchison and Palgrave did to the positions they attained, then I should have felt that I had earned Knighthood as they did, and might have accepted ; but my case is wholly different. My Science I owe to my father, ditto my Kew position. My services have been wholly under Govt., and if I am entitled to any such recognition as Knighthood at all, it is to one given for services unmistakeably. As it is, I am on the horns of a dilemma. I could be knighted for the saying I wished it to-morrow— the declining is interpreted into despising it, in preference to a riband which I am not offered. Lyell and Murchison say—'Take the Knighthood as a step to K.C.B.' This would be all very well if I really wanted the K.C.B.! though even then I do not think I could have stooped to any such dodge. Huxley and Lyell are the only persons with whom I have talked over this matter. Huxley quite understands and approves ; but then he *despises* Knighthood, which I do not.

Again, to Hooker's grim amusement, he found that in October 1871, in the height of the Ayrton troubles, Sir Charles Lyell urged Mr. Gladstone to make amends to Hooker by giving him the deferred K.C.B.—at the very time when the Prime Minister was specially exercised how to keep his unruly colleague in order without giving him further offence ! [1]

The next episode was in 1874. The K.C.M.G., a recommendation for which had been refused by the late Government, five years before, was again offered. The official link between Kew and the Colonies was attractive, but as he was now President of the Royal Society, it would have been regarded as given to P.R.S., and not to himself for Colonial services, and the Society, as appeared later, would not have approved of his taking anything less than the Bath given to both his predecessors in the chair, with far less service claims. The K.C.B. being so limited an honour, he now felt safe from Knighthood for some time to come.

He had followed his own inclination but felt some qualms

[1] See p. 168.

as to whether he was justified. The following tells the story at length.

To T. H. Huxley

March 29, 1874.

Lord Carnarvon (a stranger to me, as are all his C.O. Officials of any weight or influence) asked me to call on him two days ago, when he offered me the K.C.M.G., putting it solely on my services to the Colonies and not at all on scientific position. I declined at once. He pleaded hard in the interests of Public Service; he regretted that the Office had not recognised my services earlier—added that he hoped and wished this to be the first act of his official career—that my name would be agreeable to the Colonists, and add lustre to the Order, and so forth. I finally beat him on the point, that the Order was limited to 60 Knights, that it was instituted for the Colonies, not for outsiders, and that there were lots of men in the Colonies with unquestionably higher claims than mine on such a recognition.

I am not clear that I am right all round, and that my motives are not as egoistic as friend X.'s. Acceptance would have set the official seal to the value of Kew to the Govt. itself, in an unmistakeable way, and been a powerful handle for introducing more Science here, especially in the shape of a physiological laboratory. The refusal was not gracious to the Colonies, nor to the Service of which I am a member—and acceptance would have gilded the Board of Works as well as Kew. A host of unselfish considerations rise up to rebuke me. In fine, I may not have done right, but I have done what I liked and that is far better!

It is clear now that I cannot go on refusing to accept recognition of services for ever, and that I shall one day be placed in a fix. It is very much through my official services that I have attained the position I occupy and through official opportunities that I have made my way as a scientific man too. In short I dread Sabine's death and the vacant K.C.B. that may then possibly be offered me after this refusal. Now I recognise the duty of Public Servants to accept the Crown's recognition of these, so long as such recognitions exist, and they cannot wriggle out of acceptance! And though I look with aversion to being ' Sir

Joseph ' in any shape, I am perhaps as wrong in refusing
K.C.M.G. as if I refused a Royal Medal. Every herring
should hang by its own head.

Meanwhile I pray for Sabine's longevity and send him
flowers to prolong his existence. Every man's life has its
value to those who know it ;—how kind is Providence !

The final history of the matter is told in the following.

To Charles Darwin

June 18, 1877.

I should have told you before of K.C.S.I., but as I knew
you kindly would excuse me, I delayed. As Huxley will
tell you, I was taken completely by surprise at R.S. by
receiving a letter from Lord Salisbury informing me that
he had taken a liberty with my name, proposed it to the
Queen for K.C.S.I. and that I was virtually appointed !
It went on to imply that as I was not in the Indian Service
it was somewhat irregular, but that my Himalayan work
alone ' entitled me technically and substantially to the rank.'
It added a little about my beneficent exertions for India,
and was altogether a very ' pretty letter.' Huxley told
me that I could not refuse it if I would, and on recovering
my senses I could not but see that both the compliment and
the manner of paying it were the highest and most gracious
that could be. I have since heard that the Cabinet discussed the thing—that they could not longer allow my
services to pass unrecognised, there was no K.C.B. vacant,
and as I had refused K.C.M.G. it would be risky to ask me
to accept anything else—so they strained a point to give
me K.C.S.I., and in the handsomest manner gave it solely
for Indian work. I had always regarded the Star of India
as the most honourable of all such distinctions—it is very
limited (to 60 K.C.S.I.'s)—is never, like K.C.B., given by
favor or on personal considerations, and it has a flavor of
hard work under difficulties, of obstacles overcome, and of
brilliant deeds that is very attractive. Assuredly I would
rather go down to posterity as one of the ' Star of India '
than as of any other dignity whatever that the Crown can
offer. Of course it pales before P.R.S., but then they cannot clash. I do not know whether I told you some five years
ago application was made to the D. of Argyll to give it

me, on hearing of which I wrote to him begging him not,
as I thought so rare an honor should be confined to actual
Indian servants. He answered that he would have given it
me, but implied that the Statutes of the Order forbade it !
so I never thought anything more of the matter. It is as
you say a ' peculiar ' honor and I may well be proud of it
and of the way it came.

Is this not a jolly strain of self-gratulation and
glorification ?

Meantime the Genera Plantarum progressed slowly. He
was compelled to leave his share of it aside, ' having,' as he
tells Asa Gray (July 20, 1874),

so very few continuous days and half days to give to it, and
I cannot work it as I can Flora Indica, &c., by jerks. The
latter has given me unexpected trouble. (With two exceptions, one of whom was his Assistant, Dyer), no one has
worked well, and I had no idea how difficult it appears to
be a middling systematist even.

Some of the work, indeed, he had wholly to re-do.

Other botanical work which claims particular mention
includes the Botanical Primer for Macmillan's Science Series,
' to keep company with Huxley and Tyndall,' and researches
into carnivorous plants.

Of the former, which was published in 1876, he writes to
Darwin, February 20, 1873 :

I have no news except of my own folly. I have undertaken the Botany Primer for Macmillan which will be some
100 12mo pages of a sort [of] Introduction to the subject
of Botany—something different I think from an elementary
lesson-book, and yet the information must be definite, and
such as the recipient can be questioned about. I have
given the subject a great deal of thought and sketched out
a plan. The great difficulty is to go to the bottom of things
and yet avoid detail—or rather to keep pointing to the
bottom of things without going into it. I am afraid it will
be like the Sailor's ' Potato and Point,' which, as I daresay
you remember, consisted in a plate of potato and one
odoriferous red herring hung over the mess table. At every

mouthful of potato every man pointed it to the herring before eating it—by way of catching a flavor.

The work on Carnivorous plants was one of those which grew out of Darwin's enquiries. Hooker, as has been said, was constantly supplying his friend with plants for experiment and experimenting for him on plants which could not be kept in the small greenhouse at Down.

During this period his help lay chiefly among the sensitive and the carnivorous plants. As to the former, Hooker occasionally experimented ; thus in November, 1873,

> I was trying *Mimosa albida* in the Hot-house the other day and found it wonderfully sensitive compared to what it was in my room. I wonder if the *damp* heat kept it on the ' qui vive,' like a pig before rain ! It is in our hottest house now.

But his chief part was to send plants and answer questions, which sometimes were liable to be pushed home with searching supplementary questions for which he had perhaps a practical but not a scientific answer.

Thus in 1873 (as later in 1877) Darwin sought some explanation of the waxy coat or ' bloom ' and layer of fine hairs upon leaves. (See C.D. iii. 339, and M.L. ii. 409, 410.) One obvious effect of these equipments was to keep water off. But was this all ?

To Charles Darwin

August 14, 1873.

I have often speculated on the point you allude to, which is specially conspicuous in the *Nelumbium*. Ferns and various Cryptogams do not show it as Phaenogams do. It is not conspicuous in other water plants as in *Nelumbium*, which further holds its leaves high above the water ;—but for this association of the two means of getting out of the way as it were of injury from the water, I should have supposed that both waxy coat and hairs were connected with absorption in respiratory functions. I have lately wondered whether both may not subserve some purpose connected with Actinic or chemical rays of the sun, especially as the waxy secretion is often more conspicuous on the upper leaf surface. The prevalence of indumentum on the under surface points to transposition in some cases ; in others perhaps it is a provision against the attacks of insects, which harbor on under surfaces. I can quite fancy water impeding both the actinic and calorific effects of sunlight on the leaf. We find watering most prejudicial in the hot sun. It is a splendid subject for experiments.

Darwin immediately followed this up. He must hear more about the rationale of watering in sunlight when Hooker came on his promised visit the following week. This elicited the rejoinder :

I am aghast at the prospect of being cross-questioned on the subject of effect of watering in sunshine, and fear that no amount of ingenious wriggling will save me from the reputation of an ignorant pretender to the post of Director of Kew. (August 21.)

As regards carnivorous plants, the first reference to Darwin's researches is in a letter of January 7, 1873.

I have wandered away from *Drosera* and the question you put. In so far as I can remember it is an accepted dogma that there is no cutaneous absorption in living plants, and that glandular hairs are excretory only. I will however ask Dyer, who is away with a cold—he is translating Sachs,[1] and will be up to the latest discoveries. I will also ask Berkeley.

Your aggregation of the protoplasmic contents of the cell reminds me of the contraction of the chlorophyll contents and (?) inner cell wall of the cells under sunlight in a *Selaginella* (*serpens* I think). Have you tried *Begonia* leaves, or shall I look out for some plants with hyaline bladdery epidermal cells for you to operate upon ? Can you correlate the specific action of the Ammonia on the protoplasm of the cells, with that of its effect on the blood of animals

---

[1] A translation of Sachs's *Text-Book of Botany*, by A. W. Bennett and W. T. Thiselton Dyer, was published in 1875.

poisoned by snake bites ? Is it not the case that snake
poison affects the blood corpuscles ?

In August he writes :

> I rejoice to hear of your success with *Drosera* and long
> to hear more of the acid reaction and the retardation of the
> external digestive processes. I long to be at *Nepenthes*—
> the specimens are splendid and most inviting, but neither
> I nor Dyer have had time.

Later in the autumn there was more time for experimentation. The work undertaken to supply Darwin with more
facts began to grow, and Darwin early suggested separate
publication of the results ; a suggestion not carried out till
the following year.

Kew : October 20, 1873.

DEAR DARWIN,—A line only to say that I am at
Nepenthes, but it is a far more difficult affair than *Drosera*,
because of the thickness of the tissues. The structure of
the glands of the pouch *below* water mark is well made out
and described and consists of globose glands analogous I
take it to *Drosera's* tops of hairs, lying in a semi-circular
fold of the cuticle and half exposed. It is in these globose
glands that I must look for the action.

The water is acid : it has been most carefully described
by Voelcker and others and I have, I find, referred to many
papers on anatomy of *Nepenthes* in my account of the
species for DC Prod., which is printed, I believe, but not yet
published.

I have found rough copy and can send you a string of
authors.

The correspondence continues at length through October.
‘ Dyer is making excellent drawings and working like a horse
at it,’ but ‘ the whole investigation is fearfully difficult compared with *Drosera*.’ Similar experiments produce analogous
action to that on the cells of *Drosera*.

October 25, 1873.

We are now trying the egg process. The pieces I put
into *old* pitchers last night were unaffected this morning. Did

I understand you that the pieces should be 1/20 in. square ? I put in big lumps. We have still a great deal to do before arriving at any satisfactory results. The constant presence of insects in all *open* pitchers is a drawback, and we are going to experiment on virgin pitchers.

October 29, 1873.

What you say of the glands being secretory organs is suggestive, and may account for the pouches in which they lie pointing downwards—but I suppose they must be both digestive and secretive, as I understand *Drosera* hairs to be. The fluid of the virgin pitcher is very slightly acid. I find the cells of the glands of old pitchers (full of insects) with very aggregated contents.

I had no intention of publishing *Nepenthes,* the experiments were made solely for your eating, and I hope that you will absorb them in the *Drosera* paper. I thought of mentioning them at the Phil. Club as experiments suggested by and undertaken for you—if you did not object. If ever these and those on *Sarracenia* &c. should be worth collecting and making a paper of, it cannot be till you have done *Drosera.*

In the following spring he notes Lady Dorothy Nevill's gardener as saying that he had fed a *Dionaea* with raw meat and that it beat all others of the same age in growth and dimensions.

Pressure of work interrupted experiments till July 1874 :

I have splendid *Sarracenias* and will perform any miracle you put me up to regarding them.

I am charmed with your account of *Pinguicula* [1] : and should like to try if *Lychnis viscaria* has the same use for its viscid fluid—which I should have guessed was to prevent insects climbing up to the flower—but all things now go by contraries !

And on the 3rd :

I have been going on with *Nepenthes.* I have 3 plants set out in an inviolable place—a very sanctum—and shall

[1] The common Butterwort also turned out to be insectivorous.

make a point of now going on — all other duties social, scientific, and parental notwithstanding.

Any hints for observations most gratefully received. I note carefully what I do.

July 15.   I am at fibrin to-day.   Michael Foster suggests that coagulation of protoplasm may be diseased not digestive symptom, and advises my trying effect of citric acid in pitchers.

July 18.   Have you any objection to my giving an outline of what *is published* of your *Drosera* observations at the Belfast meeting?   I have to give an Address, and would like to make a résumé of the Pitcher plant results the back-bone of it, stating that they were wholly undertaken under your auspices and apropos of your *Drosera* experiments.

If you have the smallest objection to either *Nepenthes* or *Drosera* being described pray say so, as I would rather send you all *Nepenthes* matter for you to append or incorporate, than appear to filch.

We had such a night at the Mozart Festival at Covent Garden.   I was carried away with Albani's 'Dove sono,' and felt it up and down my back as when we were at New College Chapel, Oxford, in 1847.   I could not help my eyes watering.   I thought I had never heard anything so beautiful since Malibran in 1837.   Patti I cannot get up sympathy or enthusiasm for—she fails to satisfy me.

July 22.   I am stupefied by the trouble you have taken, and your kindness.   What you have sent was not in the least wanted apropos of Belfast, but will be enormously useful in my work on these pitcher plants, &c.

As to Belfast, all I wanted was your assurance that a mention of *what has been published* in Nature &c. of your observations on *Drosera* and *Dionaea*, would not interfere with your book, and that my giving a résumé of my *Nepenthes* observations would not look like forestalling your far more important work.

The Brit. Assn. Sections are all trying to get for Belfast Meeting more brief reports of what is doing in each branch of Science, and the direction in which research therein is tending.   I thought of making carnivorous plants my share of this work ; and giving it as my Address as Pres. of Subsection Bot. and Zool.   I thought of introducing it by a

notice of what was published of your results, and then going
on to my own, as supplementary to yours, and undertaken
apropos of yours. I do not intend to make a paper of it.
I should like *Nepenthes* &c. results either to go to you
altogether, or to form a paper for R.S., but would really
rather you took them.

August 17. I have been driven wild with work and the
Address, which I am taking down with me in an inchoate
state. We are off to-night via Stranraer.

I have been working steadily at *Nepenthes* every day
and made a good deal out—its appetite for cartilage is simply
prodigious—it reduced a lump as big as your finger nail in
48 hours to lovely jelly, and after 10 days there is not the
slightest trace of putrefaction in what of the jelly remains.
Nothing can be more lovely than to draw out the cartilage
attached to a thread after immersion, it looks like a ball of
rock crystal refracting the light most beautifully. I got
little or no action by fluid withdrawn from pitcher and kept
in a tube—nor with plants in a cold room. The digestive
fluid is evidently poured into the original liquid only after
immersion of meat. Fibrin as I think I told you goes ' like
smoke,' but not in a tube. I find copious honey-secretion
on glands of lid in all species but one, and in this one (the
only one of the genus) the lid lies horizontally back ! and it
would be prejudicial if it had honey, for it would decoy them
away from the pitcher. I have tried seeds, but results are
not satisfactory. After three days' immersion both mustard
and cress are killed—ditto in distilled water. One day's
immersion shows no difference. I must try seeds quite
differently.

I have made out a good deal of structure in *Sarracenia*,
but nothing of action—it is not easy and secretion is scarce.

As to *Cephalotus* it is a beast—it will not kill or eat,
and I am in despair about it—it does not catch insects to
any extent and I find no action in glands or cells. The
stomata in the pitcher is an exception to all these pitcher
plants and shows that this cannot depend on the secretion
much, it forms very little water indeed. I have made out
the secreting glands, seen them secrete acid fluid, but I
can't excite them to secrete. Cartilage rots very soon in
the pitcher and fibrin remains unchanged.

The Address is a sort of rambling statement of the
history of *Dionaea* and *Sarracenia* and *Drosera* up to your
time, ending with Burdon Sanderson. I shall not touch your
ground, but refer to you. I then go straight to *Sarracenia*
and *Nepenthes*, and shall just touch on *Cephalotus*, and wind
up with some generalities on absorption and nutrition by
plants in general.

Dyer has helped me *enormously*, and indeed I could not
but for him either have got through the work or done it half
so well. He catches over ideas and anticipates one's wants
wonderfully, and I really feel that he should share whatever
credit the historical part and general conclusions may get.

At the Belfast meeting of the British Association in 1874
the Hookers tried to arrange that the members of the x Club
and their wives (x's + yv's) should club together as a single
group ; but the pleasant plan could not be carried out. Hooker
did not take any active part in the hubbub that followed Prof.
Tyndall's Presidential Address, which fluttered the theological
dovecotes ; his only reference to the meeting is in a letter to
Darwin (August 22 or 29) : ' Lubbock's Lecture [1] went off
admirably, but Huxley's [2] was the *magnum opus* of the meeting.
It was a most capital meeting.'

The same letter contains a reference to protective changes
of colour in animals. His correspondent wrote from South
Africa.

The enclosed have just arrived from Mrs. Barber. Her
clever suggestion of the colour being as it were photographed
reminds me that Grove ages ago told me that he had seen
dead Fish take the colour of an adjacent object, I forget
what, but it was after the manner of a photograph.

The Papilio reminds me of my Indian Tick or Lizard,
which I have never quite persuaded myself to believe in till
now ! ! ! I remember telling you of the grasshoppers on
Mt. Lebanon which were grey on grey rocks and greener and
browner on other situations.

---

[1] On ' British Wild Flowers Considered in Relation to Insects.'
[2] On the ' Hypothesis that Animals are Automata, and its History.' Coll.
Ess. i.

CHAPTER XXXV

THE AYRTON EPISODE

THE years from 1870 to 1872 were ravaged and embittered by a personal conflict with Mr. Ayrton, the First Commissioner of Works in Mr. Gladstone's Government. Gossip suggested that he was taken into the Ministry to economise the time that would have been wasted had he been left free to heckle the Government. For he was gifted with a blistering tongue and a thick skin, with which he exploited in the Radical interest a breezy sansculottism akin to that which sent Lavoisier to the block, with the words ' the republic has no need of men of science,' but extending the phrase to cover a wider range of civilised amenities. Lord Suffield tells an amusing story of him. In 1873 he was present at a grand ball given at Stafford House in honour of the Shah of Persia, who was visiting London. The Shah desired to meet Mr. Ayrton, and a messenger was dispatched in search of him. He was found in the supper room, and being invited to come forthwith and be presented to the Shah, he bluntly responded, from a mouth full of chicken, ' I'll see the old nigger in Jericho first ! '

Kew, like the rest of the Royal Parks, fell under the administration of the First Commissioner. In his re-election speech at the Tower Hamlets in the winter of 1869 he enlarged on the popular aims of his rule, with a warning to ' architects, sculptors, and gardeners ' that they would be kept in their places.

In dealing with an official superior who had thus dotted

the i's and crossed the t's of his existing reputation, Hooker took care to walk warily. But this availed nothing. As soon as the First Commissioner was fairly in the saddle, one of his first acts was to send an official reprimand to the Director of Kew, the first in the twenty-nine years of the Hooker regime. Launched without warning given or explanation asked, it turned out to be based on a misapprehension.

This was not encouraging, but Hooker maintained a conciliatory attitude ; and indeed, while still smarting under this unprovoked reprimand, at the First Commissioner's special request devoted many nights to examining and reporting upon various books and pamphlets on the public parks of England, France, and America, for his guidance—a labour not very congenial and wholly beyond his province as Director of Kew, but furthermore undertaken in the hope that it might lead the First Commissioner to judge more generously of the acquirements and duties of some of the officers of the department under his control.

But such considerations had no meaning for Mr. Ayrton. Public economy was his watchword ; his method the contemptuous disregard of his subordinates' status and authority, with equal contempt for the scientific as apart from the popular purpose of the Gardens. His apparent aim was to drive Hooker to resign, and then convert Kew into an ordinary Park, and send science to the right about.

After a series of vexatious interferences, matters came to a head in the summer of 1871, when Hooker casually discovered from one of his subordinates that he himself had been superseded six months before in one of his most important duties—namely, the heating of the plant houses, which had, for scientific reasons, been specially assigned to the Director in 1867 (see p. 81). In reply to a courteous inquiry as to the reason of this, he received an offensively curt intimation that the change had been made, and that he ' must govern himself accordingly.' Hooker thereupon addressed a sharp remonstrance to the First Commissioner, complaining of the disregard of his office and the want of confidence with which he had been treated. In reply, Ayrton demanded particulars and dates of these

acts ; [1] and instead of answering them, ' contented himself,'
as he put it, ' with forwarding an official memorandum to the
First Lord of the Treasury.' This memorandum was doubly
disingenuous ; it complained of Hooker having ' launched
out into various topics,' whereas Ayrton had invited him to
specify these topics of complaint ; and it slurred over three
of Hooker's chief instances, while asserting that Hooker had
originally been appointed to supervise the works at Kew
(instead of the heating works only).

I do not [Hooker wrote] for a moment question the First
Commissioner's power to exercise arbitrary authority over
the Director of Kew, but I do submit that there has been
hitherto no plea whatever for such action as regards myself,
and that the repetition of such acts, and the leaving me to
be informed of them, on each occasion, by my subordinate,
constitute a grievous injury to my official position, and tend
to the subversion of all discipline in this department.

And he respectfully claimed the privilege of appealing for
redress to the First Lord of the Treasury, Mr. Gladstone.

The whole incident leading to this extreme measure justified the prophecy of *The Times* that it would prove ' another
instance of Mr. Ayrton's unfortunate tendency to carry out
what he thinks right in as unpleasant a manner as possible.'
Had he only condescended to explain his purpose, there could
have been little difficulty in effecting what in many respects
was a desirable change. The heating apparatus had, as a
whole, been very successful ; but an accident to one of the
pipes revealed some imperfect execution of details, and it was
difficult to decide who was responsible for the technical correct-
ness of such works. Hence the expediency of appointing a
specially qualified Director of Works, (Sir) Douglas S. Galton,
who should carry out all works sanctioned by the Department,
not only at Kew, but elsewhere, the requisition in each case
being made by the official, such as Hooker, in local authority.

But no explanation was vouchsafed, before or after. The

---

[1] ' I have launched a rostrate ironclad against Ayrton,' remarked Hooker
when he sent the particulars.

Secretary of the Department chose to think that ' Mr. Galton's appointment and the purposes for which he was appointed were so well known ' that any special notification to Hooker was needless ; and later, when a report on some proposals for works at Kew were endorsed by Ayrton with a note ' that such works should be carried out on the responsibility of the Director of Works in future,' he neglected to see that this was conveyed in the subsequent letter sent to Hooker. Accordingly, the Director of Kew, whose appointment made him responsible only to the First Commissioner, suddenly found himself in certain particulars responsible to his own subordinate as well.

As an isolated act, this was bad enough, but taken in conjunction with others equally unreasonable, it had every appearance of being part of a design to render Hooker's personal position intolerable.

Thus in 1870, when the plan of adding a part of Hyde Park to Kensington Gardens was under consideration, Mr. Ayrton directed that Mr. Smith, the Curator of the Royal Gardens, Kew, under the control of the Director, should undertake the superintendence of the proposed works. Dr. Hooker was ' to be informed accordingly, and to arrange for Mr. Smith attending at the Park as often as required.' In effect the Director, without consultation, was to be deprived of his most useful subordinate. He replied that he could not spare his Curator, and naturally complained that he had not been consulted. Answer, a curt Minute, stating that ' it is apparent Dr. Hooker is not aware that the exigencies of H.M. Service required the immediate assistance of Mr. Smith, in the manner directed by the First Commissioner.' Again Hooker explained in detail his inability to spare his Curator, and received for answer a still curter Minute, simply saying that his letter ' appears to have been written under a misconception,' and directing him to convey the First Commissioner's orders to Mr. Smith. In the end, Mr. Smith declared he could not combine the two duties, and the proposal dropped.

Nor was this all. Ayrton had first come down to Kew, and unknown to Hooker, had a private interview with Smith, discussing the possibility of appointing him to the superior office of Surveyor of Parks and Gardens. According to Smith's account, Ayrton concluded by asking him to keep this conversation secret. Ayrton first described this secret visit to Kew as a friendly visit to Hooker : his explanation later was that he simply said ' it was not necessary for the Curator to report, as the First Commissioner would himself communicate with Dr. Hooker on the matter.' But he unfortunately made no such communication, either at the moment, when he might have found Dr. Hooker by enquiry at his house or in the Gardens, nor by letter subsequently.

Again, a vacancy arose in the Herbarium, which had hitherto been filled up by selection of one of the young gardeners who had shown botanical aptitude. But nomination lacked the ultra-democratic touch so dear to the First Commissioner, and he had let it be known that these appointments should henceforth be made by competitive examinations under the new Civil Service rules, though the Treasury wrote to him that the exception allowed in the case of appointments requiring technical skill might well be admitted here. The men themselves saw the futility of a non-technical examination, and Hooker had to tell the First Commissioner that ' at present there is no candidate for botanical employment among the young gardeners ; lately there were several, but since it became known that this brought them under the Civil Service rules they have not come forward.' Thereupon Ayrton with an economy of words towards the central fact informed the Civil Service Commissioners that ' with reference to the proposal for selecting a candidate from among the best of the young gardeners, Dr. Hooker reports that there are at present no candidates for botanical employment.' The real point, that they desired the other form of selection, is omitted. This, says *The Times*,

is an illustration of a reckless roughness in transacting business in which Mr. Ayrton appears almost to take a pride. He boasts of his refusal to discuss matters of past complaint, of his success of confining his writing to the exigencies of public business, and of his skill in rendering his communications as brief as possible. He evidently has to learn that the way in which a thing is done is often not less important than the thing itself.

The last scene of this serious interlude was purely farcical. After due examination in outside subjects, a young man was appointed who had been employed at the Gardens two years before, and was described by the Curator as a ' dull young fellow who will not suit us at all.' But protest suddenly became unnecessary. To the First Commissioner's accountant, who came to investigate, the young man, with great good sense, frankly admitted that he was not qualified for the work, and the whole thing fell to the ground.

From the coming of Sir William Hooker to Kew in March 1841, until Ayrton's accession to power, the relations between the Director and his official superiors had always been marked by mutual respect and consideration. Now there was an unreasonable and arbitrary regime. Even the estimates for Kew were made without taking the Director into consultation, and proposals made to the Treasury for extensive and unsuitable alterations in the Museum, the full cost of which even was not foreseen.

Again there was a difficulty in regard to the remaining volumes of the Flora of Tropical Africa. The book had been sanctioned by the Treasury in 1864, the cost to appear in the estimates of the Stationery Office. The first volume appeared in 1868, the second in 1871, the arrangement with regard to the former being that their distribution should, for scientific reasons, be in the Director's hands. Now, without enquiry, the remainder of vol. i. was withdrawn from him, and sent with vol. ii. to the Stationery Office for sale. Hooker pointed out the objections to this procedure, adding that in his opinion the copies ought not to be sold. In spite of a letter from the head of the Stationery Department, afterwards printed in the Parliamentary papers, pointing out that Hooker was right, and that the new arrangement would have been inconsistent with the contract made with the publishers, Messrs. Reeve, and would have amounted to a breach of faith, the First Commissioner overrode his protests, telling him officially that the Treasury had decided against him—and this in the summer of 1872, when an 'amicable settlement' had been suggested. Small wonder that Hooker writes to Bentham (February 2, 1872) : ' My life has become utterly detestable and I do *long* to throw up the Directorship. What can be more *humiliating* than two years of wrangling with such a creature ! '

As long as the attack on Kew appeared to be Ayrton's only, he was prepared to resign if Ayrton were not removed or Kew placed under another department. But when he found the Government had known of his views and had not checked them—whether or no they favoured them—he at once changed front, and determined to hold on till turned out, if so they dared. The attack was on Science, and his scientific friends rallied round the cause. Huxley, who was away in Scotland, writes on August 23 :

From T. H. Huxley

Ardlui, Arrochar : August 23, 1871.

MY DEAR HOOKER,—I heartily wish I could have been within tongue-reach of you and have aided and abetted in the cooking of your new kettle of fish. I hope you have not made the fire too hot, which is what one generally does if left to one's own devices—at least, *I* do. As for your resigning Kew, that's out of the question. Ayrton has made such a brute of himself in all quarters, that the fact of your rebelling against him will be a strong *prima facie* argument in your favour in the minds of all men—and we shall make common cause and shew him that he has caught a Tartar in presuming to meddle with Science. Only let not thy soul be vexed by that Amalekite to the verge of losing sleep, Morpheus being the god of temper and patience.

I like what I have seen of Thomson much. He is, mentally, like the scene which lies before my windows, grand and massive but much encumbered with mist— which adds to his picturesqueness but not to his intelligibility. Tait [1] worships him with the fidelity of a large dog

---

[1] Peter Guthrie Tait (1831–1901) was educated at Edinburgh and Cambridge, where he was Senior Wrangler and First Smith's Prizeman. From 1854–60 he was Professor of Mathematics at Queen's College, Belfast, and thereafter Professor of Natural Philosophy at Edinburgh. Besides important works on mathematics and physics, he wrote in conjunction with Prof. Balfour Stewart, *The Unseen Universe.*

—which noble beast he much resembles in other ways. I
cannot say I greatly admired the address. It wants cohesion
and resembles a flash of his own aerolite more than anything
else—bright points in the midst of much nebulosity.

We have come over here to spend a few days with some
old friends but we shall [be] back at St. Andrews on Monday.
Let me hear of any new incidents in the fight. Wife unites
with me in best love.

Ever yours,
T. H. HUXLEY.

Mr. Gladstone, being appealed to, forwarded Hooker's
complaint to the First Commissioner and obtained a reply
which Hooker found unsatisfactory on one point, inaccurate on
the second, and avoiding reference to the three other points.
Accordingly, to avoid departmental embarrassment and unnecessary publicity, Hooker asked to be put in communication with Mr. Gladstone's Private Secretary, Mr. (afterwards
Sir) Algernon West, to whom, on October 10, 1871, he offered
all explanations of his case, summarising it the same evening
in a letter :

> I am at a loss what to say as to my future position under
> a Minister whom I accuse of evasion, misrepresentation and
> misstatements in his communications to the First Minister
> of the Crown, whose conduct to myself I regard as ungracious
> and offensive, and whose acts I consider to be injurious to
> the public service, and tending to the subversion of discipline.
> : . . Granting that the functions of a Director are restored
> to me, how am I to act when ordered to undertake works
> that involve wasteful expenditure, or are otherwise detri-
> mental ? I should be thankful for Mr. Gladstone's instruc-
> tions on this head.

By this time the matter had become a political one,
and quite apart from the merits of the case, he felt he had
political strength in the number of his friends who would be
troublesome ! [1]

[1] Mr. Ayrton and his supporters had had no idea of the halo of popularity
with which Sir William Hooker had surrounded Kew; they now began to be
alarmed at the outcry about ' Kew in danger.'

Knowing this [he tells Darwin, October 20, 1871], I am determined that my voice shall not be withered. . . . I should lose caste altogether if I did not stand up to fight. I am putting all this in plain language to Mr. Gladstone. I quite feel that I should hold on here, and that it is my duty to do so, and that I ought not even to hint at resignation. On the contrary, my cue is to treat *my* being turned out as a ridiculous idea. Moreover, to threaten even to resign would be a dishonourable ruse. But I shall let Mr. Gladstone know that I continue in Office under protest, and Mr. Ayrton's office subordinates, no less than my own here, shall know this, and that there is no sort of compromise of principle in doing my duty under such circumstances.

To C. Darwin.

Kew : October 31, 1871.

DEAR DARWIN,—I think that you should see enclosed.

I have at last driven Mr. Gladstone into a corner, and obliged him to take up my grievances. I told you that he had forwarded my complaints against Mr. Ayrton to the latter to be answered, and he has sent me Mr. Ayrton's in the form of a paper of explanations, and allowed me an opportunity of discussing them with his private Secretary as his representative. I have unhesitatingly pronounced Mr. Ayrton's ' explanations ' to be ' a tissue of evasions, misstatements and misrepresentations,' and I further charge him with telling the Prime Minister a direct falsehood. I then proceeded to show how all but impossible it is, that I should hold office under a Minister of whom I entertain and express these sentiments, and whose conduct to me has been so ungracious and offensive, and whose acts I regard as so detrimental and subversive of discipline in this establishment. I further appeal to Mr. Gladstone as the First Minister of the Crown, *by whom Mr. Ayrton was set to rule over me!* to direct the latter to restore to me that authority and those functions of a Director that his Minister has taken away.

' So you see,' he continues a few days later, ' I am enjoying a good shriek at my Lords and Masters, and I rather enjoy tossing my horns against the Sun and Moon '—the more so because circumstances introduced a touch of ironic comedy into the business.

The Duke of Argyll has been most kind about it. But
the best part of the fun is—only for God's sake don't let
it escape you—that Lyell has written to Lord Granville,
asking him to get Gladstone to confer Murchison's K.C.B.
on me! I'd give a hundred pounds to see Gladstone's
face when this ' obus ' was dropped into the embroglio—
it gives infinite zest to the whole proceeding. Of course
Gladstone would rather give it to Ayrton than to such a
pestilent fellow as I am, who have continuously worried
him for three whole months. But how good of dear Lyell,
and how like him to cleave to an old friend and seek his
honor when *in extremis*. I am immensely touched by it,
and for his sake (not for my own, God knows) would have
wished him success. As it is, that incubus is now put off
sine die. Lyell has, I know, done it by way of influencing
Gladstone to my side in this pitiful quarrel with Ayrton,
and in that respect the application cannot but have immense
effect—in fact, if I beat Ayrton I shall rank Lyell's shot
as in the bull's-eye—bless him.

Ever your affectionate,

J. D. Hooker.

While no exception was taken by Mr. Gladstone or his
Secretary to Hooker's written or spoken statements, the
situation was a troublesome one for the Cabinet, who naturally
wished to avoid anything like a public scandal. A couple of
months passing without any steps being taken, Hooker wrote
again to Mr. West, who replied that a plan was under consideration that would materially alter his position in regard to the
Office of Works.

Writing to Sir John Lubbock, February 9, 1872, he notes
that such steps will involve an Act of Parliament, but the
Ministers have not the faintest notion about the working of
Kew. Can he go to Mr. Lowe and point out the necessity of
first consulting himself or some other naturalists ? Two of
his friends who knew the Government's intentions had been
bound over not to tell him a word of what they were.

Kew [he writes] is what my father and I have made it
by our sole unaided efforts ; and the Ministers have for
three months or more been considering a scheme for funda-

mentally altering its constitution and my position, without consulting me either directly or indirectly in the matter. I say nothing and try to think as little as possible of their utter disregard of my experience and position. I have no wish to throw up my post, but I must do so if matters go on thus.

However, this scheme to move Kew from the Board of Works and set Hooker ' on his legs again ' fell through in the third week of February ; as did another essay immediately after, for which Hooker waited with some hope, for ' Mr. West assures me that my grievance is *thoroughly appreciated* and that there is a determination to support me.' (Feb. 23). His own opinion, however, was

> that matters are not ripe for any great scheme, but that a beginning may be made by removing Kew to the ' Lords of the Committee,' where it will be in the same category as Jermyn Street and South Kensington. Kew has come to be classed with, and as one of, the Parks, most erroneously, and out of that it must be hauled forthwith, if possible. (To Lubbock, Feb. 19, 1872.)

This scheme being dropped, Mr. Gladstone placed the matter in the hands of a Committee of the Cabinet, consisting of Lord Ripon, Lord Halifax, and Mr. Cardwell,[1] to whom, on March 21, Hooker handed in a memorial stating the points wherein his relations to the Government required definition and correction.

The upshot was a communication from Mr. Gladstone through Lord Ripon's secretary, that ' Mr. Ayrton has been told that Dr. Hooker should in all respects be treated as the head of the local establishment at Kew ; of course in subordination to the First Commissioner of Works.'

This thoroughly inconclusive statement provoked a vigorous answer from Hooker. The issues raised were not met by such

---

[1] Edward Cardwell (1813–86), cr. viscount on retirement from politics in 1874, was a first-rate administrator, best remembered for his Merchant Shipping Act of 1854, and the reform of the Army when Secretary for War under Mr. Gladstone after 1868. Others will recall that it was he who defeated Thackeray at the Oxford bye-election in 1857.

a curt and vague announcement. The First Commissioner, to whom alone he was responsible, had officially subordinated him to the Secretary of the Board and the Director of Works in London. In the eight months which had passed since the appeal was made to Mr. Gladstone, four of them under the assurance that a measure of effective relief was under consideration, the aggression had continued, and the Director's position had become not better, but worse.

This answer was inconveniently uncompromising in reply to a message which he had been told was official and final, but which was later defined as a private and friendly communication. As an official answer was now to come from the Treasury, Hooker was begged to let this letter be regarded as *non avenue*. Since, however, he had shown it to his friends and counsellors, he felt that Mr. Gladstone should see what had been seen by others, and the letter remained as part of the correspondence.

The Treasury's official reply, dated April 25, merely repeated the message, with the addition that

they anticipate no difficulty in the regulation of the relations of that important establishment (Kew) to the office of the Board of Works, in which the duties and powers of management are vested by statute.

The vagueness of this statement is only equalled by that of the final paragraph.

The present form of estimate for Kew Gardens laid by their Lordships before the House of Commons cannot now be altered, but it will be acted upon, and will in future be framed in accordance with this letter.

Hooker therefore (on May 1) begged an interpretation of these generalities, without which he could not understand his position, in regard to the original points at issue. But to this letter no answer was sent.

So far the Government had admitted the essential justice of Hooker's case by trying to effect his release from an injudicious superior. If this could not be effected at once, he must be patient. Kew, politically, was of negligible importance; Cabinet solidarity could not be imperilled on its account. It was the politician's instinct to look at the affair as a personal matter. Granted that the First Commissioner's action had been rough, even incorrect—to take offence was to be too thin-skinned. Now that complaint had been made and considered, all could be smoothed over by a general expression of confidence that in the future the rules would be observed.

But it was a great deal more than a matter of personal offence. True that the attack was on Hooker's personal position; but in his person were attacked Kew itself, and science and administrative fair dealing. It remained to make Kew less negligible politically; to send the Prime Minister an expression of the weightiest scientific opinion, and finally, to lay the matter before Parliament, through Sir John Lubbock, the natural representative of science in the House of Commons.

Accordingly a full statement of the case was drawn up over the signatures of Sir Charles Lyell and Darwin, of George Bentham, Sir Henry Holland, George Burrows, George Busk, and H. C. Rawlinson, Presidents respectively of the Linnean Society, the Royal Institution, the Colleges of Physicians and Surgeons, and the Geographical Society, of Sir James Paget (the surgeon), William Spottiswoode (afterwards President of the Royal Society), and Professors Huxley and Tyndall.[1] This recited the history of Kew, its debt to the two Hookers, and the overbearing acts of Mr. Ayrton. The concluding paragraphs run as follows:

It but rarely falls in either with our duties or our desires to meddle in public questions; and not until we found Dr. Hooker maimed as regards his scientific usefulness—not until we saw the noble establishment of which he has hitherto been the living head in peril of losing services which it would be absolutely impossible to replace; not, indeed, until we had observed a hesitation upon your part which

[1] Tyndall, writing to Huxley on April 27, remarks that the Government lacks 'inner fibre of rectitude sufficiently strong to resist Ayrton, so the only plan is to lift up the hands of Joshua by external aid. What a smashing memorial could be written on this correspondence.'

we believe could only arise from lack of information—did
the thought of interference in this controversy occur to us.
Knowing how difficult it must be for one engrossed in the
duties of your high position to learn the real merits of a
conflict like that originated by the First Commissioner of
Works, we venture to hope that you will not look with
disfavour on an attempt to place a clear and succinct statement of the case before you.

That statement invites you, respectfully, to decide
whether Kew Gardens are, or are not, to lose the supervision
of a man of whose scientific labours any nation might be
proud ; in whom natural capacity for the post he occupies
has been developed by a culture unexampled in variety and
extent ; a man honoured for his integrity, beloved for his
courtesy and kindliness of heart ; and who has spent in the
public service, not only a stainless, but an illustrious life.
The resignation of Dr. Hooker under the circumstances here
set forth would, we declare, be a calamity to English science
and a scandal to the English Government. With the power
to avert this in your hands, we appeal to your justice to
do so. The difficulty of removing the Directorship of Kew
from the Department of Works cannot surely be insuperable ;
or, if it be, it must be possible to give such a position to the
Director and such definition to his duties, as shall in future
shield him from the exercise of authority which has been so
wantonly abused.

Little as the Government desired to give battle on behalf
of so unpopular a representative, in a cause which could not
possibly do them any good, conflict became inevitable when,
early in July, the question began to be discussed in the public
press. The *Spectator* of July 13 had a strong article, based
on the memorial of the men of science, declaring that if by such
treatment Hooker were compelled to resign, it would be a
great and very real calamity to the nation. An article in the
Daily Telegraph of the 15th suggested the line that would be
taken up by the First Commissioner. The trouble, it was
alleged, arose from his zeal for retrenchment, cutting down the
tropical exuberance of the Kew estimates. The plain fact
was precisely opposite to this. The economies effected were

due to Hooker's management ; the additional estimates were those he had protested against. Accordingly Lubbock gave notice that he would raise the question on the 30th, and called for papers to be laid before the House. On the 25th the papers were tabled. These 128 folio pages were remarkable as combining redundancy with incompleteness ; redundancy in irrelevant matters of ancient history, of which *The Times* remarks that ' so scrupulous an economist as Mr. Ayrton might have been expected to save the country the expense of printing such trivialities ' ; incompleteness in the omission, save for one memorandum, of the correspondence between Mr. Ayrton, Mr. Gladstone, and Dr. Hooker, on which the latter's charges of ' evasions, misrepresentation and misstatements in his communications to the First Minister of the Crown ' were based.

Additional papers, however, were presented to the House of Lords, and formed the basis of some discussion on the 29th. They gave fifty pages of correspondence between the Treasury, the Board of Works, the Civil Service Commissioners, and Dr. Hooker, showing that in the ordinary course of business the Treasury ' intimated to Mr. Ayrton a very decided opinion ' that ' he has failed to treat Dr. Hooker with proper considera- tion,' while ' in addition to this, the Return closes with an important Treasury Minute, dated July 24, which deals generally with the whole controversy, and with ample con- sideration for Mr. Ayrton, admits substantially the justice of Dr. Hooker's remonstrance.' It was very plain speaking to say that ' the Lords of the Treasury are not surprised that in various cases Dr. Hooker should have thought that he had just cause of complaint,' and ' they direct so decidedly that in all matters connected with the scientific branch of the Gardens Dr. Hooker's opinion should be followed, subject only to the consideration of expense, and lay down so distinctly his right to be consulted in all matters relating to the manage- ment of the establishment, that there can hardly be room in future for substantial disagreement.'

The most unpardonable feature, however, of the Return laid before the House was the publication of an official report

on Kew and its management which had never been submitted
to the Director for answer or comment. Ayrton had caused
it to be written by Prof. Owen, who was notoriously hostile
to Kew and to its Director for his evidence before the Science
Commissioners, and Owen had employed all his great dexterity
to belittle Kew and its applications of systematic botany, to
urge the transfer of its collections to the British Museum,
where they would come under his own government, and to
insinuate a bitter personal attack on both the Hookers.

Nothing could so rouse Hooker as an attack on the memory
of his father. He insisted on adding to the parliamentary
papers a vigorous and dignified reply. It was easy enough to
expose the long string of misrepresentations as to the aims of
a Botanic Garden, the actual arrangement of the plants, the
need of a first-class herbarium and library at hand, and a disingenuous comparison with the old-time practice of identifying plants—to challenge the perversion of Hooker's known
views as to the Herbaria at Kew and the British Museum
—to dispose of the approved sneer at systematic botany in
the herbarium, whose ' net result ' was ' attaching barbarous
binomials to dried foreign weeds '—to repel the false innuendoes as to the labours and rewards of Hooker and his
fellow workers. But his greatest satisfaction was to pulverise
the attacks upon his father—not on the strength of his own
assertions, but by citing the Treasury Minute which followed
the Report. Did Owen make sidelong appeal for an official
enquiry into the benefits received from Kew by the leading
gardeners ? He was answered in advance by the Royal
Horticultural Society's address to the Premier, by the meetings
of the botanists and the consensus of the gardening papers.
Did he impute neglect to introduce new, rare, and beautiful
plants ? For five-and-twenty years, from the beginning to
the end of Sir William's Directorate, his Botanical Magazine
was full of descriptions and illustrations of these. Eagerness
to find any handle for attack overreached itself no less signally
in specific charges of mismanaging the trees at Kew. Especially
there was the great *Araucaria*. Look how inferior it was
to its coeval at Dropmore. Sir William was in fault ! Sir

William's son must have had particular satisfaction in giving the actual facts. Here was no matter of opinion. The mere recital of facts was damning to the accuser who was in an official position to know them. The tree in question had come to Kew in 1796 ; it had been planted in poor soil and cossetted till its growth was checked and its strength failed ! Sir William, forty-four years later, found it moribund : he left it a strong tree 30 feet high and 90 in spread. The better appearance of the Dropmore tree was due to its having been planted at once in more favourable soil and better atmosphere.

The debate in the House of Lords on July 29 was marked by one piece of disingenuous tactics that was immediately exposed. One speaker was put up to say, *inter alia*, that the memorial of the men of science in support of Hooker was to be discounted because some of the signatories were moved not by sympathy for Dr. Hooker, nor illwill to Mr. Ayrton, but opposition to Owen's views on a central botanical museum, and other differences.

This gave Prof. Huxley, as one of the signatories, occasion to write to *The Times* (July 31, 1872), pointing out that the scientific memorial was in Mr. Gladstone's hands a month before anyone knew of Owen's participation in the affair, and that no divergence of opinion as to the botanical value of Kew touched the points at issue.[1]

Before the debate in the Commons, on August 8, various negotiations took place, through Sir John Lubbock. Mr. Gladstone wrote to him that Ayrton had no intention of giving offence. Hooker responded in kind. Then Mr. Gladstone

---

[1] 'I am not aware that there is anything (except its strong infusion of hostility towards Dr. Hooker) in the paper presented two months ago by Professor Owen to the First Commissioner which the memorialists might not accept without incurring the risk of a charge of inconsistency. For example, if I thought it a wise way of convincing people of my fitness to express an opinion upon a botanical question, I might speak of the great science of systematic botany as a process of " attaching barbarous binomials to foreign weeds," or I might advocate the conversion of Kew into a sort of colossal kitchen garden, and the transference of its vast collection to the British Museum, without being thereby estopped from entertaining any of the opinions which have been expressed by the memorialists as to the justice or propriety of the dealings of the First Commissioner of Works with Dr. Hooker.'

Owen's reply in *The Times* of August 8 may be read as an example of ineffective ingenuity.

desired withdrawal of the charge of evasion—a sign, apparently, that the case was too strong to be successfully defended. Hooker agreed on condition that a substantial reply to Owen's attack be placed at once before the House. This was effected by a Treasury Minute, whereupon Hooker, in accordance with the Prime Minister's wish, withdrew any imputations which may be regarded as of a personal character in his letter to Mr. West of October 30, 1871, at the same time requesting permission, in justice to his father's memory, to place on record in the Office of Works his reply to Owen's report. He refused, however, to withdraw the letter itself, the basis of all his charges for the last ten months, lest this should be used against him as a withdrawal of the charges themselves.

Lubbock's speech in the House was effective in its studied moderation, backed by the force of the Treasury's official rebuke to Ayrton's roughness and the incompleteness of his presentation of the case. Ayrton, in a defence 'forty times as able as his written memorandum,' as the *Spectator* describes it, ' exerted his whole capacity in developing this thesis, that when, as Justice Maule said, " God Almighty was addressing a black beetle," He could not be expected to choose His words.' The whole drift of his reply was that he had not injured Dr. Hooker, and that Dr. Hooker was far too low an official to have a right to raise questions of manner with a Minister of the Crown. He was a mere subordinate spending £12,000 a year, while the ' departments I control spend £1,200,000.' It was a great thing for a Minister of the Crown to take such ' trouble to satisfy a person occupying so subordinate a position.' Dr. Hooker ought to have called on the Secretary, if he had anything to complain of, ' like anyone else who was one of a number of subordinates.' His scientific friends had written a scurrilous libel on him (Mr. Ayrton) secretly, though they only knew ' about organic and inorganic matter,' while he knew something ' far higher,' the science of the law. Evasions ! Those were ' *errors used by a slave to escape from the anger of his master, but which a master, conscious of his power, was not in the habit of using against a slave.*'

The House was so taken aback by the strong man's repre-

sentation of himself as the ' weak and helpless victim of a
scientific tyrant ' that it allowed Mr. Gladstone to wind up the
debate without expressing its opinion on what it had heard.
The Government could congratulate itself on escape from a
most unpromising situation. It had avoided producing the
crushing correspondence which, though Ayrton had declared
it should not be produced if asked for, Mr. Gladstone had not
refused. Hooker, though still eager to substantiate his charges
to the hilt, bowed to Mr. Gladstone's wish, and wrote Ayrton
a formal withdrawal of imputations that might be regarded
as personal or incompatible with official subordination. Ayrton
rejected this qualified withdrawal as being less than he had
demanded in the debate. The truce, however, remained unbroken. At last, in August 1874, Mr. Gladstone transferred
Mr. Ayrton from the Board of Works to the resuscitated
office of Judge Advocate General. With the resignation of the
Ministry in 1874 his political career came to an end, as he twice
failed to secure re-election to Parliament. But till then Hooker
lived in perpetual uncertainty as to the next move, and exclaims
to Maw (November 2, 1872) :

> How I long for your liberty of life. You cannot con-
> ceive the depressing effect of working under a chief in whom
> you have less than no confidence. I dread opening every
> letter from the Board, lest it should contain something
> offensive, and I suspect every unusual communication.

Sir Algernon West, in his ' Recollections, ' recalls the part
he played as mediator in the quarrel, and says that he found
Ayrton the more reasonable man to deal with. I think this
is highly probable, for Ayrton had no reason to stand out for
redress of grievances and was quite ready to accept an act
of oblivion and indemnity—for his own indiscretions—and to
promise official correctitude—if he might be judge of what was
officially correct.

CHAPTER XXXVI

LIFE AND FRIENDSHIP AT KEW

A VISITOR to Kew about this period would have found the Director always busy, though never hustling. Entering the house from Kew Green, to which it turned an old-world front of brick half covered with ivy, the visitor would be shown along a passage hung with old engravings to the Director's studies, two simple rooms lined with books. In the wide south window that filled up most of the farther end and looked out upon a green lawn backed with trees, stood a table and microscope. On the right stretched tables and a long desk covered with letters and reports, perhaps a pot of strange flowers and several coloured drawings of rare plants. Over the mantelpiece hung a medallion of Sir John Franklin and a portrait of Darwin, and on the walls various portraits of botanical worthies, including his father and Lindley, as well as some of the beloved Wedgwood portrait medallions framed.

Somehow he would generally be able to steal time from his long day to show his visitor something of the beauty and the scientific worth of the Gardens, for he was proud of both. He was eager to stir interest in Kew for its own sake; well-informed public opinion would resist its possible starvation by a penny-wise Government.

When he sallied forth, it would not be in the conventional silk hat and black coat always worn by Bentham and Oliver in the Gardens. Much travel had confirmed his liking for comfortable clothes; he appeared in the freedom and ease of a light suit and a flat topped felt hat or occasionally a white 'topper.' It is recorded, however, that once being cornered in the Herbarium by distinguished visitors, he dashed into Oliver's room and borrowed a black coat in place of his working jacket.

Each corner of the Gardens would suggest a particular aspect of Kew's activities : travel and discovery ; special modes of cultivating tropical plants, which at last made even the languishing plants flourish and the flourishing ones expand beyond their enforced limits to the veritable splendour of their own homes ; and not least, the abounding benefit arising from the practical side of economic botany.

He always rose early and worked before breakfast. As to his ordinary routine, the day's work had its share of outdoor movement in the morning round of the houses and gardens. This, the gardening side of his work, though he enjoyed it, was not his special *métier* in the same way as botanical science. He was not a gardener steeped in the empirical treatment so different from the plants' natural conditions with which travel and travellers had made him familiar. The story runs that on his round of the houses he marked a particular plant and gave an order : ' Don't water this ; it is in Nature's three months' drought.' The foreman followed just after with a nod and a whisper : ' Never mind what Sir Joseph says ! ' All the same, cultivation reached a high level under the Hookers' regime, though with supplies pared down to the barest minimum, it was a struggle to maintain things adequately. The palmiest days for this side of the Gardens could only be when the Department was ruled by a minister who had personal appreciation of such work and helped it with a liberal hand.

The largest part of the day's work, however, lay in the correspondence. Letters poured in every day from Europe and Asia, Africa, Australia and America, with enquiries about plants large and small. In the Herbarium curator and assistants would be busy naming plants from the most out-of-the-way parts of the world. These were generally sent in duplicate; one specimen going to swell the Kew collection in return for naming plants which the senders could not identify. Correspondence, much of it in Hooker's own hand, was maintained with the directors of botanical gardens elsewhere, and with collectors and unofficial correspondents. The raising of useful plants from seeds and cuttings and sending them to new countries was a vast undertaking in itself. Reports on the ordinary work and on the special subjects referred to Kew had to be written, the Botanical Magazine and the Icones Plantarum to be published, work to be done on the Colonial Floras that were being issued in connexion with Kew. It was only after official work was over that he could turn to his own original work, and official work did not necessarily end with official hours.

Yet all this never cut short his scientific work; the Botanist was never swallowed up in the Official, though he kept in the closest touch with the details of administration. In all this he looked well after his subordinates. He never lost a chance of picking up a promising young man to whom he could give work in the Gardens till he was fully trained and thus exempted from Civil Service examination before being added to the staff. Often he found excellent places for them in Colonial botanic gardens where they could best serve science and keep in close touch with Kew. His personal interest here is illustrated by a cheery letter to the elder Oliver in 1865, telling how he had bidden one of the other Garden officials restrict his correspondence with a gardening paper to time outside the official hours of 8.30 to 5. At the same time he warned him about excessive smoking and his habit of rushing back to work immediately after meals—'which you should be told of too! He suffers from dyspepsia (no wonder).' Prof. F. W. Oliver remembers his father obediently resting three-quarters of an hour after lunch. He was always fearful of assuming the position of scientific mentor over his subordinates, especially at first. Still, in case of need, he would lose no opportunity of keeping their work on the right lines, till they found their way for themselves.

Bad workmanship and waste of time were his abhorrence,

and he would condemn them emphatically.[1] To give must always be to give of the best. Judge of his horror when once he found Crump, the Herbarium man, ' picking out the worst specimens from the borders for von Mueller, and then '—what was almost worse than such misplaced parsimony—' making them up into shocking bad parcels,' for he himself was excellent at making up parcels, and often sent away plants with his own hands. This was how he was occupied when Prof. Daniel Oliver first set eyes upon him in 1858, in the little room to the right of the Herbarium door.

Mean motives were even more hateful to him. To protect the Gardens from the dust and dirt that came from the increased traffic outside, it became necessary to raise the wall along the Richmond Road. He would have preferred simple railings, for they would have added to the amenities of the district, only they would have ruined the Gardens. Thus he was regretfully compelled to resist the local property owners, who desired the railings so that a nice view might be opened up for their houses overlooking the Gardens. But this regret was mitigated when he found that the nice view thus obtained was to be a ground for raising their tenants' rent.

In general he was outspoken and downright, but he could be beautifully diplomatic. Prof. Oliver was with him when he interviewed Disraeli about a pension for Fitch, the admirable botanical draughtsman. Disraeli was rather unwilling, but Hooker played on his Imperialist feelings by showing him drawings by Fitch of the Victoria Regia and suchlike high-sounding names, and succeeded. The auditor was greatly tickled.

For young people he had a great liking ; an unquenchable touch of boyishness kept his spirit from ageing. Prof. F. W. Oliver tells how as a boy he had one day climbed up an oak tree in the Gardens. Hooker at the same time was moved to ascend the Pagoda close by, and spotting the boy by the move-

---

[1] In his last letter to the Secretary of the Linnean Society he begs to have the leaves of their Journal cut, after the good example of the Geological, Royal, and Statistical Societies. He wastes so much time and temper over cutting the leaves of the books which must be read, that he fears they will be registered against him aloft. It will be a mercy to him to have the pages cut.

ment in the branches, hailed him, and they exchanged greetings
and fun. As they walked back, they were met by one of the
Hooker boys. Hooker told him of the adventure ; the spirit
of rivalry was stirred, whereupon he set them on a challenge
climb up the cables of the great flagstaff. The result has
faded from memory ; the picture that remains is of the central
figure ordering them down when honour was satisfied. So,
too, he always looked in on the children's parties ; Prof. Oliver
remembers his appearing from under the table as a lion, and
a very fine lion he made.

It was this same buoyancy of spirit that made it so difficult
to induce him to talk of the past. In a moment he was back
in the present and the future, the things that were being done,
the things that still might be accomplished.

His unceasing interest in the education of his sons, as they
reached school age, is reflected in his letters both to Darwin,
whose sons were past that age, and to Huxley, whose cares
in that direction were beginning. In 1872 Charles Hooker was
at the International College at Isleworth, under Dr. Leonhard
Schmitz, a modern school, where the main stress was laid on
science and modern languages. Brian was at a preparatory
school. It was disappointing to find that at a scientific school
science was not yet emancipated from bookish methods, while
at a literary school the headmaster did not know what was
inside his books of literature.

To T. H. Huxley

(Christmas 1872.)

I am disgusted with the so-called Science teaching at the
International, and have written a sharp remonstrance to
Schmitz : it is an utter sham, worse by far than nothing,
and calculated to bring the thing into contempt.

Per contra, Brian brings from Weybridge, as a school
prize, a copy of Chaucer, with all its obscenity, verbatim
and literatim, reproduced,—a sweet thing for an ingenuous
youth of 12 ! So I send a shell into that camp, and am
answered that it is a mistake, and that the Master (a Rev.
D.D.) never read Chaucer and got it as a prize for another
boy who had ' been examined in the " Faery Queene." ' I

don't see the connection myself. Perhaps the D.D. thinks that Chaucer wrote the latter. I have prescribed for him a course of the ' Miller's ' and ' Reeve's ' tales with analyses.

A pleasanter echo came from the Himalayas after more than twenty years.

To Charles Darwin

November 1873.

I am in a state of temporary inflation—a book just published on the military operations in Sikkim says of my Travels : ' Never was the officer commanding a force favoured with a fuller, more able, or more lucid report of a country and its inhabitants than I was by the study of Dr. Hooker.' I wonder whether Leonard [1] will ever display such military sagacity and acumen as this Commander-in-Chief ; and he has his reward by being made ' Keeper of Crown Jewels,' a sort of Lady's Maid Extraordinary, you will say.

The serious illness of his friend Huxley gave occasion for drawing the links of friendship yet closer. Consistent overwork had led to a breakdown, aggravated by the black misery of acute dyspepsia. In 1872 a trip to Egypt improved matters, but much of the good was undone by renewed overwork, coupled with the worry of a wholly gratuitous lawsuit. On his move into a new house, a rascally neighbour pretended that his property was damaged by certain building operations. His efforts at blackmail were contemptuously thrown out in the Courts, but, as he was a man of straw, no costs could be recovered from him, and Huxley found himself heavily mulcted for being in the right.

His friends were deeply concerned at the threat of a renewed breakdown ; and Darwin, in whom generosity and delicacy went hand in hand, organised a joint gift from eighteen such friends, as ' to an honoured and much loved brother,' which should enable him to rest, free from every care, until he had won back to health. (See ' Life of T. H. Huxley,' chap. xxvi.)

[1] His third son, now Major Darwin, R.E.

To Charles Darwin

Monday (March or April, 1873).

I write by return of post. Come on Wednesday. I am so happy to think you can. I have never liked to worry you by asking—thank goodness you know that.

Fanny called on you the other day with some such proposal on the tip of her tongue. She had suggested to me the paying to Huxley's Banker the amount of his law expenses (to be raised by you, I, and a couple or so more). I asked Tyndall the night before F. called on you, and he thought the affair too small, and that H. would not like the 'stealing a march upon him.' So we agreed that F. should say nothing to you about it. The matter has however never left our minds.

My impression is that the thing (raising £3000 or so) should be done, if we can assure ourselves that H. will have it, but I feel sure that this will be a difficulty. He and I have been railing at the testimonial system within the last few weeks; [1] and as a public testimonial I feel sure he would not accept. I have beaten my brains to find out if we could practise a pious fraud, and hand it over to him as a legacy from a defunct friend—but he is a deal too sharp, and no one *could* be other than open with Huxley. I can't conceive deception, however innocent, in his presence. I have no time to say more, but do, I pray you, cudgel your brains. I will come to lunch with you to-morrow, Tuesday, lest weather or health should prevent your coming here on Wednesday.

The gift was accepted in the spirit in which it was offered. On April 25, Hooker tells Darwin:

I am charmed with Huxley's noble-minded letter. We had a walk and talk together yesterday, but no allusion passed—but he said he was determined on a long holiday and was very doubtful whether to give up his summer Lectures to Schoolmasters or no. He asked whether I would go with him in July to Auvergne and Germany;

[1] Apropos of certain memorials and monuments to men of science, to which as P.R.S. he felt bound to subscribe (' these are the luxuries of Presidentship '), Hooker had recently told Darwin he need not contribute to one of them: ' I see no call *whatever*, and disapprove this eternal touting for dead bones.'

I promised that I would if I did not go to America, of which I have heard nothing more.

This is modestly put ; in plain fact he had to wrestle hard to overcome the invalid's stubborn desire to ' carry on.' A special inducement was this visit to the volcanic region of the Auvergne with Scrope's [1] classical volume, which they both knew and admired, as a guide-book.

They began their month's trip on July 2. ' I will take great care of Huxley,' he wrote to Darwin. He was loaded with doctor's orders as to what his friend should eat and drink and avoid, how much to sleep and rest, how little to talk and walk, orders

> that would have made the expedition a perpetual burden had I not believed that I knew enough of my friend's disposition and ailments to be convinced that not only health but happiness would be our companions throughout.

And so it was. After the first few days, depression was lightened ; mental recreation was found by picking up at a bookstall a ' History of the Miracles of Lourdes,' which were then exciting the religious fervour of France, and the interest of her scientific public. He followed this up with keen interest, getting together all accessible treatises on the subject, favourable or the reverse, and forming a very definite opinion as to the nature of the original ' vision ' from which the rest followed.

By the end of another week, he was equal to any expedition they cared to make in the still primitive conditions of Central France and its rural districts. Geology was an unfailing lure ; and near La Tour on the Pic de Sancy they made what they thought was a new discovery—namely, evidence of glacial action in Central France ; striated stones, a seemingly glaciated valley and huge perched blocks. (See ' Nature,' xiii. 31, 166.)

---

[1] George Julius Poulett Scrope, F.R.S. (1797–1876), a pioneer, with his friend Lyell, of modern geology, though after the Reform Act of 1832 he devoted himself principally to social reform. His two most important works were on *Volcanos* and on the *Geology and Extinct Volcanos of Central France*, which ' is still carefully read by every geologist who visits Auvergne.' He was awarded the Wollaston Medal in 1867.

It turned out, however, that these had been already observed by Sir William Guise in 1870 and von Lasaulx[1] in 1872.

Geology, too, offered a special interest at Le Puy in the skeleton of a pre-historic man found in a cave with the remains of a rhinoceros, elephant, and other extinct mammals. This Huxley carefully examined and sketched.

Then, to quote Hooker once more,

after leaving the Ardèche, with no Scrope to lead or follow, our scientific ardours collapsed. We had vague views as to future travel. Whatever one proposed was unhesitatingly acceded to by the other. A more happy-go-lucky pair of idlers never joined company.

So they wandered to Dauphiny ; then, driven from Grenoble by heat and maleficent drains, to the Black Forest.

By August 2 he was home, ready to attend the meeting of the British Association at Bradford, having parted from Huxley at Baden Baden, ' really remarkably well,' as he reports to Darwin.

On his return he found a surprise for himself and his friend.

Kew : August 4 (probably 5), 1873.

Taking up the *Times* last night, over a ' Jamaica ' in the kitchen, I find it announced that you, Tyndall, Airy, and self, are all made Knights of the Polar Star !

[His own insignia had arrived]—a great nuisance, having to send it back with a long yarn. I shall get into the black books of all the crowned Heads, and think of putting up a notice warning off all such.

For acceptance was formally barred by the rule that ' the Queen neither authorises in her officers and Civil Servants, the acceptance, nor allows the wearing of these orders,' in accordance with which Hooker had previously refused decorations offered by the Tsar and the Emperor of Brazil.

A common line of action had to be arranged, for even one who, like Tyndall, held no official position, was not free in the matter.

---

[1] A. von Lasaulx, a Sicilian, published *Ein geographisches Charakterbild*, 1879, and *Der Aetna* in 2 vols. in 1880, from the MSS. of the late Dr. Wolfgang Sartorius.

To T. H. Huxley

August 8, 1873.

As to the North Star—it is only given for Scientific or Literary merit, is very limited (to 50 I think), and a very great mark. All which is 0 to the purpose. The Queen's Order in Council is absolute. ' No subject of H.M. shall accept a Foreign Order, or wear the Insignia thereof, without previously obtaining H.M.'s permission to that effect.'

It goes on to say that no permission can be granted except for ' active and distinguished service before the Enemy or except the person shall have been actually and entirely employed beyond H.M.'s dominions, in the service of the Foreign Sovereign.'

This would equally apply to the Order ' pour le mérite,' which I should decline on the same grounds. Now comes the hitch. Little Sweden is very proud of this Order, which is sent to us at the instigation of the Swedish Academy without a doubt—and who but such would single out Airy, Tyndall and you and me. Such concentrated wisdom is not of Courts and Camps—and it appears to me that hasty action on the part of the representative men (of R.S.) might give offence. So I took the liberty of suggesting to Mrs. H. to acknowledge receipt of communication in your name, and add, that owing to your absence, it would be some time before you could write in person. Then I wrote to Sec^y. of Embassy, telling him I had refused similar Orders and must this, but that if the sending back the Brevet and Decoration would give offence, I would make further application to the Foreign Office for instructions. His first answer was evident bewilderment, which was followed by next post by another very nice letter, to the effect that should permission to accept be refused as, he added, it no doubt would, he hoped that I would still not refuse to retain the brevet and decoration. I took this to the F.O., and was advised by the head of the Treaty &c. Department not to send these back, as it would clearly give offence, but to let my refusal to accept stand. The position being this, that neither I nor the Queen can prevent the King of Sweden *naming* me one of his Ritters, whether I accept or no, but that this cannot absolve from the duty of declining to accept. Then I had to revolve in my mighty mind what to say to H.M.

of Sweden—which I settled by asking H.E. Baron Hoch-schild, through Count Steenbock, to ' convey to H.M. the K. of S. and N., my grateful sense of the honour he has conferred upon me, in holding me worthy of being created a Knight of so illustrious an Order as the North Star, which is distinguished for the high Literary and Scientific attainments of its members.' My previous refusal to accept, being official, holds good, and I retain the badge and brevet to please them. If this course approves itself to you, you might write to Count Steenbock, expressing your regret that H.M. Orders in Council peremptorily preclude the formal acceptance and wearing of the Order, and thank him in the sense I have (i.e. for his intentions) and either say nothing about badge and brevet, or that you retain them as a pledge of H.M.'s gracious message. Address, Count Otto Steenbock, 2 Great Cumberland Place, W.

The death of Lady Lyell in 1873 broke one of the links with olden days. She was the eldest daughter of Leonard Horner, the geologist, education reformer and public worker.[1] Among his friends were the elder generation of Hookers, Lyells, and Darwins, through his friend S. T. Galton, who had married Dr. Darwin's sister.

Mary Horner, eight years Joseph Hooker's senior, was first of a group of sisters distinguished for beauty and charm, which touched the artist in him as well as the friend.

To Charles Darwin

April 25, 1873.

Lady Lyell's death is a complete upset. I called to-day and had a long talk with poor Mrs. Lyell [2] and saw (at her wish) for the last time that most lovable face shrouded in flowers in the coffin—looking so calm and beautiful. Amid a flood of later memories my mind rushed back to long years ago, when quite a boy, I felt rather than thought, that I

---

[1] 1785–1864. He was the first Warden of University College in London, a founder of the Edinburgh School of Arts and the Academy, the well-known boys' school, President of the Geological Society, and afterwards for twenty-five years a Factory Inspector under the new Factory Act.

[2] Katharine Horner (1817–1914), the fourth sister, married Charles' brother, Col. Henry Lyell.

never could look at it without emotion—I used to dream
of it as a child. I have no morbid or other liking for seeing
the faces of the dead, but am glad I have seen this ; it was
so beautiful—and I should not have liked my last thoughts
of her to have been coupled with a face worn by sickness.

The happy but strenuous tenor of his life was soon to be
broken in upon by a grievous and unexpected blow. This was
the death of Mrs. Hooker. The year had opened with brightness
as well as shadows. The Presidency of the Royal Society
was the crown of his scientific career. In connexion with a
Botanical Congress at Florence during the spring, he made a
charming tour with his wife in Northern Italy. He was happily
able, as his letter of July 18 tells, to take once more a deep
draught of the music he loved so well. But the physical
strain was more and more tense, and was aggravated by a bout
of whooping cough early in the year, caught from his children
who brought it back from school. A sign of fatigue in
September was the recurrence of ear-trouble, while, as a last
straw, Prof. Dyer was unable to continue as his private secretary.

I am in the depths of despair [he tells Darwin]. He is
quite right—he ought to be at original work, and I am only
too glad to think that he will now settle to good work, though
to me the loss of his hour a day is dreadful.

As has been told already, his request to the Office of Works
for the assistance so imperatively necessary with the expansion of the functions of Kew, was coolly shelved until in the
following January he appealed direct to the Treasury. This
time the Treasury officials showed no unwise parsimony. The
Office of Works was invited to do its duty ; and when, being
internally at sixes and sevens, it was again in default, the Prime
Minister himself, Mr. Disraeli, intervened. Prof. Dyer was
appointed Assistant Director in the summer of 1875, and in
August the most obnoxious official was politely retired.

It was in the midst of this wearing strain that the blow
fell. Mrs. Hooker died quite suddenly on November 13, 1874.
She had lived to see her husband reach the highest scientific
fame and the highest position in the scientific world. The

daughter of a botanist and a considerable botanist herself, her active interests marched with his. As a good writer of English, she constantly aided him in his writing and correction of proofs, where he relied greatly on her judgment. Her knowledge of foreign languages enabled her to play her part whether as hostess receiving the many foreign visitors to Kew, or as guest with her husband on his official visits abroad. Her skill was also shown in the translation of Le Maout and Decaisne's ' General System of Botany,' which was published in 1873, with additions by Dr. Hooker. An attachment buttressed by mutual affinity, by a share of common interests and pursuits, by the same measured firmness in ensuring the same ideals of family and social life, had taken into its fabric the common joys and sorrows of three-and-twenty years.

Of the six surviving children, three still required care. The inevitable problems of the home and education added their undivided anxieties to Hooker's workaday burdens. For the blow had fallen at a moment when not only the honours but the labours of the time had been heaped up.

His elder daughter was able to be a great help to her father, and fortunately his aunt, Mrs. Dawson Turner,[1] and her daughter (afterwards Mrs. Calverley Bewicke), were able to join his family circle and were of great assistance to him. He especially delighted in his cousin Mrs. Bewicke's beautiful voice and cultured singing—a voice afterwards devoted for many years and until, indeed, the present time, to the enjoyment of the sick and wounded in Westminster Hospital.

Two brief references may be permitted to his sense of loss. One is in a letter to Huxley, a fortnight after the event.

December 7, 1874.

As for me, barring fits of depression, I am getting on. I am still in a sort of trance—my memory of the immediate past is blurred, and I have difficulty in recalling her features. I think of her mostly as the girl I so long and so dearly loved

---

[1] The wife of Dawson William Turner (1815–85), philanthropist and educational writer; son of Dawson Turner; demy of Magdalen College, Oxford; M.A. 1840, D.C.L. 1862; for some years Headmaster of the Royal Institution at Liverpool, and a most generous benefactor to the London Hospitals.

25 years ago, and feel as if I had never returned from the East to marry her,—and never shall now. And yet I am perpetually stumbling into pitfalls of recollections of the immediate past.

The other is in a letter to Darwin eighteen months later; from Nuneham, near Oxford.

I am here on a two days' visit to a place I had not seen since I was here with Fanny Henslow in 1847 ! I cannot tell you how depressed I feel at times. She, you and Oxford are burnt into my memory.

The following recollections, contributed by Mrs. Bewicke, date mainly from this period, when she and her mother came to live at Kew after Mrs. Hooker's death.

From my earliest childhood to the close of Sir Joseph Hooker's long life, I remember ' Cousin Joseph,' as he liked us to call him, as the best and kindest of friends to my mother and myself. His kindness was especially shown at a period of great trouble and anxiety in our lives. It was during this time that I had an opportunity of knowing Sir Joseph well, and appreciating his truly lovable and noble nature.

My father was ill and had been ordered a long rest and a voyage to the Antipodes. My mother and I were in great trouble, when Cousin Joseph, with the thoughtful kindness so characteristic of him, proposed that we should go and live with him at Kew. He would take no denial, and made us feel it was all for his benefit, when under the circumstances it was entirely for ours. He made Kew a real home to us, and I think my mother was a help to him with his children, while I thoroughly enjoyed the companionship of his daughter Harriet, my contemporary in age. My brother too, then quite a boy, was always a welcome guest, and Cousin Joseph took the greatest interest in his work, helping him in every way he could.

Nothing gave Cousin Joseph greater joy than the progress of his children, and, if one of them brought a good report from school or answered correctly any of the many questions he asked them at meals, it would make him proud and happy for hours. There never was a father more appreciative of the

good points of his children, and my mother and I often said
to each other that he was either ' up in a balloon ' or ' down
in a diving bell ' according as the children's reports were good
or bad.

About many little things he was particular.   For instance,
I soon learned that I must put down my knife and fork very
quietly at table, and, if he asked for half or quarter of a cup of
tea, I must not give him more.   Again, I must not leave a
door open, especially my bedroom door.

I had been brought up to think it a virtue to go to bed
early, and was greatly astonished to find that Cousin Joseph
looked on it in quite a different light and was really shocked
if I proposed going to bed as early as ten o'clock.   He himself, though a very early riser, used often to sit up until
two o'clock in the morning writing.

He was very fond of music.   His eldest son, Willie,
played the violin, and every evening I used to sing to him
for about an hour.   About nine o'clock, a big bundle of
newspapers tucked under his arm, he would come up to the
drawing-room for a little recreation.   I well remember how
he would stretch himself out in an armchair, his head thrown
back, his eyes closed, and, with a sigh of relief, would say,
' Now sing to me.'

His favourite songs were, ' Angels ever bright and fair,'
' Old Robin Gray,' ' Robin Adair,' and Blumenthal's ' Love
the Pilgrim.'   The first of these songs is associated in my
memories of Cousin Joseph with a visit we made in his
company to Mr. Spottiswoode, the Queen's printer, and Mrs.
Spottiswoode.   We were invited for a few days, and there
was a large house party of distinguished people, one of them
Henry Irving, the actor.   The first evening was devoted to
science, Mr. Spottiswoode, a keen student of science, giving
us and some of his tenants an after-dinner *causerie* on
spectrum analysis, telling us how we could be sure from
the spectrum what metals there were in the sun and the
planets.

The next night one of the visitors, a great lady,
monopolised the piano in the music room.   Cousin Joseph,
as was usual on Sunday evenings, wanted me to sing his
favourite song, so when the piano became vacant, he helped
me on to the platform, and, though very frightened, I sang

'Angels ever bright and fair.' When I had finished, the aforesaid great lady remarked, ' Don't know whether you are aware of it, Miss Dawson Turner, but you sang that song quite out of time.'

Very much embarrassed, I was beginning to apologise, when Sir Henry, or rather Mr. Henry Irving, as he was then, jumped on to the platform, and said for the benefit of the audience, ' No doubt her ladyship is right, but in my opinion that song requires the time broken for the expression.'

' That may be very well, Mr. Irving, with an ordinary composer, but it does not do with Handel.'

' Ah, when I'm reciting *Shakespeare* doubtless you would like me to count four at the full stops and two at the commas, but for my part I prefer the time broken for the expression.'

It was very kind of Mr. Irving to defend such a young girl as I was then, and Cousin Joseph was so pleased that he at once asked my champion to come down to Kew and spend a day at the earliest opportunity.

Our life at Kew was as simple as it was happy. With regard to meals, as in everything else, Cousin Joseph was abstemious. There was breakfast at half-past eight, luncheon, or rather dinner, at one o'clock—two courses as a rule, and some very light wine, and no afternoon tea unless there were visitors down from London, and he thought they would like it. His own evening meal, consisting of tea and cold meat of some kind, was at seven, with nothing else afterwards.

Cousin Joseph liked us young people to talk with him at meals and other times, but to talk subjects, not people. I remember that I would often read a newspaper or book with the special purpose of finding some topic likely to please and interest him when we met at table. When we drove into London with him, he would tell us the names of the big houses and their owners, and then expect us to know them as we drove back. If Reggie, his youngest son, was in the carriage, he would tell him to count the different trees we passed, his idea being, of course, to teach us to be observant. He would always take us himself to see the pictures at the Royal Academy, and he was, I remember, a great admirer of Leader's landscapes, Hook's seapieces, and Poynter's work. He was less appreciative of the after-dinner speeches at Royal Academy dinners, and I remember his saying apropos of one of them, 'I could not sleep the night before for thinking what I should say, nor the night after for thinking of all the good things I might have said.' Though he encouraged us to express our views, again on *subjects*, not *people*, Cousin Joseph would often tease us, ending an argument with his favourite expression, ' What you say true, my dear, is perfectly correct.' He would parry our questions with grave humour. For instance, on one occasion when I asked what the doctor, whom he had been consulting, had said about his health, he replied, ' I had to be very firm with him, very firm indeed, or he would have stopped all the things I liked.'

We young people often used to smile at the way everyone spoke of our distinguished scientist and to him as ' dear Joseph.' I have never known a man more genuinely beloved, and deservedly so ; so childlike and ingenuous he was and so really modest, always putting others before himself, always unconscious of his own importance. One of his most delightful traits was his tender-hearted affection for his friends ; and I shall never forget how overcome he was with grief when any of his old friends died, and how anxious he was to do everything to help those they left behind to mourn them.

Many of his friends were distinguished scientists, like himself, Herbert Spencer, Tyndall, Huxley among the number. I remember Herbert Spencer coming to lunch one day and my mother, who was a great student and admirer of his books, asking him what he thought on a certain subject.

' I forget what I think on that subject,' was his reply, ' but you will find it in such and such a book of mine.'

I remember Dr. Tyndall coming down at Christmas and taking Cousin Joseph and all of us on to the lake in Kew Gardens, where there was some skating. Dr. Tyndall began making experiments in sound when, as if in the special interest of these experiments, a thick London fog came on, and, if I remember rightly, he was able to prove to us that sound travelled more quickly in a fog than in clear air.

Cousin Joseph was interested in all the sciences, not only in those he had made his special study. He would often regret that he did not know much about astronomy. He liked my mother to point him out the stars, often going out in the evening on purpose to learn them. I have heard him express regret at the neglect of scientific knowledge in the army ; and I remember he would say, ' Oh, how much suffering might have been spared on that expedition, if only the men had had a little scientific knowledge.'

We often went with him to scientific parties at his friends' houses or the Royal Institution. I remember one party at Dr. Busk's in Harley Street, where he was greatly interested in an experiment shown us, a pith ball moved by light— proving, I suppose, that light is a mode of motion. He took the greatest trouble to explain the phenomenon to his daughter Harriet and myself. I also remember going with him to the Royal Institution, and Dr. Tyndall showing us a ray of light in which he held a bottle which did not impede the light at all. He then made a vacuum in the bottle, put it again in the ray of light, and showed the inside of the bottle in unillumined darkness—proving, I imagine, that light in itself is invisible.

A great friend of his in the medical world was Sir James Paget, the well-known surgeon. We often went to lunch with him ; and I recollect Cousin Joseph's naïve enjoyment of the Norfolk biscuits that always figured on the menu, making a prelude to pleasant reminiscences of his mother's native county.

Many royal visitors came to Kew to see the famous gardens. I forget the date of the Emperor of Brazil's visit, but I remember Harriet Hooker and I were sleeping in a room looking on to Kew Green, when we were awakened by the noise of a carriage being driven round the Green at six o'clock in the morning. There were four gentlemen in the carriage, and when the front door bell rang, we guessed it was the Emperor and his suite, as we had been told he was a very early riser.

Cousin Joseph went down as quickly as he could, but not before the bell had rung more than once, and a Brazilian parrot we had in the hall had given the Emperor a warm, if hardly polite welcome. Then he accompanied the Emperor and his party round the gardens, while we waited their return for breakfast.

To our surprise the Emperor took the head of the table,

and asked us in French if we would sit down and take coffee. I had been teaching little Reggie Hooker French, and to his father's delight, when the Emperor, addressing the little boy, asked 'Parlez-vous français?' my pupil promptly replied, 'un peu.' The Emperor had hit upon the one sentence Master Reggie happened to know.

The Emperor, who I remember was a very fine, tall; good-looking man with a charming manner, had the royal gift of never forgetting a face. Meeting me two years later at a garden party given by Canon Duckworth in the Dean's Yard, he at once recognised me.

Another recollection connected with a royal personage is of being taken by Cousin Joseph to Buckingham Palace, shortly before a visit that the Shah of Persia made to England, and of his saying, 'How foolish to tell me to bring all these plants over from Kew and bed them in Buckingham Palace Gardens! The Shah will think they grow out of doors in England.'

A charming group of royal visitors to Kew were the Princess Alice of Hesse and her sisters. They came with two or three ladies-in-waiting, and Harriet and I took them over the gardens. They were really very good-looking girls with charming manners, expressing so gracefully their thanks for our escort and hoping they were not tiring us. Cousin Joseph admired these young Princesses very much.

The close of our visit to Kew was marked by an event of great importance, Cousin Joseph's second marriage. If there is anything in a name, it seemed most appropriate that Dr. Hooker, a botanist, should marry a lady of the name of Hyacinth Jardine.[1]

Willie Hooker and I were present at the wedding, which took place very quietly at Hereford. Afterwards I joined Dr. and Mrs. Hooker at a meeting of the British Association, where he received the congratulations of all his scientific friends and we had a most interesting time.

Later, when Dr. and Mrs. Hooker went on a little tour to Oban and the Isle of Skye, I was invited to accompany them, Cousin Joseph encouraging me to sketch all the time. Mr. Arthur Lyell also went with us, and kept all our accounts for us during the journey. It was a case of history repeating

[1] See p. 202.

itself, for, as my cousin pointed out, sixty-two years before, in 1814, a Lyell, a Hooker, and a Miss Dawson Turner with her parents and a younger sister had made a tour together through Normandy to France, and on that occasion too, the Lyell, afterwards Sir Charles Lyell, the distinguished geologist, kept the accounts for the party. A diary written by the younger ladies of the party, Maria and Elizabeth Dawson Turner, with many beautiful sketches of the churches seen on the way to Paris, still exists, an interesting record of a tour that had its part in Dr. Hooker's family history. For the Hooker of the earlier journey was William, later Sir William Hooker, Sir Joseph's father ; and Maria Dawson Turner, the older of the two sisters, was later to become Sir William's wife and the mother of Sir Joseph. The fifteen-year-old Elizabeth was the future Lady Palgrave, wife of Sir Francis Palgrave.[1]

[1] See p. 203 ; the journey repeated, p. 341.

CHAPTER XXXVII

LOSS AND GAIN

THE one anodyne for loss and sorrow lay in the remedy he prescribed for others—application to work, especially official work with its impersonal necessities. Kew, single-handed, was growing an intolerable burden ; intolerable, also, as has been recounted, the immobility of his official superiors ; the prospect of a new official ' row' used up the last of his fighting spirit. In all this ' the Royal Society is my " great consolation "—everything there is smooth and pleasant so far.' (To C. D., February 24, 1875.)

Success in this struggle, with the appointment of an Assistant Director in June 1875, brought relief ; but the strain told. Months before, the new labours brought by the Presidency of the Royal Society made him long for the ' peace and quiet and sound sleep' of a week-end at Down. Now with his ' arrears of work pressing and Bentham craving for Gen. Plant.,' he could not break off till after the Royal Society soirée on April 16, when he joined his eldest son and daughter at Algiers for a month. Moreover all through the year from February to September he was troubled with headaches and dyspepsia, varied by attacks of lumbago and bronchitis, ' one off, and t'other on,' and on October 11 reported himself to Huxley (who was still feeling the effects of his recent breakdown) as well again after a horrid bout of rheumatism and ear trouble ; he adds : ' And if I did not get as hipped of a morning as a Huxley, I should be all right.'

As he wrote afterwards to Darwin (March 15, 1879) :

I cannot but think that a little public duty is an excellent thing for any man who has health, energy, and acquirements

enough to perform it, and I think I am not wrong in sur-
mising that in X.'s case such a duty would be eminently
beneficial. I well remember my own extreme aversion to
undertake public duties, and your affectionate encouragement on very many occasions when I would fain have held
back. I now know how good it has been for me, and how
grateful I am to you for your encouragement I only know.

Now, too, another link with the past was broken, another
lifelong friendship ended. Already in 1871 he had written
apropos of 'dear old' Murchison's death and Sedgwick's
retirement from lecturing : 'After a year or two there will
have been a regular clearing out of the old philosophers, all
dying at a ripe old age.' Chief among these, Sir Charles Lyell,
who had for some time been failing, died on February 22, 1875.

This is another black week [he tells Asa Gray on February
26]. Dear Lyell is gone. . . . Stanley had as good as offered
[Westminster Abbey for his burial], and there we shall lay
the grand old Philosopher,—the kind friend and sympathiser
in all my ups and downs. He was indeed great ; so truthful,
so fearlessly honest, such a hater of everything mean, small,
or doubtful. To me the loss is very great. I loved him
so, as I did his wife.

To Charles Darwin

February 24, 1875.

I feel Lyell's loss most keenly, he was father and brother
to me ; and except yourself, no one took that lively, generous,
hearty, deep, and warm interest in my welfare that he did.
I cannot tell you how lonely I begin to feel, how desolate,
and how heavily the days, and worse still, the nights, hang
on my mind and body. Well! it is all for the best, i.e.
the best that man is born to, poor lot as that may be, it is
one that no one really wishes to exchange for an unknown
one ; and we are hence logically driven to the conclusion
that the sum of life is more happiness than the reverse.
Assuredly the sum of happiness derived from having known
and loved Lyell is greatly in excess of the pain felt at his
loss : the gap he filled has to be compared with the chink
his mere absence for the rest of life opens.

I have arranged for his burial in Westminster Abbey. On Monday I got up a petition signed by some 50 Fellows of the Royal, Geological, and Linnean, and at Stanley's suggestion and promise that it should be attended to (communicated to Spottiswoode) I sent it in yesterday. It was by mere accident I went to Town on the Monday to vote at the Athenæum, heard of Lyell's death, and was able to secure so many voters to sign the petition.

As to any other testimonial, I think that this is so incomparably beyond any other that none need be thought of —any other would in my eyes dim the lustre of his memory —his Principles *must* live for ever—they will no more be forgotten than Plato's or Faraday's works : they will always be classical. The idea of a testimonial being in any way required seems to me rather an underrating of the durability of his works.

The choice of an epitaph involved much consideration.

To Charles Darwin

June 20, 1876.

Mrs. Lyell has asked me to help her with an inscription for Lyell's slab in Westminster Abbey—such as Stanley may approve (I have fainted away twice).

She sends me two, neither of which I like. I enclose them.

I have asked for some days to consider, and the longer I do so the more awful the task appears. How would it do to suggest something of this sort :

His long life was devoted to searching for Truths and to reasoning on their Teaching ; and he gave to the Public the results of his labours in a memorable work of enduring scientific value—' The Principles of Geology.'

The epitaph took final form as follows :

Throughout a long and laborious life he sought the means of deciphering the fragmentary records of the earth's history in the patient investigation of the present order of Nature, enlarging the boundaries of knowledge and leaving on scientific thought an enduring influence.

Two later letters to Darwin of 1878 and 1881 touch on the publication of Lyell's Life and Letters.

Mrs. Lyell has been consulting me confidentially as to what is best done with Sir Charles' correspondence, and I have her permission to ask what you think, and if you would kindly help her with an opinion.

I have read a great many of the letters, to Horner and others, and am greatly taken with them—they are so full of matter, so pleasant, lucid, and tell so much of his unwearied labours and of the progress of Geology during its comparative infancy. Then too they are so full of feeling to, and ready recognition of, the labours of others. They are full of local colouring as regards the places (often very obscure) that he visited for the purpose of verifying statements and collecting facts ; and full of little notices of admirable local collectors and Museums that are worthy of being remembered.

Mrs. Lyell has a mind to put all in print for private distribution, after revision and cutting out all passages that could hurt any one (of which I have seen no trace), and afterwards publish a selection as a contribution to his life. My idea is that the number will prove too great for printing, but this must depend on their value. I suggested her picking out a dozen by chance (without looking at them) of the bundle I have perused and sending them to you for your opinion, as to their value to Science. I am a partial witness I know, and so would you be, but that must be taken into account.

Mrs. Lyell has riches and is devoted to Lyell's memory, and if good can be done by the printing now is the time.—
Ever affectionately yours,

JOS. D. HOOKER.

The value of the work would be : (1) The history it is of the progress of Geology ; (2) the evidence of the ease with which Lyell sifted facts and evidence, and the interest attached to the facts.

[1881.] I have been reading Lyell's Life with great interest. It is a great pity that it was not cut down to one volume, but as it is I am only too glad to get it in any shape. I

really think that Mrs. Lyell has given us a very important
contribution to the history of Science—and it does make
one 'warm to' Lyell himself. The accounts of the early
history of the Geological, its dinners, &c., are most entertaining and instructive ; so too is the substance of many
of his journeys, in which he chronicles the labours of many
good men whose names deserve to be remembered. The
account of Cuvier and his way of working is most curious.
The letters to Herschel are the best, they are evidently
very careful compositions.

Do you observe certain passages that seem to *prove*
that he never expected to come into the Kinnordy property
on his father's death ? and that on the contrary he looked
from an early age to providing himself with a modest com-
petency for his latter days.

So the months passed till a fresh thread of happy and sus-
taining companionship was woven into the broken fabric of
his life. ' No one can have an idea who has not experienced
it, what a house of six children is without a female guide—let
the children behave ever so well !' At the end of August
1876 he married Hyacinth, daughter of the Rev. W. S. Symonds,
Rector of Pendock, Worcestershire, and widow of Sir William
Jardine. Her friendship with the Hookers dated from 1864,
when her father brought her to the Bath meeting of the British
Association ; it was drawn closer from 1869, when they fore-
gathered at Sir Charles Lyell's, and visits were frequently
exchanged between Kew and Pendock. To native personality,
education and environment was added a community of general
interests with her husband. Her lines had been cast amidst
science and letters. Her father (1818–87) was a considerable
geologist and a writer of merit [1]; Sir William Jardine (1800–74)
a lifelong student and writer on Natural History, especially
ornithology.[2]

---

[1] Among his scientific books were *Stories of the Valley*, 1858 ; *Old Bones*,
1859 and 1884 ; *Records of the Rocks*, 1872. He also wrote two novels, *Malvern
Chase* and *Hanley Castle*.
[2] He published with Prideaux Selby *Illustrations of Ornithology*, 1830, edited
the *Naturalists' Library*, 1833–45, contributing sections on birds and fish,
founded the *Annals and Magazine of Natural History*, and for a time was joint
editor of the *Edinburgh Philosophical Journal*. In 1860 he was appointed a
Commissioner on Salmon Fishing.

Part of the honeymoon was spent in North Wales. It was on this occasion that, going up Snowdon, they were caught ' in a storm such as I have not seen in mountains since I left the Himalaya,' and had ' the *top* of an umbrella, incautiously raised, blown *in*.'

Then to Glasgow, for the British Association, where they stayed with his niece, Mrs. Campbell, and with the old friend of the family, Miss Smith, of Jordan Hill. Thence for a week to Skye, with a contingent of Lyells and Mr. Symonds ; [1] then to other friends in Perthshire and Fife and to the Hodgsons' in Gloucestershire, thus picking up the strands of many friendships.

One scientific note of this journey is in a letter to Prof. Oliver (September 25) :

> The Geology of all that part of Scotland [from Loch Maree to Dingwall] seemed to me wonderfully complicated, and gave me a new impression of the labours of Macculloch and Murchison. Skye Geology too impressed me much ; the island resembled some of the Antarctic ones in many particulars ; and though volcanic on the whole, it contains beds representative of most of all the British formations from the Laurentians upwards ! and I could not help wondering if future discoveries, say in Kerguelen's Land, may not throw as much light on the Geology of the Antarctic regions, as Skye alone would have done in respect of Northern Europe. Perhaps the Fossil wood of Kerguelen's Land may be the nucleus of a great light.

To Darwin he adds : ' Were you aware that Dickie of Aberdeen had examined the earth beneath the Glen Roy roads and found them to contain *Fresh water* diatoms ? '

But in the midst of this happiness he was deeply moved by the sudden death of a young friend's wife. They had not been married a year. It seemed to open up his own too recent loss and to depress him utterly :

> They were so happy and she so lovable—how I envied them a few months ago. . . . Give the dear fellow my most affectionate sympathy. . . . Oh dear, oh dear, what a weary world it is, and yet I should be the last to complain.

[1] See p. 197.

And remembering the kind words about re-marriage that
had come to him from his friend's mother, he adds with a
look from his own history towards his friend's future: ' I
can but hold the prospect for him deep down in a far off corner
of my heart.'

The new era at home was rendered yet more happy by
the engagement of his elder daughter in the following spring
to Professor Dyer, now for two years established as Assistant
Director at Kew. As he confided to Darwin, the only objection
(a crumpled roseleaf) was

> that it is ridiculously apropos, i.e. commonplace, and reminds
> me of Hogarth's industrious apprentice. I never had any
> ambitious desires for my sons and daughters, and a good
> scientific man though poor (if otherwise honest, as Sydney
> Smith ? said of a poor man) is the best of all matches in
> my eyes. . . . What especially pleases me is that he is just
> the brother-in-law I should like my sons to have.

The birth of a son (Joseph Symonds) in 1877 opened another
happy phase in his varied life. That the infant son of the
President of the Royal Society should visit Burlington House
was an unusual episode.

To Mrs. Hodgson

March 10, 1878.

Hyacinth and I went to the Old Masters the other day,
the first dissipation ! either of us have had since my return.
She took the baby and left it at the Royal Society's rooms
with the Porter's wife, who has a baby of her own, and ' the
President's baby ' created quite a sensation in the House !
What a ' rum world ' it is—the more I think of my own life
and career, the more unintelligible it appears to me. I feel
as if I had been divested of my individuality every ten years
or so of my life, and then given quite another body and
mind.

And though at times in these years memory could not but
cast lingering looks behind and catch the shadows of past
sorrows, still he could say ' nothing can be brighter than my
visible future, little as I now dare to trust to it.' (1878.)

CHAPTER XXXVIII

THE outstanding event of 1877 was the long looked for visit to the United States. This had been half-planned for the last four years. America had a two-fold call for Hooker, in the problem of the North American Flora and the friendship with Asa Gray. These represented the personal and impersonal sides of the same impulse. Their common interest in the same question, approached from different sides, had initiated an unbroken correspondence which deepened in personal and scientific interest with their united appreciation of Darwin. Gray had already visited England four times, and was urgent for Hooker to come over and join him in a personal study of the complexities of botanical distribution in the States.

Two features in the problem which cried most loudly for explanation were the remarkable connexion between the plants of the Eastern States and those of Eastern Asia and Japan—with no living intermediate connexions—and the hard line of division between the Arctic floras of America and Greenland. Independently of each other, Gray had investigated the former [1] and Hooker the latter.[2] Both came back to a common cause in the Glacial Period and the earlier land connexion with an Arctic continent.

But why had not the Glacial Period produced the same

---

[1] 'Observations upon the Relations of the Japanese Flora to that of North America, and of other parts of the North Temperate Zone,' *Memoirs of the North American Academy of Sciences,* vol. vi. p. 377. Read December 14, 1858, and January 11, 1859.

[2] 'Outlines of the Distribution of Arctic Plants.' Read before the Linn. Soc., June 21, 1860. *Trans. Linn. Soc.,* xxiii. p. 257.

results in the great mountain chains of the West, where now only a few botanical 'pockets' of East Asiatic type exist, among plants of Mexican and more southern types ? Unsupported suggestions had been advanced of contemporary submergence of these high lands or of recent unsuitable climate ; actual investigation was to show that these 'pockets' extended over specially favoured areas. Considering that the high mountains would have kept the glacial cap long after it had retired from the other levels of North America, the plants of East Asiatic type could have got no foothold there save in these favoured areas, and by the time that the general change of climate had melted this belated ice-cap, it would also have affected the now treeless prairie district, exterminating these plants and leaving the survivors isolated in the more congenial forest district of the Eastern States, with no possibility of re-invading the Rocky Mountain area, which was thus left open to the plants advancing from the Mexican highlands until they met, not temperate, but Boreal forms.

The Polar problem in its relation to the whole question of distribution was constantly before Hooker's eyes, and it has been noted that in 1873 he was working in this connexion at the flora of North-West America.

The journey this time, though often beyond beaten tracks, could not class with the wholly adventurous trips which had so strongly appealed to his spirit in earlier days, and which he had renounced after his Marocco expedition. Still there was a flavour of the elemental joy and labour of the wild that might not have been welcome to every man of sixty. The two elderly botanists were indefatigable, and Hooker especially, who never carried a superfluous ounce on his bones, astonished the rest of the party by his activity, though his own remark is, ' Gray is a man of extraordinary energy, and though 5 or 6 years my senior is the younger of the two ! '

After seeing his daughter married on June 23, Hooker sailed for New York on the 28th. Professor Dyer was to return in a week to 'carry on' until his chief came back, when he would be free to take the rest of his honeymoon.

With him went his old friend Major-General (Sir) Richard

Strachey, R.E., and his wife. Strachey, like himself, was a Himalayan traveller, who had surveyed the Kumaon valley, and was both a geologist and botanist.

At the invitation of Professor Hayden, chief of the Topographical and Geological Surveys of the United States, they joined the official surveying party which was at work in Colorado and Utah, Nevada and California, whose formal report, it was arranged, was to utilise their general botanical results, especially in regard to the character and distribution of the forest trees.

On the *Parthia* there were only some thirty-five cabin passengers, and he had a state cabin to himself. Finding some excellent books on board, he had occupation in Macaulay, Evelyn's ' Diary,' Keyes' ' Lives of Eminent Indians,' Longfellow's Poems, and one volume of Lyell's ' Travels in North America,' to beguile the tedious voyage, for the *Parthia* was a slow boat, so that from the first, with the prevailing west winds, he despaired of a ten days' passage.

Moreover he found the motion of the screw so unpleasant as contrasted with the rhythmical beat of a paddle-boat's engines, to which he was better accustomed, that he grew more weary of this voyage than of any other he had taken. Boston was reached on the night of July 8. One day was spent with the Asa Grays ; three with Professor Sargent, curator of the Botanical Garden at Harvard and of a magnificent park, the Arnold Arboretum, which was not yet laid out, but was to be the Kew of Boston. He was ' up to the eyes in trees, flowers and shrubs.'

Boston, with its charm of openness and good upkeep, the cleanness and comfort of the labouring classes, where ' coachmen and railway guards look and speak like gentlemen, and in the market the butcher is as clean as the grocer, betraying no disagreeable features of his trade in apron, hands or head,' represents the best of long established culture, in contrast to the grime of London streets. Comment on the comfort and variety of the public conveyances rouses no astonishment in those who can recall the frowsy discomfort of our omnibuses and 'growlers' of the seventies. Everyone seemed to be well educated. A visit to the Museum and Natural History Institute endowed by Peabody at Salem, where the Professor was teaching Zoology to a mixed class of school teachers, for the most part, on the lines of Huxley's courses at S. Kensington, prefaces the remark, ' The thirst for knowledge in this State is most wonderful ' ; and the sight of Wellesley College, a rich man's gift to the State for the education of female school teachers, prompts the reflection, ' Education is the rage here ; wealthy people do not know what to do with their money.'

The journey was broken for a day at Cincinnati and at St. Louis, where the party was joined by Dr. Lambourne, Professor Leidy, ' the very great zoologist whom Huxley swears by, who wants to explore the minute animals, Diatoms, Rhizopods, &c., of the Colorado waters,' further north in Wisconsin, his wife and adopted daughter, Mr. Hayden, head of the Geological Survey, and Captain Stevenson, his chief assistant.

Then followed two nights and nearly two days on the newly made railway to Pueblo across the prairies along the Arkansas river. At Pueblo the Leidys went north ; the others to Cañon City. Then Hooker and the Stracheys went by wagon over the hills to La Veta, visiting on the way Dr. Bell, an Englishman who had settled there, and was President of the local railroad. The main survey party went direct to La Veta by rail, and established a camp at 9000 feet as a centre for botanical work.

> The facilities [he notes] of getting about this world's end of a country are wonderful, but travelling is very fatiguing, as you have to go great distances and there is so much to learn and see by the way, and everything is rough and hard.

Again,

> the education, intelligence and general prosperity of the people still impresses me very agreeably. Here at this wretched collection of scattered ' balloon ' cabins [i.e. rough pine planks] and adobe huts I find *eight or ten* journals and newspapers sold, and of the very latest date, and there are several ' balloon ' churches.

ENCAMPED IN THE ROCKY MOUNTAINS, 1877.

Sir J. D. Hooker. Mrs. Strachey. Mrs. Gray. Dr. Lambourne. Dr. Hayden.
Dr. Asa Gray. Gen. Strachey.

The camp consisted of five tents pitched at the edge of the great pine forest, one for the Grays, one for the Stracheys, one for Hooker and Dr. Lambourne, one for the cook and black man and one for the mess. On the 26th they proceeded to Fort Garland, a lonely post in the midst of a vast plain, garrisoned by five officers and fifty soldiers. There were no Red Indians left within fifty miles or more : no skirmishes save at distant outposts : the chief duty that of escorting stores. In this monotonous existence the travellers' visit was a most welcome break, especially to the four ladies who had accompanied their husbands to the Fort. From this they ascended the Sierra Blanca, said to be the highest of the Rockies, 14,500 feet, a very fatiguing ascent, for to pass the timber line they had to force their way for five hours through thickets of aspen, then through forests of pine, the fallen branches of which encumbered the ground.   They slept at 13,000 feet under thick blankets on the ground by a huge fire very comfortably, ' though my breath turned to frost all round my head.' After a day's botanising on the heights, they returned to the Fort, very tired and in rags, to rise at four next morning to return to La Veta, and proceed beyond Colorado Springs to the neighbourhood of Pike's Peak.   Two days of botanising there, and they reached Denver on August 1, leaving that on the 2nd for Georgetown and Gray's Peak, returning on the 5th to Denver on the way to Salt Lake City, two days and a night by train, for a botanical excursion to the Wahsatch Mountains. Salt Lake City was left on the 9th, and from Ogden, where to Hooker's great regret the Stracheys went home, the arid mining region, with its astounding mushroom cities, was wearily crossed. This portion of the journey began with twenty-nine hours in the train to Reno and Carson City ; then by Silver City and on for ten days by wagon across the Sierra Nevada to the Yosemite and Calaveras Groves, winding up at San Francisco. Hooker's botanical work was to end with the Forest region of the Pacific Coast. With his very large collection of plants and a good general idea of the Flora of the whole continent from East to West, it would be, he felt, a splendid achievement in Geographical Botany, but a very laborious one. ' I am so sick,' he says feelingly, ' of railway cars and perpetual packing of traps, drying plants, writing notes and seeing endless people and things.' This was more trying than the dust and dirt, though these were ' horrible ' on the railway and in driving and riding. In the dry Californian climate not only the roads, but even the mountain paths were loaded with dust, and a light camping outfit did not provide much in the way of change and comfort. What troubled him least, though he had passed his sixtieth birthday, was the fact of being never in bed till midnight and up at five or six.

A sketch of a mining city, Georgetown, in Colorado, and of a visit to Brigham Young, deserve quotation. Colorado had but recently transformed from a Territory into a State which should take care of itself with no questions asked by the Central Government as to how criminals are punished and how laws are evaded. Public organisation was inefficient, and the better disposed were still compelled to keep order by lynching the incorrigible. But this voluntaryism was often very efficacious.

> Here, at this little town at the extreme finger-end of civilisation [Georgetown] the streets are watered better than at Kew, people sleep without locks to their doors, the fire-engines are well manned and in capital order, and of food there is *no end*, though it is too high to raise vegetables or any garden produce !—all is brought up by train from Denver to within a few miles of the City. The smallpox has been raging in a neighbouring mining village, i.e. city, to this, and the authorities sent the beds and bedding of the sick to the Capital City (about 50 houses) to be stored there for the casual poor. The citizens sent a vigorous remonstrance to the said authorities, who paid no heed, upon which they coolly set fire to the building. The alarm bells were rung, and the fire brigade refused to turn out, and so infection was stamped out by ' lynch law ' ! This is the sort of way matters go on, quite illegally, but in the right direction and in the interests of the community. (To his Wife, August 5, 1877.)

August 8, 1877.

To-day we called on Brigham Young and had a chat with him. He is about 70, stout, well dressed, and with rather a refined countenance. He reminded me more of a stout, elderly and *thoroughly respectable* butler, than anything else. In person and conversation he is less of a Yankee than $\frac{9}{10}$ of the gentlemen I have been introduced to. Of course he is an arrant impostor, but nothing in speech, look or manner differs from those of a quiet well-bred English gentleman. I talked a good deal with him about the climate, history and productions of his country, and found him communicative and intelligent. He gave us iced water and ' God blessed us ! ' when we left ! His missionaries are bringing in converts from all quarters, especially Wales, Sweden and Prussia —of course from the most profoundly ignorant classes, but once arrived here, they get plenty of work, good food, comforts and domestic happiness—for a plurality of wives, which few care for and fewer can afford, is the only sin that B.Y. allows, and for that he quotes the Testament. All the school children are brought up to believe in him and in a lot of Scripture history as useless and idle as that taught in our schools, and the religious teaching is altogether contemptible. The Gentile ladies hold no intercourse with their Mormonite sisters ; nor is it likely they should. Educated U.S. ladies would not care to associate with the ignorant class to which the Mormonite ladies belong. In short as far as I can make out, the system of polygamy is that of making young female servants your wives. They are servants without pay who cannot run away ! and a well-to-do man here with large farms, cattle, vegetables and other produce of all sorts for distant and near markets, has plenty for many wives to do, if he will take the trouble to teach *and then rule* them.

The following letters give some impressions of his tour :

To Professor Oliver

August 8, 1877.

I should have written to you ere this, as my work has been always and altogether Herbarium work, over and above travelling, since I landed in this country. . . .

On the steamer between Providence and New York we picked up Thurber, a big, stout, very intelligent man, of a rather leucophlegmatic temperament and curly hair. He is fond of grasses, and knows them ; is in bad health. I liked what I saw of him very much indeed ; he reminded

me of Berkeley a little. He took us about the streets of
New York for the two hours we spent there, which city did
not delight me. It is just like Liverpool. The sea, islands,
and shipping, and especially the gigantic Ferry and coasting
steamers all white, with Saloons piled one over another,
and paddles 40 ft. diam., are all extraordinary sights.

Thence we took rail *via* Philadelphia to Cincinnati,
where we staid a night, and then on, sleeping in Pullman
cars, to St. Louis, where I saw a great deal of Engelmann,
who is still hot on Pines, Oaks, Yuccas, and Euphorbias.
It is astonishing what a lot of information one picks up of
trees and shrubs, especially in travelling with such a man
as Gray, and both at Cincinnati and St. Louis (and with
Sargent at Boston) I saw much of the native tree- and
shrub-floras East of the Mississippi. The number of associated trees struck me as most curious ; a few yards walk
in the forest would introduce you to perhaps 20 different
forest trees and of more than half as many genera.

The Herbaceous Flora, and especially the Compositae,
were equally numerous. I visited several Gentlemen's
seats, where the native trees were carefully preserved and
replanted.

From St. Louis West was over the gradually ascending,
weary prairies with *Helianthus* rampant, also here and
there the Compass plant with its leaves in a vertical plane
N. and S. tolerably conspicuously. Of the many big yellow-
flowered Compos. some certainly open towards the sun,
but do not appear to me to follow it ; they wriggle about
afterwards according to the wind or their own inclinations.

Buffalo and savages are all gone from the prairie on our
line of rail which was South of the main one and struck the
State of Colorado about the middle, at Pueblo, whence we
went into the Mts. by the cañon of the Arkansas. The
R. Mts. are not a range of Mts., but a multitude of rocky
ridges rising to 12 and 14,000 ft. over a huge elevated pro-
tuberance some 300 miles broad. Their ridges are rocky
and rather bare except of Pines and Aspen. They are
usually craggy and sparkled with snow patches (not perpetual
except in hollows). Between these ridges are vast open
downs 6–9000 ft. above the sea, with grass and herbs but
few or no trees. The Forest Zone extends to and above

2000 ft., and the Alpine Zone above it is not rich. By going into the range from three different points, La Veta, near New Mexico, Pueblo (the Arkansas), and Colorado Springs (a feeder of the Arkansas), and Georgetown (the Platte) on the North, we had a fair general view of the Colorado vegetation from 4000 to 14,000 ft., yielding me altogether about 500 species I think, of which I have dried small specimens.

After these explorations we came up to Cheyenne on the main line of rail from the Mississippi to the West, and took it across the R. Mts. to Ogden in Utah, whence a branch 36 miles long runs between the Wahsatch Mts. (12,000 ft., with rocky peaks and patches of snow) and Salt Lake, to Salt Lake City, where we arrived last night and whence I write.

You expressed an interest in hearing about the manners, social &c. of the Americans as they impressed me. Of course all such impressions must touch principally upon very superficial observation. The New Englanders are most like us in language, speech, and habits, and have least of the nasal twang, which is simply obtrusive and detestable. As a rule I find the Americans too loquacious, for ever praising themselves and introducing you to *most* remarkable men. They think the curious things of their country have no parallel with us, and forget how ' Colonial ' they appear to us. Their high sharp voices, and of the women especially, is the most grating feature of their life to us. In other respects they are superior to us, as in education, civility, great desire to oblige and take trouble for you,—decent, cleanly manners, clean shirts and a far superior condition and manner of the official and subofficial classes attached to public conveyances and to Hotels, &c. These people are most universally well conducted, civil and obliging to all, far more so than with us. Meal hours are at very irregular times. At Hotels you pay so much a day for everything whether you eat or no. The food is most abundant, wastefully so, and I do not like the little messes of endless meats, breadstuffs, and vegetables that are served to each at all meals. Each individual is surrounded by a constellation of little thick white plates, which the waiters throw down and about like quoits, making a *dreadful* clatter all through the huge or small dining-halls. Few drink at meals anything but tea and coffee, or iced do., iced water or milk in tumblers —of the latter you get any quantity—and this and the bread and oatmeal porridge (admirably made) at breakfast are my great supports. Of course quality of beef, &c., varies in different States. The Americans are great and promiscuous eaters, and are too fond of talking of their foods. Their fruits are simply contemptible, except the Californian pears, which are splendid. (The apple season is not in.) The peaches are great coarse things with a big mamilla at the top, and decidedly compressed sides. The apricots, though large, are also flattened and poor. The plums shrunk up my tongue to the size of a snail.

There are many *Rubi* and *Vaccinia*, all very poor. I have seen nothing equal to a good *V. Myrtillus*, or wild strawberry or raspberry. The *Rubi* are greatly eaten at all meals, raw, stewed, or in tarts called pies. A good account of the native and cultivated native fruit is greatly wanted.

Beds are remarkably clean and good but the pillows too soft. Even in out-of-the-way mining districts we got good food, clean beds, and civil service. The scale of living amongst all classes West here is enormously high. Here (I am now at Carson City, Nevada, Aug. 23rd) gold is the currency, paper loses 6 per cent. ; and nothing less than a silver 10 cent piece is taken. The miners' and labor wage generally here is four dollars a day, and the men live like fighting cocks, as to eating and drinking especially.

Gray and I took a trip into the Wahsatch Mts. (E. of Salt Lake), and at 10 or 12,000 ft., in a hut of seven miners with some women and children, we found a dinner fit for a prince preparing—clean plates, knives, forks, castors, &c., &c., &c. Before every cabin door are heaps of empty tin cans (of fruit, vegetables, and luxuries), clean unbroken bottles, good new empty casks, and often hundreds of playing cards—and this over whole States. Wealth is largely distributed. The poor are starved out and seek the towns. Here I am in the centre of the greatest gold and silver producing mines in the world. Virginia City, 6000 ft. above the sea, is built at the mines 21 miles from here. We went there yesterday. The country is a hideous desert, but full of good plants : lots of *Erigone, Grayia, Ephedra, Pinus monophylla, Juniperus*, scattered with tufts of *Artemisia, Bige-*

lowia, Argemone, and Compositae, all over the face of the bare, yellow and red, hilly landscape.

The mines are marvellous, yielding daily thousands of pounds of gold and silver. The machinery for working the mines, crushing, amalgamating and smelting and assaying, &c., is quite superb, gigantic in short, and all so well kept, clean, and complete. I have seen nothing like it elsewhere. Then too the Banks, places of business, Hotels, &c., at these mining cities are as perfect in fittings, carpets, pictures, clocks, &c., &c., as the trimmest house in Lombard Street. Everything is done regardless of expense, and yet efficiently. You see no makeshifts anywhere. The houses are small and wooden (no one cares to invest in such things, they are too often burnt down, and the more lucrative lodes of silver and gold may be worked out at a day's notice) but look very neat inside, windows always clean ; outside, i.e. about the house generally, they are very slovenly, no, or few, little gardens or grass plots.

Since I began this I have added very largely to my collection from the desert region, though the season is too late. We are now starting for the Yosemite, by some interior route across the Sierra Nevada which rises in rocky ridges with *P. ponderosa* to the West of us, while all to the East are hideous, bare, rounded, stony hills with a grey tufted vegetation as mentioned above, and only *Pinus monophylla,* which finds both its N. and its W. limits here and continues E. only to W. Utah and S. to Arizona ; it is a very small species, a round bunchy thing, and appears to me to represent *P. edulis* of the lower zone of the Rocky Mts. I am getting a great deal of information about forest trees, and have learnt an *enormous* deal of botany in a month (it is just 31 days since we left New York for the West). The Stracheys have gone home. Dr. and Mrs. Gray and Prof. Hayden go on with me to California. We shall be ten days in getting to San Francisco by this route. I miss letters most terribly. If forwarded from Boston we do not get them, and I have no news since July 16 (from Dyer).

With kindest regards to Mrs. Oliver and the Bakers,

Ever affectionately yours,

Jos. D. Hooker.

P.S.—The weather has been on the whole most cool and

propitious till we entered this desert region, which extends
from Salt Lake westwards, and is very low, dusty, and very
trying.   It is awfully hard to keep up, travelling by rail,
collecting, drying plants, writing notes and Journal in such
heat and drought.   I am getting very sick of it, but to-morrow
we shall be, I hope, over the Sierra.   I cannot fancy any
route over which a European would get more accessible
Botany new to him than a railroad trip across N. America.
The Floras of the E. and W. are of two continents !

To Charles Darwin
October 19, 1877.

I have indeed had a splendid journey ; and thanks to
A. Gray a most profitable one—nothing could or can ever
reach his unwearied exertions to make me master of all I saw
throughout the breadth and not a little of the length of the
U. States.   The Geographical Distribution of the Flora is
wonderfully interesting, and its very outlines are not yet
drawn.   We have material for a most interesting Essay.   I
have brought home upwards of 1000 species of dried specimens for comparison of the Rocky and Sierra Nevada and
Coast Range Floras, an investigation of which should give
the key to the American Flora migrations.

As usual with me when at sea I caught the Equinoctials,
and we had the longest Eastward voyage that the Captain
had ever known!   Thirteen days of heavy contrary gales and
a high sea continuously from Boston Harbour to Cork.

Dyer has done uncommonly well in my absence, and
goes for the last three-quarters of his honeymoon on
Monday.   Crowds of people asked for you in America, so
pray accept the national greetings through me, for I can't
individualize.

To George Maw
November 16, 1877.

I had not the ghost of an adventure in America, where
I saw a prodigious deal and learnt much.   California was
burnt up with nine months' drought, which obliterated the
herbaceous vegetation and allowed me full time for the
arboreous and fruticose.   I was charmed with New England,
disappointed with the Rocky Mts. as a range, and have no

love for California, but all are full of great interest, and
wonderful resources.   Niagara did not disappoint me nor did
the big trees.   I travelled 7–8000 miles by rail and never
but once missed a connection! but I did not like the cars.
The people I found to be wonderfully nice, and A. Gray is
a trump in all senses.

To the Same

June 16, 1878.

I was quite as much struck as you were with the ' mature
aspect ' of Boston ; and not of Boston only, but of many
towns, and even villages in New England.   Nothing impressed
me more with my ignorance of America ; and long before I
left I rallied the Americans on thinking themselves so young
a people when they were long past youth.   It is only when
one comes to look close and finds the absence of any old
dwellings of a poor class, that one realises that all must be
new after all.

A first conspectus of the general scientific bearings of the
expedition was given by Hooker in a lecture at the Royal
Institution, April 12, 1878.

To Asa Gray

February 25, 1878.

Well done your hypothesis ! it is splendid.   It fits in
splendidly to a Friday evening Lecture on our work which
I am to give to the Royal Institution on April 12th, entitled
' On the Distribution of Plants in N. America.'   You may
not know that the Friday evenings are reserved for the
single Lectures of *Swells* ! The Committee, who for years
have given me all the privileges of the Institution gratis,
had over and over again besought me to lecture, and I had
steadily refused ; but this time I could do so no longer.
I have made Meridional Distribution my principal theme,
and had intended to treat of Pliocene Flora, &c., and the
effect of the Alps as compared with the American Mts.,
in the latter directing the course of migration, and in the
former favoring the extinction of N. Pliocene forms ; but
I had not come to the formulating of the subject as you
have done.   I did not think so much of the Mediterranean

as a barrier, regarding the dry extreme climate of the South
shore as sufficient to kill any Northern Pliocene that might
have arrived there.   Also who knows what may turn up in
the Pliocene of N. Africa ?

I have been comparing E. States Flora with Californian,
and am more than ever amazed at the difference even in
such an order as *Caryophylleae*.

I hope that you will indicate to me any views or papers
of yours that you think I may have overlooked or am likely
to overlook.   I intend to show, first how your researches
on the Japan Flora and mine on the Arctic each come in,
and are foundations upon which we meet in theory (one
of us in England, the other in America), and how we coalesce
as to results in our present labors after travelling together.
How ever I shall get the Lecture finished, i.e. the subject
properly elaborated, I do not see, for I am really busier
than ever !

Talking of the E. and W. Floras of N. America, I am
surprised to find so many Asiatic types in W. America that
are not in East ; and the Western American representatives
of Asia seem to belong to a different type from the Eastern
American representatives.   Can both (the East and the
West Asiatic types) have branched off from one Asiatic
migration into N. America ?   or were there two migrations
at very different periods, one into East, the other into West ?
if so which first ?   I have not read up this matter ; please
tell me where to look.

The full botanical results were worked over when Asa Gray,
on his visit to Europe in 1881–2, spent a couple of months at
Kew, thereafter joining the Hookers on a spring trip to the
Continent.   The work was issued conjointly in the Bulletin
of the U.S. Geological Survey for 1882, vol. vi, pp. 1–62, under
the title of ' The Vegetation of the Rocky Mountain Region
and a Comparison with that of other Parts of the World,' by
J. D. Hooker and Asa Gray, albeit Hooker, enquiring after
progress on August 2, 1879, had written : ' Dare I ask what
has become of our report ?   I will do anything you like but
have my name put before yours.'

The following letters deal with the progress of the work.

To Asa Gray

August 22, 1878.

Here I am with my wife immured in a little inn to which we came yesterday, to view the beauties of Killarney, which have ever since been obscured by torrents of rain accompanied by a furious gale, which for aught I know has blown here ever since I sighted the S. of Ireland last October. It is a beastly climate. How different from Nevada!

How stupid of me to forget the Miocene Flora of Iceland! which I knew of. It is well my letters to you are not publications!

Our difference as to ' gouging out ' of Yosemite is probably verbal only. I never intended it to be understood ' that the Glaciers had initiated the valley,' but I think that the mass of material has been removed by glaciers, and that they have given its sides their configurations to a much greater extent than you do. A glacier enduring for ages in such a valley must have carried away an incredible amount of stuff, and not merely ' scraped the sides.' That sort of granite offers *no* resistance to ice, such as Limestone, Porphyry, Slate, and other Metamorphic rocks do.

Just regard the amount of solid rock on the lateral and medial moraines of any glacier at any one time—add the grooved detrition of sides and bottom and sum up the annual loss of material—it is stupendous. You say you see no proof of anything more than ' Glaciers smoothing the sides a little.' I saw proof of enormous removal of stuff, in moraines everywhere, and no doubt had we gone down the valley we should have carried old glacial detritus to its mouth. As to your ' seeing no proof,' I do not see the force of this unless you have made a study of glacier action ; a non-botanist ' sees no proof ' of *Ruscus* being allied to *Asparagus*! You have not lived alongside glaciers for months and watched day by day what they do and what they *must* have done.

I am very stubborn about Greenland and have asked Dyer to review the subject. Certainly *Platanthera hyperborea* is European, if books are to be trusted. You make no allowance for the great rarity in America of so many Greenlandic plants. I mean such as just cross Baffin's Bay, or turn up in one or two places in America.

Now for yours of July 15th. I am glad to hear of the
Chilian types in West N. America—but may retort upon
your proverb apropos of Greenland : ' Ponderandum non
numerandum, &c.'

I deny that the Equator is the Grandfather of climate
—it is the Grandmother—the Poles are the Grandfathers ;
i.e. it is the alternate heating and cooling of the most extra-
tropical areas that ' kicks up the bobbery.'

Assuredly you should try for an English market for
your Introduction to Morphology and classification. It is
much wanted—but all the world is mad after Physiology
and Histology, and Morphology *pure* and classifications
are despised on the Continent, and Britain is fast following
suit.

To *Charles Darwin*

October 4, 1878.

I am very busy at Gray's and my joint paper on the
Botany of Colorado in relation to the rest of America, and
the Universe I suppose. It has I find curious relations with
Altai which I hope to show are not shared by the Floras of
either Eastern or Western America, but these comparisons
are very laborious.

To *the Same*

October 7, 1878.

I am working hard at the Rocky Mountain Flora, and
find that it contains many Old World genera and species
not found in the equally lofty Sierra Nevada which runs
parallel to it for so many hundred miles, and I am excessively
interested about it. One would suppose that the migration
along the American meridional ridges from the North South-
wards and back again was the simplest thing in the world,
but it has not been so I am sure. The Rocky Mountain
Flora will stand a very fair comparison with the Altai,
which the Sierra Nevada will not.

To *Asa Gray*

August 2, 1879.

What a splendid time you had of it on the Alleghanies.
I should indeed like to have been with you both.

So next year you are normally due in England. Come

then ! [1] I have thought much of my next trip to America, and of my great obstacle, which is Bentham. If I do not go, and he continues as active as now (and I really see no dimming of intellect or cessation of power of work), we really should get on well with Monocots. Except yourself there is no one who can work like him. I have been closely observing all he has been doing with the genera of *Coniferae* and can only marvel. Now that I am rid of R.S. we see more of one another, and I of his work and he of mine. With the Gen. Plant. on hand I cannot think I ought to leave him.

We are dreadfully impatient for the continuation of Watson's Bibliography. Nothing short of it will mitigate the curse that hangs over American Botanists. You can have no idea of the labor you cost *all hands* at the Herbarium when we revise a sheet of Gen. Plant. for checking the references.

To Charles Darwin

November 29, 1879.

We are still thinking over our conjoint work on the Geographical Distribution of American Flora. I have sent him (Asa Gray) a comparison between the Rocky Mt. Flora and that of Altai, which present many curious points of affinity, as in rarity or absence of Oaks, Nuts and other cupulifera which abound all round both areas. He now wants my Lecture to Royal Institution in a modified form, and a comparison of the European and Asiatic Floras, which might be very interesting in reference to America. I have a notion that the E. Asiatic and W. European Temperate and Subtropical Floras are very distinct, but not so distinct as both are from the intermediate Area, and that the Himalaya is the bridge between them, crossing the intermediate area.

Further the Himalaya contains a mingling of European types with others typical of both Eastern and Western America.

Three years later, at the York meeting of the British Association in 1881, he delivered before the Geographical Section an address ' On Geographical Distribution,' a survey of

[1] Gray duly came in 1880, and when he left Hooker wrote (October 19, 1880) : ' We missed you *awfully* for a full week, and then put off mourning.'

the whole question to which he had himself contributed so much exact material from every quarter of the globe. It had been his earliest as it was his most constant interest. Like Darwin, he saw that the secret of Distribution must throw light on the origin of existing plants and animals ; it was reciprocally true that a valid theory of the Origin must reflect light on the dark places of Distribution.

When Humboldt developed Linnæus' analysis of the habitats of plants into a richly furnished Botanical Geography, and further strengthened this with numerical data, he founded the science of Geographical Distribution, of which Forbes was to be the reformer and Darwin 'its latest and greatest lawgiver ' ; but he, no more than his predecessors who noted the most obvious instance of Distribution in the succession of plants in ascending the mountains, could realise its full bearing. It was left for later investigators to show ' that the parallelism between the floras of mountains and latitudes was the result of community of descent of the plants composing the floras, brought about by physical causes.' Not otherwise could the existence of representative types, so utterly perplexing to earlier naturalists, be explained.

Historically it had first to be shown (and this was Lyell's work) that our continents and oceans had experienced great changes of surface and climate since the introduction of the existing assemblages of plants and animals ; that there had been a glacial period, as betokened by the Arctic survivals and fossils found in N. temperate lands ; and, long before, a warm Arctic period, attested by the abundant fossils brought back by Arctic travellers, of plants belonging to a warm temperate zone.

Stirred by results so largely due to himself and Asa Gray (though these, to be sure, were barely touched upon in the address) new research and new interpretations had pushed the history still further. As Forbes had traced successive plant immigrations into the British Isles, so Blytt traced in the Norwegian peat bogs the succession of five different waves of plants following wet or dry periods. And as for the world-tides of migration so clearly worked out in the Northern hemisphere, not only were they paralleled in the Southern, and from similar causes, but the work of Saporta and Dyer had gone on to make it probable that all plant life had originated in the polar regions, and radiated thence to be differentiated in different regions.

Kew : August 4, 1881.

DEAR DARWIN,—I am groaning over my Address for York after a fashion with which I have more than once bored you awfully.   Now do believe me when I say that it is an unspeakable relief to me to groan towards you ;—and I will have done.

I am trying to formulate my ideas on the subject of the several stages or discoveries or ideas by which the Geog. Distrib. (of plants) has been brought up to be a science and to its present level, and showing that these stages have all been erected on ideas first entertained by great voyagers or travellers, thus ' hitching ' myself on to the sympathies of a geographical audience ! something in this following sort of way :

1. Tournefort's enunciation of the likeness between the vegetation of successive elevations and degrees of latitude : the true bearings of which have come out only now that we know that said vegetations are affiliated in fact as well as in appearance.

2. Humboldt's showing that great Natural Orders, *Gramineae, Leguminosae, Compositae,* &c., are subject to certain laws of increase or decrease relatively to other plants, in going polewards (in both hemispheres) and skywards. I should also refer parenthetically to his construction of the isothermals as so great an engine towards the advancement of Geog. Bot.

Now will you give me your idea as to whether I should be right in calling Humboldt the greatest of scientific travellers, or only the most accomplished,—or most prolific ?   It is the custom to disparage Humboldt now as a shallow man, but when I think of what he did through his own observations during travel, for Geographical distribution of plants, for Meteorology, for Magnetism, for Topography, for Physical Geography and Hydrography, for Ethnology, for political history of Spanish America and for Antiquity of Mexico—

besides the truth and picturesqueness of his descriptions of
scenery and all else—I am constrained to regard him as
the first of scientific travellers ; do you ? This is however
a digression.

3. Lyell's showing that distribution is not a thing of the
present only or of the present condition of climates and
present outline and contours of lands, and Forbes' Essay on
the British Flora.

4. The establishment of the permanence since the Silurian
period of the present continents and oceans. Were you not
the first to insist on this, or at least point this out ? Do
you not think that Wallace's summing up of the proof of it is
good ? (I know I once disputed the doctrine, or rather could
not take it in—but let that pass !)

5. The Evolution theory.

6. The discovery of fossil warm plants in high Northern
regions, leading to exact ideas as to effect of glacial period
as shown by Gray's Essay.

7. I must wind up with the doctrine of general distribution being primarily from North to South and always along
existing continents, with no similar general flow from S.
to N.—thus supporting the doctrine which has its last
expression in Dyer's Essay read before the Geog. Soc., and
referred to in my last R.S. Address [1879, p. 15]. Now if
this is accepted, we may not too hastily throw overboard
Saporta's doctrine of the boreal origination of the main
types of vegetation ; and if this again is accepted we cannot
altogether neglect Buffon's argument that vegetation should
have commenced where the cooling globe was first cold enough
to support it, i.e. at a pole ; and lastly, if this is accepted I
must bring in Buffon's speculation in its proper chronological
order, and put it as No. 2 of the stages that have led up to our
present state of knowledge. But I am disposed to regard
Saporta's and Buffon's views as too speculative for that and
to introduce them at the end. What do you think of this
point, and of it all ?

It is not even on paper, and how I am to get it all in
shape before the end of the month passes my limited powers
of prevision.

I have to take some part in this Congress,[1] and *by request*

---

[1] The International Medical Congress held in London, in August 1881.

give a Garden Party on Saturday—it will be a dreadful ordeal I fear (except it rains !).

Kew : August 11, 1881.

MY DEAR DARWIN,—Your letter and memos have been unspeakable comforts—for I was beginning to despair of making my Address anything but a budget of snippets of facts and ideas, and you have both helped and encouraged me to give one part of it at any rate a consecutive and scientific character.

Then too the revival of our scientific correspondence and interchange of ideas is extraordinarily pleasing to me, who regard myself as your pupil.

I am indeed glad that your old appreciation of Humboldt is no more dimmed than is mine. I have been re-reading all his Geog. Bot. Essays, and it is impossible to deny their supreme *ability* and approach to originality. I wish I had time to write, and space to give to all I think of them—his ' Distributio Arithmetices ' of the great groups, expressed in definite proportions, is a stroke of originality, if not of genius, and I have called it a sort of parallel (?) (I can't find a good word !) to his Isothermal lines.

I cannot find a reference to the permanence of continents in your ' Coral Reefs '—a book by the way that shook my confidence in that theory more than all others put together, and the effect of which it has required years of thought to eliminate or rather to overlay. I thought the idea was first published in your ' Geological Observations,' of which I cannot find my copy (but shall). Any of Dana's works must have been long after both. Where does he ' reclaim,' and where does J. Mellard Reade publish his views ? [1]

I may have to allude to this subject from the Chair at York in view of the papers to be read on the progress of Geog. discovery in the great Continents. In respect of them I have long cogitated over the fact that the main water parting of Asia is not coincident with the greater

---

[1] Under date of August 20 he writes : ' I find that Dana was the first (of all I have yet found) who broached the doctrine of permanence of position of existing continents. You somewhere do the same for existing oceans, and I read it lately but for the life of me cannot turn the passage up. Also in the *Origin* you imply this. But I do not know of any one except Wallace who has summed up all the arguments for it, and marshalled them with convincing force.'

elevations of that continent but runs obliquely from S.W.
to N.E., and is sometimes determined by huge sedimentary
deposits as in Upper India, at others by very low mountains—
does this not imply vast oscillations over an already formed
land of continental extension ?

I am doubtful about going into the Flora of past ages,
beyond the Tertiary. I quite believe in the sudden development of the mass of Phanerogams being due to the introduction of flower-feeding insects, though we must not forget
that insects occur in the coal and may have been flower-
feeding too.

I have dealt with Saporta's view of the polar origin
of Floras in my last R.S. Address.

I hope we may talk over them and many other such
matters when too late for my Address !

It appears to me that the great Botanical question
to settle is, whether the main endemic Southern temperate
types originated there and spread Northwards, or whether
they originated in the North and have only just reached
the South, and have increased and multiplied there (to be
turned out in time by the Northern perhaps). The balance
of evidence seems to favor the latter view, and if Palae-
ontologists are to be believed in crediting our tertiaries
(even polar ones ?) with Proteaceae, it would tend to confirm this view, as do the Cycadeae, now about extinct in
the N. Hemisphere and swarming in the South.

Buffon's and Saporta's views of life originating at a
pole, because a pole must have first cooled low enough to
admit of it, is perhaps more ingenious than true—but is
there any reason opposed to it ? If conceded, the question
arises, did life originate at both Poles or one only ? or if
at both was it simultaneously ?—but this is the deepest
abyss of idle speculation. Ever yours affectly.

J. D. HOOKER.

To the Same

September 9, 1881.

Your criticism anent Southern Glacial Epoch is just—
my loose statement was due to hasty condensation of matter.
What I should have said, and did originally in MS., was,
that from the appearance of Antarctic plants on mountains

north of their home, a glacial period might be *inferred*, as *proved* on astronomical and geological grounds, or something to that effect.

Yes, I do hope to live to work out the relations of the Southern Temperate Flora. I do wish I could throw off my official duties here ; I am getting so weary of them, and Dyer does them so well ; but I could not nearly afford it yet.

CHAPTER XXXIX

END OF THE PRESIDENTIAL TERM (1877–1878)

MEANTIME the building of the new Herbarium was a fresh landmark in the history of Kew. But it was not accomplished without considerable friction, in great part the aftermath of the old trouble with the Office of Works, for Lord Henry Lennox continued the Ayrtonian tradition of supercilious officialdom towards science and glorified gardeners, and Ayrton's right-hand man was not retired till August 1875. The one helpful person in the office was Mr. Bertram Mitford (the late Lord Redesdale), and his chief was at loggerheads with him!

When the last personal obstruction was removed, Hooker could exclaim, " Thank goodness I have all the Office and the Treasury at my back and beck," and continues :

To Charles Darwin

August 16, 1875.

I only hope that now my Lord will find himself unsupported, he will retire from active interference in the Office. Meanwhile he is moving heaven and earth with the people about the Queen to prevent the Herbarium being kept in the Queen's private grounds, for a small piece of which I have asked (as a site for the new building). He insists on my finding a site for it in the public part of the Gardens ! which I absolutely refuse to do, except the Queen refuses a corner of the ground where the Herbarium now is.

To Asa Gray

March 12, 1876.

After a great deal of worry, lasting over nine months, the Herbarium building is in a fair way of commencing. The

—— family, who had an eye to the ground and house, were bitterly opposed to it, and got over my present chief, who, after the Queen had given the site, continued throwing obstacles in the way. When Lo ! by a stroke of luck, it turned out, when preparing for a legal transfer of the site, that the present Herb., House and grounds all belonged to us !—that old scamp, George IV., having sold it for £84,000 to pay his debts, in 1824 ! There was no legal conveyance, but the receipt of the money is to the fore ! Thus both William IV and Victoria have for half a century been giving to others (the King of Hanover and the Herbarium) a house not their own.

I shall retain the present building for the Library and Working rooms, render them sufficiently fire-proof, and throw out a Herb. Hall at the back in the same style of architecture that suits the site and surroundings. I am all the more glad of this, as George III had given the building originally for Library and Herbarium, and Banks had begun to have it fitted up as such, when his death stopped it all, and it reverted to the King's use.

The result was thoroughly satisfactory : ' I wish,' he tells Asa Gray, August 2, 1879, ' you could see our Herbarium and Library arrangements before you begin to build, for which I quite saw the need.'

Hooker had hoped to combine with this needful addition to Kew the physiological laboratory offered by Mr. Phillips Jodrell. The Royal Commission on Scientific Instruction and the Advancement of Science had stated in its fourth report (1874) that ' it is highly desirable that opportunities for the pursuit of investigations in Physiological Botany should be afforded at Kew to those persons who may be inclined to follow that branch of science.'

Hooker was not primarily a physiological botanist, but he well understood the value of this branch of his science. This was being vigorously taken up in Germany under the influence of Sachs, and should be no longer neglected in England, which in olden days had led the way in this direction, before systematic botany, expanding with the discoveries of our expanding empire, had swamped other lines of research.

Beside Hooker, moreover, now stood William Thiselton-Dyer, a leading representative of the new school of physiologists, whose ardour had received inspiration and direction from Huxley. He was Professor of Botany to the Royal Horticultural Society, and was then aiding the Director of Kew as Private Secretary. Hooker already had him in mind for Permanent Secretary, or better still Assistant Director, as soon as the Government should sanction the appointment, which took place in 1875. Under his care the Physiological Laboratory could find its true development, while it would also afford him immediate scope for his own branch of work.

Hooker was therefore well advised when he persuaded Mr. Jodrell to make his benefaction to botanical science in the form of this Physiological Laboratory.   Private munificence thus outstripping a laggard Government, the Jodrell laboratory was built, equipped and in working order by 1876, before the new Herbarium was well under way. It is interesting to note that the first research made in the laboratory was by Professor Tyndall on the organisms of putrefaction.

The immediate need and scope of such a laboratory are illustrated by a letter of 1874 arising out of Darwin's correspondence with G. J. Romanes,[1] who was experimenting to raise seedlings from graft hybrids. If the seminal offspring of plants hybridised by grafting should show their hybrid character, it would be striking evidence in favour of pangenesis. This experiment, however, did not succeed. (See M.L. i. 280 and i. 359.)

To Charles Darwin

Kew: December 22, 1874.

By all means let Mr. Romanes come here and I will do what I can. Our best grafters &c. get such good places abroad that we cannot keep them, but he shall have the

---

[1] George John Romanes (1848–94), a student under Michael Foster and Burdon Sanderson, proceeded from physiological work to wider scientific problems, especially on the development of the intelligence in animals and in man : while in Darwinian and post-Darwinian theory, he put forward a theory of Physiological Selection. Elected F.R.S. 1879.

best aid and advice that we can give. Why should he not experiment at Kew himself ? I would put plants and all appliances at his service. The only thing is, that he must himself daily inspect his own work. I cannot get anything of the kind done for myself even, with any approach to skill and care—but I have plants and appliances to any amount.

I am now writing to the Board about a Physiological Laboratory, which Mr. Phillips Jodrell offers to build, and which I hope we may get as an adjunct to the new Herbarium building. Mr. Romanes's is just the work which should be conducted in a laboratory, which should be at the service of such men as Mr. Romanes, on payment of a small fee for materials, &c., which should be had from Govt. Grant or other funds.

This is the sort of encouragement that I think Govt. should give to original research. Let Govt. find the appliances and buildings and Colleges, Universities, &c., and private enterprise find the workers and funds when they require it for their support. The R.S. will have abundance to repay workers at present, and I am not sure but that it would be well, if the Gilchrist works well, to have a similar one raised by subscription.—Ever your affectionate

J. D. HOOKER.

The last piece of work to be recorded for 1878 is the long delayed publication of the ' Journal of a Tour in Marocco and the Great Atlas.' In July he was busy correcting proofs with John Ball, to whom had fallen the writing of the greater part of the book. The seven years' delay since the journey itself in 1871 (see p. 90) scarcely prejudiced the scientific or historical value of the book, dealing as it did with an unchanging country ; but it may have contributed to lessen its popularity. ' It will, I think, be interesting,' was his opinion ; but it was an expensive book to produce, and to his disappointment the sales were far from recouping the cost and left a deficit of over £100.

The following letters touch on these and other occupations about this time.

To T. H. Huxley

[He was about to attend the Dublin meeting of the British Association.]

Kew: July 30, 1878.

DEAR HUXLEY,—There is a talk of giving me a sub-section of D., which I make no objection to : but I have heard nothing more of it. I shall certainly give no address if I am called upon to act as President in any capacity. I have too long resisted Satan to make it worth the old gentleman's while to tempt me in that line.—Ever yours,

J. D. HOOKER.

To Charles Darwin

July 31, 1878.

Huxley tells me he will give no address to his section and I applaud his resolution. I think that even he will soon find that the power of giving addresses is exhaustible, and that he will be reduced to a state of nudity—the address becoming no dress. I am at my wits' end for a subject for the Anniversary of Royal.

To the Same

October 4, 1878.

Ball's and my Marocco Journals are nearly out, they await a brief Essay from me on the comparison of the Floras of Marocco and the Canaries—the differences are marvellous and quite unexpected. There are no islands in the world so near the mainland with such a difference in their vegeta-tion—they beat the Galapagos in certain respects, but then the separate islands do not differ much.

I must clear the American and Marocco works off before I begin my Address : happily the matter of these is in my head. Then I must go to Paris on the 18th to be present at the Prize giving of the Exhibition, which is to be my only duty as a Royal Commissioner ! I have shirked every other without exception and cannot have the impudence to decline this, though I do hate it.

I am still looking out for a country cottage within easy distance of Kew to retire to on Sundays and perhaps in the

end for weeks, months, years of Sundays, for between you
and me I am getting giddy with science in all shapes, and
with the worry of social, scientific, and official life, and I
long for rest and nothing but the Library and Herbarium
to busy myself with.

This is the best and most sensible growl you have had
from me for a long time.—Ever your affectionate

J. D. HOOKER.

To the Same

[Darwin had sent for his criticism a paper he had received
on the flora and insects of St. Helena.]

October 7, 1878.

DEAR DARWIN,—I had already read [the] paper and
corresponded with [the author] about his conclusions.
Unfortunately the Botany is all dead against him. There
is no relationship whatever between the N. Atlantic Island
Flora and that of St. Helena.

You have marked a passage to the effect that ' one or
two genera of plants common to St. Helena and S. Africa
are strongly suggestive of a Palaearctic origin, and dis-
persion by the influence of a Glacial epoch ; for example
Sium, which has an endemic representative in St. Helena,
and the very characteristic Cape genus *Pelargonium*, which
has a straggler in Syria.'

Now the *Sium* which I first described, I have stated
to be closely allied to the *S. Thunbergia* of the Cape, which
is no Palaearctic form ; and how *Pelargonium* is to be classed
as Palaearctic because one species grows in Syria, whilst
hundreds are confined to the Cape, which is its headquarters,
passes my comprehension.

I have come to the conclusion that the Flora of St. Helena
is very S. African and not in the least North Atlantic, and
as the plants must have got to St. Helena before the insects,
these must, if they came from the North, indicate that the
Flora has survived the Glacial epoch, i.e. had come from the
Cape before it.

The difficulty of attributing to the Flora a Miocene age
or origin is, the absence of any old types, such as Conifers
and Cycads or examples of exceedingly limited (i.e. dying
out) Natural Orders. If I remember aright, most or all

the plants belong to large and very cosmopolitan Orders,
well represented in S. Africa. Ascension does not help ;
its only shrubs are of South African affinity and St. Helena,
and these are, if I remember aright, its only flowering plants
(except tropical weeds). St. Helena has affinities with Tristan
d'Acunha. If we could only make the insects antedate the
plants I would understand the argument. Is the Entomology
of the S. African Mountains known ? especially of those
Mts. of the W. coast.—Ever yours affectionately,

JOS. D. HOOKER.

To the Same

January 18, 1877.

Wheat brought by Nares from Smith's Sound, where the
Polaris left it some five years ago, has germinated splendidly.
I am now planting a lot of various seeds which I sent out
and which have been exposed to cold of 60°–70°. A grain
of maize that was with the Polaris wheat has also grown ;
this being properly a tropical plant is remarkable.

What a rum thing living protoplasm must be, so quickly
to decompose in some seeds and resist change in others.
That the freezing of its watery constituent (if it, water, is
a constituent) should not affect its vitality is very wonderful.
A good man might make a splendid thesis on ' vitality '
in the abstract. Jas. Salter [1] has been writing to me about
another series of experiments on burying seeds, but I do
not think he is prepared to carry it out. I should be disposed
to attack the problem another way—viz., to experiment on
means of prolonging vitality of seeds which are notoriously
short lived.

It seems an age since I heard of you all.—Ever
affectionately yours,

JOS. D. HOOKER.

[1] Samuel James Augustus Salter (1825–97), or James Salter as he called
himself, received his medical education at King's College, London, and after
graduating as M.B. at the University of London, became partner with his
uncle Thomas Bell, dental surgeon. He was elected F.L.S. 1853, and F.R.S.
1863, and on his retirement from his profession occupied himself with horticulture at Basingstoke. He was the first to make the remarkable observation
of perfectly formed pollen-grains in the nucellus of the ovary (see *Trans. Linn.
Soc.* 1863, xxiv. p. 143). He published his *Dental Pathology and Surgery* in
1874.

He was immensely interested to hear that Mr. Anthony
Rich, the antiquarian, having no other kith and kin but his
sister, with whom he lived at Worthing, had resolved to
bequeath his fortune ultimately to Darwin, in token of his
admiration for the man and his work.

To the Same

December 13, 1878.

Well, I shall dream of that blessed old couple at Worthing
—it was indeed a curious thing, and I have no doubt that it is
the precursor of many such acts ; as knowledge increases,
so must appreciation of the people and institutions to whom
we owe it. Govt. may do much, but it must always be
under such vexatious restrictions that it tries a man's temper
and patience, let his patriotism be what it will, to undertake
the expenditure of what Government gives, and I fear it
ever must be so. Between ourselves I think there will be
a wretched outcome of the Govt. Fund (the £4000 per annum).
I am sure that if I had the uncontrolled selection of persons
to grant it to, and was free to use my authority over them,
I could have got ten times more done with the money. I
shirked the subject with my address.—Ever your affectionate
rejoicer,

J. D. HOOKER.

There was an old couple at Worthing
Who resolved to reward the deserving ;
And with wise resolution
Pitched upon Evolution,
That pecunious old couple of Worthing.

Kew: February 8, 1877.

MY DEAR GRAY,—I have not yet wished you a happy
New Year and many of them,—but like Martha, I am
' troubled with much serving.' Now too I have a new edition
of my Student's British Flora on hand, anent which nothing
strikes me as so curious as the contrast with your Manual in
respect of the limits of species. Will you ever be bothered
with the subspecies and varieties that drive me frantic,

and in my view are not worth the time they take to elucidate ? [1]

What I wish now to consult you about is the position of Gymnosperms, whether to make of them a sub-class of Dicotyledons, or a group equal to all other Phaenogams: i.e. should it be

1. Monocots.
2. Dicots.
    a. Angiosp.
    b. Gymnosp.

or Phanerogams :

1. Angiosperms.
    Monocot.
    Dicot.
2. Gymnosperms.

I see that you and Decaisne, and I (in Decaisne and Maout) have adopted the first course, and I still incline to it. Oliver is disposed to go in for the second with Dyer. No one could weigh the evidence on both sides so well as you could. Much should depend on the structure of *Gnetum* embryo, sacs, &c., and I think *Gnetum* is quite overlooked by the Physiologues in removing Gymnosperms from Dicots.

I have just sent to Press the corrected Primer,[2] a work which has cost me immense labor. I feel terribly the want of that facility for writing such a book as lecturing would have given me.

---

[1] Writing to Mr. Bolus on November 20, 1883, he calls the accurate determination of them 'a most fidgetty affair, and most unsatisfactory.' To complete the new edition it was further necessary to collate the 'new edition' of Nyman (presumably the *Conspectus Florae Europaeae*, 1878–82, the *Sylloge Florae Europaeae* having appeared in 1854–5), to rearrange many Orders by the *Gen. .Plant.*, and revise distribution by the last edition of Watson's *Topographical Botany*.

[2] I.e. the third 10,000.

CHAPTER XL

THE years that followed the end of the Presidential term were
still full of incessant activity ; but it was the activity that
centred in Kew and the systematic botany to be completed at
home. Work on the Councils of the Royal Society and Linnean
Society continued into the early eighties ; but the best of his
active life being past he refused to think of new presidential
duties, whether at the Linnean or Geographical Societies, even
for the sake of carrying through desirable reforms. The days
of camp and field work were over ; the old explorer could
only respond to the call of the wild through others. For these
his sympathy, his experience, his advice, were unfailingly
ready—more especially for Antarctic explorers, such as Dr.
Bruce of the *Scotia*, and Captain Scott, who twice revisited
the southernmost land of which he himself had been one of
the original discoverers. Past also were the days when he
had travelled with Darwin as a pioneer in speculative regions
more difficult and more perplexing than the unmapped in-
tricacies of the Himalayan passes. The joyous pains of the
long wrestle with Nature, the rapture of finding a way
through the maze, the first great conflict with a hostile
organisation, all these also were now of the past, and the
paths so laboriously broken had become the common highway.
Thus the picturesque element grows less though the solid
work moves on in the business of Kew, its constantly improved
organisation, the completion of the Genera Plantarum, and
the yet greater burden of the Indian Flora and the undertaking
of Darwin's last great gift to science, the Index Kewensis.

This gift had a personal as well as a public aspect. Darwin
owed a debt of ' happiness and fame to the natural-history
sciences which had been the solace of what might have been
a painful existence.' He was moved by admiration for the
work done at Kew and gratitude for the incalculable aid which
for so many years he had received from the Director and his
Staff. At the same time his botanical work had shown him
the importance of a complete index to the names and authors
of the genera and species of plants known to botanists, together
with their native countries. The plants he received from all
sorts of sources were often incorrectly named, and without
precise means of identification by other workers, his own researches would be misleading.

Steudel's ' Nomenclator ' had partly fulfilled this purpose ;
but it had been published in 1840, and in the next forty years
the number of described plants had been doubled. At Kew
the list had been kept posted up by means of an interleaved
copy, by the help of funds supplied by private generosity.
But this was unpublished. Wishing definitely then, ' to aid in
some way the scientific work carried on at the Royal Gardens,'
he set aside a considerable sum to complete and publish the
Kew ' Nomenclator ' under a scheme drawn up by Hooker at
the end of 1881, with the help of the Kew Staff and Bentham,
and carried out in detail by Dr. B. Daydon Jackson, the
Secretary to the Linnean Society, for ' we should of course
all help.' As the work proceeded, Darwin's original idea of
producing a modern edition of Steudel was practically abandoned, and the aim kept in view was rather to construct a
list of genera and species (with references) founded on Bentham
and Hooker's ' Genera Plantarum.'

This ' Nomenclator Botanicus Darwinianus,' or more briefly
Index Kewensis, was brought out in four quarto volumes
between 1892-5. It was no trifling work which proceeded for
fourteen years under Hooker's supervision. In 1887, Sir F.
Darwin notes, the MS. of the Index was estimated to weigh
more than a ton,[1] and the completed work, all the proofs of
which Hooker read and criticised, comprised 2500 pages, each

[1] See C.D., iii. 351-54, on which I have largely drawn.

with three columns containing some fifty names each, a total of about 375,000 entries.

The officialdom of the period was still characteristically left cold by this rich gift to a national home of science.

To Charles Darwin

January 19, 1882.

DEAR DARWIN,—The enclosed requires no answer. The history of it is this. I, as a matter of course, informed the Board of your munificent offer, showing what a grand aid it would be to our own work, as well as to science in general, and how honorable to Kew. The First Commissioner (one of your d—d Liberals) wrote a characteristically illiberal and ill-bred minute on it, addressed to me, in effect warning me against your putting the Board to any expense !—and this though I expressly stated that ' your offer involved the Board in no expense or other responsibility whatever.' I flared up at this, and told the Secretary, whom I saw on the subject, that the F.C., rather than send me such a minute, should have written a letter of thanks to you. I suppose that this shamed him, and he has taken me at my word, though I did not seriously contemplate such action.

In the administration of Kew the years brought no relaxation to the Director. The general lines for the development of the Gardens had long been laid down ; the same operations went on, only on an annually increasing scale ; development from within proceeded unceasingly, while correspondence with collectors and gardens overseas grew with the central importance of Kew. ' The ordinary correspondence, etc.' he assures his friend Maw (March 24, 1882), ' gets more extraordinary every year.' At that moment, for instance, all the Cinchona papers relating to the cultivation of both kinds, and the policy of making both quinine and the febrifuge in India, were coming to Hooker to be reported on. The making of new sections, the re-arrangement of the old, went on busily, for the progress of science, inverting the familiar proverb, makes the better the enemy of the good, and leaves the excellences

of the past out of date. But the burdensomeness of this
inevitable work could be sadly increased by the spasmodic
action of Government departments.

To Asa Gray

I wrote you the other day and have no further news.
Sargent wants any amount of the Indian woods, &c., of which
and other things there are 36 tons measurement coming to
Kew from the India Store department, and I cannot tell you
how many tons we have already disposed of—the accumulation of 30 years' extravagant collecting in India without
judgment or regard to cost, and of utter mismanagement,
indolence, and caprice on the part of the India Museum
authorities here. I suppose there never was such a revelation of the sort (in the Museum way). Many many thousands
of pounds must have been spent in India upon the collecting
duplicates on duplicates, put up in the most expensive manner,
to be destroyed unopened by rust, dust, rats, and insects.
There are, I am told, *cases* of Cashmere shawls riddled by
vermin, sent for exhibition (these of course were not coming
with my 36 tons ! of vegetable produce), and a silver Elephant
Howdah. I need not say we are tremendously worked.
Dyer gets through work most wonderfully, and is a very
skilful manager.

The Indian Government gives us £2000 to add to the big
Museum, and I have screwed £450 out of the Treasury to add
to the little one (that my Father inaugurated), so we shall
have space enough ; but it will cost us the re-arrangement
of both Museums ' au fond,' and as poor Dyer has just completed that operation, we do growl at the job.

A far more agreeable feature of the time was the inauguration of the regular garden parties, which became a social link
with the many personal and official friends in London. A letter
written by an old friend in December 1915 gives a capital
description of these.

No one privileged to be present at one of the famous
Royal Gardens' summer parties during the last years of Sir
Joseph Hooker's Directorship could ever forget the especially

unique enjoyment which distinguished them. The small
garden belonging to his official residence used to be thronged
with people, famous, not only in science (a natural product
there) but in Politics, Art, Literature, etc., etc. The Archbishop (Tait), the Prime Minister (Lord Salisbury), Travellers,
Literary men, Artists, Journalists, came in crowds, and among
them, genial, radiant, happy, moved the eminent host,
enjoying himself to the full and delighting in being surrounded by so many friendly faces.

These gatherings made Kew not only a well-managed
Government Department, but a meeting-place for the best
social life of London. All were gratified by an invitation
from the President of the Royal Society, and it was doubtless
his position as head of the Scientific World that suggested
Kew's possibilities as a Universal host. No set entertain-
ment nor amusement, not even a band, was ever found there,
none was wanted. The crown and joy of it all was *just*
the pleasure of meeting everybody, of seeing and hearing the
celebrities of the time, listening to the busy hum of talk,
enjoying the boundless hospitality, and then passing at will
into the wider Public Gardens. Yet the exact secret of the
charm of these gatherings was, I think, something above and
beyond all these factors. It was simply because the host
and hostess themselves enjoyed the gathering together of
their friends, and never regarded it as a duty to be discharged,
or a burden to be borne. Memory reports that these parties
were generally favoured as to weather, but if rain fell, who
cared if it did pour without, for within that spacious house
were treasures innumerable, Wedgwoods, Indian curiosities,
pictures, books, etc., and the fragrance of exquisitely arranged
flowers, while host and hostess excelled themselves in their
welcome, warmer and more cordial because the weather was
unkind.

His heart's desire now was to throw off the trammels of
official life and get back to pure science. First of all he was
eager to finish the ' Genera Plantarum,' for Bentham, born
in 1800, was now growing old and frail, though he came daily
to the Herbarium for four or five hours, and his memory,
judgment, and power of botanical work seemed unimpaired.
And in the second place, he wanted to deal finally with the

Indian Flora. But with a family to maintain, this was not yet practical politics. He could only hope to economise time by retiring from the active work of the learned societies which he had served steadily for so many years. As he tells Asa Gray (October 28, 1880) :

The R. S. Councils begin to-day. Happily this is my last year of them (*for the present*) after 10 continuous, and 16 in all ! just half my period of membership. I suppose I must have been useful ! or else have been an egregious impostor—a little of both, we may conclude logically. As it is, I feel a loathing to all that sort of work.

How I wish that we could join you in Spain, but it is impossible. We cannot leave home now, even if my duties allowed of it, and I must get three months Bot. Magazine off my hands before I go anywhere.[1] Bentham too is commenting on my slow progress at the Palms.

[1] It may be of interest, in alluding to Sir Joseph Hooker's editorship of the *Botanical Magazine*, to give a brief account of its beginning and uninterrupted existence of 129 years. Since 1787 it has appeared regularly every month without fail, and probably no other serial or periodical can show such a record, due in great part to the energy and resourcefulness of its editors, in bearing it safely through more than one great crisis. For seventy-seven years of this time it had two editors only, Sir William and Sir Joseph Hooker, and for seventy-one years (from the time lithography was first employed in 1845, up to the present volume) the plates were put on stone by two lithographers only, W. H. Fitch and his nephew J. N. Fitch—verily two further ' records ' ! while, except for a brief space, when W. H. Fitch retired in 1878, he and the present artist have been the only artists employed on it for seventy-eight years, from 1837 to 1916—vols. 64 to 141 !

The founder of the *Botanical Magazine* was William Curtis, of the Society of Friends, who was born in 1746, and, after some years as an apothecary, was appointed Botanical Demonstrator to the Apothecaries Society at Chelsea. He started the magazine, a serial of octavo size with coloured illustrations, in 1787 the first number containing only three plates. On his early death in 1799, Dr. John Sims became editor. He was born in 1749, and studied medicine at Leyden and in Edinburgh, and finally settled as a physician in London. He retired from the editorship in 1826, and was succeeded by Sir William Hooker, who was artist to the magazine as well as editor for some ten years. His work has a delicate finished style of its own, and he introduced the plan of giving full analyses of his plants. In 1834, or even sooner, finding his other botanical work too exacting, he began the training of a young man, Walter H. Fitch, who in a very short time became one of our most famous botanical artists. In vol. 61 (1834) two plates are signed by him, and his work must have appeared, unsigned, for some time before that, though only in 1837 (vol. 64) does he seem to have become the regular, recognised artist to the magazine. No one has ever excelled him in the life and vigour he put into his portraits of plants, the plant itself lives before us, and his drawing was so accurate and yet so free, that Sir Joseph once said delightedly, ' I don't think Fitch *could* make a mistake in his perspective and outline, not even if he tried ! ' while in his sketch of his father's

Subsequently, also, Hooker refused to be put forward for the presidency of the Linnean and the Geographical Societies, even for the sake of battling for desirable reforms, especially where his efforts were likely to produce no more effect than 'pinching a pillow.' That form of work belonged to the past.

But his experience and authority were still in demand, and he found it impossible to escape from the ' endless Councils and Committees in London of which I am heartily tired,' as he tells Asa Gray, May 26, 1885. ' The grasshopper becometh a burden at last,' he confesses to Huxley (1884), even in regard to ' The Club,' that ' mixed lot of savans and swells, all very agreeable as far as my experience goes.' Meantime he threw himself into his official work at Kew as energetically as ever, ably seconded by his skilful lieutenant, only indulging in a ' growl ' to his intimates now and again. There was no prospect

life he writes of him as ' One who by his artistic talents contributed to the value of my Father's work.'

On Sir William's death in 1865, Sir Joseph became editor, and remained so until 1904, when, owing to his living at a distance from Kew, he felt himself unable to continue his duties. Under him and Walter Fitch the magazine attained its greatest height of excellence, and some very interesting and remarkable plants were figured. Mr. J. G. Baker, Assistant Keeper, and later Keeper of the Herbarium, supplied the letterpress for the Monocots, but all the rest were described by Sir Joseph himself, until some few years before his withdrawal, when Dr. W. B. Hemsley, Keeper of the Herbarium, gave him such great assistance that both their names occur on the title-page of Sir Joseph's last volume (130th).

Even in 1898, when dedicating the 124th volume to Dr. Hemsley, Sir Joseph says, the dedication ' is offered as a record of the interest you have shown in this work, and an acknowledgment of the valuable aid I have received from you in conducting it.'

One great crisis was safely surmounted in 1878, when Fitch gave up his connection with the magazine, which was left with little or no stock of drawings to fill its monthly issue. But Sir Joseph's energy was equal to the occasion ; he secured the continuance of J. N. Fitch's services as lithographer, and succeeded in getting various artists to contribute. His daughter, Mrs. (now Lady) Thiselton-Dyer, and his sister-in-law, Mrs. Barnard, daughter of Prof. Henslow, both contributed drawings for some five or six years (in part), with most praiseworthy and admirable results, considering the ' Master ' they followed, while Miss Eleanor Ormerod, the entomologist, and others helped to supply a few plates. No less than nine different names appear in that volume (105th). It was then that the editor undertook to train the present artist, Miss Smith, and thanks to his skilled instruction and high standard of artistic feeling, she was able to be of use at once, and finally to become the sole artist with J. N. Fitch as lithographer, and both continue to fill these posts up to the present time. On Sir Joseph's withdrawal in 1904, Sir William Thiselton-Dyer edited two volumes, and at the end of 1906 the present Director (Sir David Prain) succeeded to the editorship.

of time to experiment on further species of pitcher plants, and at sight of the ' hopeless pile of literature to glance at ' accumulating on his desk, he murmurs : ' The " intellectual activity " of the age is horrid.' The release he desired was not from work, but from uncongenial work, which would allow him to take up the wider interests always to the fore in his correspondence with Darwin.

To Charles Darwin

November 24, 1881.

I must just thank you for the ' Movements,' which seems a most capital production, and I am so pleased to see Frank's name associated with yours in it. I have read only two chapters, 7 and 8, and they are splendid, but I hate the zigzags ! *Bauhinia* leaf-closing is a curious case ; does it not show that said leaf consists of two leaflets ?

The fact that for good action the leaves want a good illumination during the preceding day is very suggestive of experiments with the electric light. They are like the new paint that shines only by night after sun-light by day. There are heaps of points I should like to know more about.

Dyer and Baker are taken aback by the keel of the *Cucurbita* seed ; which keel was a wonderful discovery in *Welwitschia* ! ! !

I have had no time to read more than the two chapters as yet, for I have a stock of half-read books on hand and no time for any of them. I am only two-thirds through Wallace and it is splendid. What a number of cobwebs he has swept away. That such a man should be a Spiritualist is more wonderful than all the movements of all the plants. He has done great things towards the explanation of the New Zealand Flora and Australian, but marred it by assuming a pre-existent S.W. Australian Flora—I am sure that the Australian Flora is very modern in the main, and the S.W. peculiarities are exaggerations due to long isolation during the severance of the West from the East by the inland sea or straits that occupied the continent from Carpentaria to the Gt. Bight. I live in hopes of showing by an analysis (botanical) of the Australian types, that they are all derived from the Asiatic continent. Meanwhile I have no chance of tackling problems—I must grind away at the Garden, the Bot. Mag., and Indian Flora, which I cannot afford to give up, and Gen. Plant., which alone I delight in. I am at Palms, a most difficult task, and sometimes weeks elapse and not a stroke of work done ! I am getting very weary of ' working for a living;' and am beginning to covet rest and leisure in a way I never did before. But I must first look out for the education of three sons,—all hopeful I am glad to say, but one still an infant !

Have you read Paget's Lecture [1] on plant diseases ? it is very suggestive and a wonderful specimen of style aiding in giving great importance to possibly very superficial resemblances between animal and vegetable malformations. Still there must be a great deal in the subject to be investigated.

Paget has started the idea of a Vegetable Pathologist for Kew, and I have asked him to corkscrew Gladstone about it.

To the Same

June 12, 1881.

I am groaning as usual,—now under the incubus of the Sectional Presidency of the B.A. for York (Geography) [2]— which I was ass enough to accept—because of Lubbock.

Kew is becoming more toilsome than ever, and I can rarely get an hour for ' Genera Plantarum,' for which I have been doing the Palms for 16 months at least ; the most difficult task I ever undertook. They are evidently a very ancient group and much dislocated, structurally and geographically.

The Palms, with ' near 120 genera, many very imperfectly known,' took over three years to finish : stigmatised again not only as the most difficult job he ever undertook, but ' perhaps the most unsatisfactory.' Indeed ' when the Gen. Plant. is off my hands, I shall be a happy man—I hope.'

To the Same

June 18, 1881.

I quite understand your misery at finding yourself where you have ' all play ' offered you, and no work to fall back

---

[1] 'Disease in Plants,' by Sir James Paget. See *Gardeners' Chronicle*, 1880.
[2] See *ante*, p. 223.

upon.'[1]   I should be as bad, but then I know not the
condition.   When I go away I have work that I can always
take with me, official and other : and my misery is the lots
accumulating at home.   I cannot tell you how I long to
throw off the trammels of official life and do like Bentham.

Now at last he ' struck out ' for a more liberal treatment of
Kew, for the pay of the principal officials was far below that of
similar positions elsewhere ; a clerk was needed for the Director ;
a proper office and office keeper to relieve the Curator of work
after office hours, besides accommodation for a couple of good
gardeners and provision of further labour.

While the Assistant Director took over most of the Garden
work, Hooker's special share from 1882 onwards was the
Arboretum, of which he was making ' a noble thing ' with
groups of species of trees, &c., in systematic sequence, no easy
task.

In the spring of 1882, also, a new rock garden had to be
provided to receive a large collection of rock plants, the generous
bequest of Mr. G. C. Joad of Wimbledon, and the present long
rock garden, 160 yards in length, was made ' in the form of
a winding valley with rocks and tree trunks on both sides
rising 8–12 feet.'   As a consequence of these developments,
it became necessary to prepare new guides for the Garden,
Arboretum, and Museum.   The latter promised to be the most
complete summary of economic plants as yet published.

Thus in the letter of November 20, 1883, to Mr. Bolus,
already quoted, he sums up the situation by saying :

Everything goes on here much as when you were with
us, only on an annually increasing scale, the correspondence
especially waxes more formidable every year.   The old staff
I am happy to say holds on, and we have no changes, only
additional assistants in the Herbarium.   In the Garden we
have greatly improved the cultivation by obtaining better
pay for really good foremen as growers and propagators ;
so that now even the Orchids are praised.   The Palm House,

[1] Charles Darwin was staying at Patterdale, and had written despond-
ently, ' I have not the heart or strength to begin any investigation lasting
years, which is the only thing which I enjoy, and I have no little jobs which
I can do.'   (June 15.)

which is Dyer's special pet, is magnificent, and he has gone in for Cycads, and by correspondence all over the world got together a wonderful collection of them.

What with his writing and his administrative work, he finds himself so busy that he never gets to the Linnean and very rarely to the Royal Society. In short, as he reiterates to Hodgson (May 28, 1884) : ' The Kew work does not decrease ; the contrast between what the Directorship now is and that when I took it is enormous, and I shall not be sorry to be relieved of my share.'

Most of his spare time went to the Flora of British India. Progress was slow ; the number of species had exceeded his estimate, and at the desire of many Indian botanists he had expanded quotations and references far beyond his original intention. At the same time the actual examination of the specimens was exceedingly laborious. ' Polygonum,' he remarks, ' was a hard task, but I think I have squared them all up.' The Peppers, when he came to them, were ' by far the worst genus I have ever had to do with, and I shall have to assign lots of Miquel's and De Candolle's species to the limbo of the unknowable.' A few weeks later : ' I am groaning over Myristica ! The species are indeterminable without ♀ and ♂ flowers and fruit—all three ! and the specimens of Manigey and others play the deuce with Thomson's descriptions in Flora Indica.' In the Laurels he had perforce to follow his old grouping, but would have done them differently if he had had time. One genus brings up a situation familiar in his early systematic work : ' Cyanadaphne is I am sure a myth, and its one species is in two or three other genera as well ! ' Finally, the bibliographical details and the readjustments in the Herbarium involved much drudgery, and he responds to Asa Gray's sympathy on September 26, 1885 :

Yes, the references to volumes, pages and plates are horrible plagues. I groan over them daily. Nine-tenths of them are not worth verifying. I am now bored with revising the reprint of Gen. Plant., Part I. I sympathise with your supplement work, but oh, how I wish you could

have been persuaded to throw all other work aside for the
ten last years for Flor. Bor. Amer. !

We have much accelerated the work of intercalation at
the Herbarium, and shall do more yet. I am putting away
my Indian accessories and am more than ever impressed
with the huge waste of labour in putting away a few at a
time. We have adopted the supplementary pigeon holes
at the ends of large genera and small orders, &c. It is the
gigantic accessories that now weigh on us, and to decide when
and where to stop gluing down more specimens is distracting.

Other letters of the period again illustrate the sustained
interest in the botanical work that was being done in India and
the colonies. Now it is a matter of detail. Mr. Duthie is going
into the Himalayas ; he is bidden make special pilgrimage in
search of a strange plant with a strange history.

To Mr. Duthie

August 12, 1882.

Strachey and Winterbotham found in Kumaon a most
remarkable little plant, which neither Brown nor any other
botanist could refer to any natural order. I made a careful
drawing of it, with full analyses for the Linnean Society,
some 28 years ago. This and the specimens were lost
soon after, and nothing more was known or heard of the
plant till Maximovicz sent us a specimen from N. China as
an unintelligible nondescript, and he has since published it
as *Circaeaster agrestis*, Max. in Bull. Acad. Petrop. xi., 345,
and adopted an idea of Oliver's which I hardly share, that it
is *Chloranthaceous*.

Now I think perhaps you may be able to re-find this
pigmy, so I send you a rude sketch of it ; and I have asked
Strachey, who remembers it well, for the exact locality,
which is at ' Sába Udiyár, on an overhanging rock, at or
near the halting place below Rálam, on the road from the
Gori valley—Jalat or Munshári, elevation ? 8000 feet.'

The plant is worth a pilgrimage, for I know nothing in
the least like it.

A year later he identifies various plants in Mr. Duthie's collection from the Alpine Himalayan region, and adds a postscript

delight at the revival of Sikkim memories. And in a similar letter on October 27, 1884, he strikes the same note : ' *Abies dumosa* seeds will be most acceptable, as we have lost it at Kew.   Only to-day, I gave Masters a pencil drawing I made of it in 1848, for *Gardeners' Chronicle*.'

November 18, 1882. Mr. Duthie has been planning an extension of the Mussoori Gardens for European plants :

> I am glad to find that the ' powers that be ' support you so well, and that your European Garden is likely to be carried out, upon a scale that will necessitate a European gardener ; you must never rest till you get this, for I quite agree with you, that without a European it will not work.

> November 20, 1883 :—I am glad to hear of your being utilised for the Lahore Garden. Botanical matters want drawing together in India very badly.

On this point a note of June 22, 1884, to Hodgson records :

> King [1] of the Calcutta Garden and Mr. Duthie [1] of the Saharunpur too, are both home now and I have been at the India Office with them about getting a better botanical organisation in India.

In the spring of 1883 Mr. Bolus sent from South Africa a paper on the Cape Orchids and a quantity of specimens ; all was put into shape by one of the Kew assistants, and the paper duly read at the Linnean. Hooker's suggestion of a botanising trip to the north-west bore immediate fruit.

To Mr. Bolus

April 18, 1883.

What you say of the richness of S. Africa in terrestrial Orchids is truly surprising and a curious contrast to Australia where the Order abounds in specimens but where there are no great number of species. The way the *Caladenias, Prasophyllum, Diuris* and *Pterostylis* come up through the grass everywhere in Tasmania is enchanting, one gets 20 or 30 species at once, but extending the area adds but

[1] See p. 275 and p. 281 respectively,

slowly to the number of species. I hope that your *Cymbidium tabulare* will turn out the right thing.

To the Same
May 29, 1883.

It will indeed be a good work done to get the Cape Orchids into order ; at present they are in a little confusion with regard to many genera. Pray get figures however rude.

I wish that you could manage a trip to the N. W. frontier, there must be many curious things there.

To the Same
November 20, 1883.

Your letter has interested me extraordinarily, for I know very little of Namaqualand beyond what I had picked up from old travellers and Baines's paintings. Your account of the geographical and botanical features of the country is very instructive. I had a mind to communicate some of the contents of your letter to the Linnean, but was not sure that you might like it ; I do hope you will send in a paper on the subject. The brilliant display of flowers you mention reminds me of California in spring—as it now is— but where, as they cut down the forests of the Sierra Nevada ranges they may expect droughts like those you describe, and only an occasional flush of vegetation.

The following to Mr. W. Hancock, of the Chinese Customs, illustrates the way in which a fresh correspondent would be enlisted for Kew.

Kew : June 17, 1883.

MY DEAR SIR,—The perusal of a most interesting and instructive article from your pen about Tamsui, in the Chinese Customs reports, emboldens me to address you, in the hopes of enlisting your kind offices in behalf of the Royal Gardens at Kew.

We are deeply interested in Chinese products and most anxious to know more of its vegetable resources, whether by Herb. specimens or seeds ; or, especially, objects illustrative of the uses to mankind of the members of the vegetable kingdom : of these we are incredibly ignorant. I may

mention that it is within the last very few years only that
we have ascertained the origin of Chinese Cassia, of the
Star Anise, and of the Coffin wood,—all through the energy
of Mr. Ford of the Hong Kong Botanic Gardens : and the
more we hear of China, the more persuaded we are of the
inexhaustible riches of her vegetation in the way of utilised
plants for textiles, drugs, dyes, gums, waxes, &c., of which
nothing is known except by vague reports.

It is a singular fact, that none of the late expeditions
into the interior of China, Giles's, Baber's, or Colquhoun's,
has added materially to our knowledge of its vegetation
and native products.   Not that I would speak disparagingly
of their labors or works, for those of Baber especially were
quite admirable, but those explorers have not the knowledge
which you possess and which renders your Tamsui Journal
so deeply interesting to botanists, and which makes Kew so
desirous of obtaining collections from you.

It would afford me sincere pleasure if I could aid you in
any way. If you were to send us Herbarium specimens
numbered, I would guarantee a speedy return of the names
of the plants so numbered, as far as we could determine
them, and our means of doing this are unrivalled.

I have been studying the Palms lately and we have not
a single one, known to me, from Formosa ! The seeds of
any China Palm would be most acceptable.

I take the liberty of forwarding a copy of the latest
Kew Report, and trusting that you will forgive my intrusion,
Believe me very faithfully yours,

Jos. D. Hooker.

To turn to more personal interests, in 1881 a trip to
Italy was taken in company with Prof. and Mrs. Asa Gray,
' most agreeable and most considerate travelling companions,'
and lasted from the beginning of March to May 12.   They
began with the Naples district ; ascended Vesuvius on the one
fine day of their stay ; and worked northward by Rome and
the Umbrian cities to Florence, and finally to Venice and the
Italian Lakes.

As to botany, he confesses he did very little, and indeed the
spring was too backward for much.   He visited various Italian
botanists, and the Forest school at Vallombrosa.   At Florence

he did what he could ' to stop the *insane* project of moving the
Herbarium and Garden from their present unique position to
the horrible stables at the other end of the town.' At Rome,
he tells Prof. Oliver, the collections and herbaria were good, but

the new Botanic Garden is a complete fiasco ; a dry
rounded pile of brick rubbish on the tip top of the ' Mons
Viminalis ' without shade or water ! There is however a
scheme for making a new one on a capital site on the baths
of Caracalla and to unite with it a Zoological and acclima-
titative one. The Italians are a wonderfully go-ahead people ;
but they have a great deal of leeway to make up yet. Mr.
Newton, whom I met at Rome, and who had attended the
meetings of the Lyncei Academy and of the Senate, was
much impressed with the practical businesslike way in
which they conducted affairs in both.

The impressions of so great a traveller as regards scenery
are worth recording. In the same letter to Prof. Oliver (April
15, 1881) dated Florence, he writes of his delight in Cortona
and Orvieto, and their fine situation ; adding,

though as far as scenery is concerned, Italy is, I think,
far behind many others parts of Europe ; and but for its
atmosphere would be considered quite third-rate ; of the
picturesque there is no end, but little of the grand or beautiful in so far as I have seen. Naples I thought greatly
overrated.

As regards Roman sightseeing, he remarks to Bentham :

Rome as a whole, in its antiquarian aspect, is a headache
and a nightmare ; the comparatively modern Churches
are a great relief ; one is thankful that the ancient Romans
were not Christians to have burthened these with their
' ancient history,' their endless gods, Censors, Senators,
Kings and Emperors. The history of Rome is too convulsive.
History, it is said, repeats itself ! any schoolboy will tell you
the contrary, were it so it would be easily learnt ! History is
the curse of modern Education ; it not only doubles itself
as time goes on, as population increases and as people segre-
gate, but not content with this it burrows in the past for
new (and best forgotten) facts for boys to be crammed with.

His general impressions appear most clearly from the following.

To Brian Hodgson

April 10, 1881.

We have been now for the best part of a fortnight in this august city, and I know you will be kindly glad to hear of our impressions thereupon.

I can easily give my wife's ; she would prefer green fields and flowers and insects a thousand times over to the Ruins, Sculptures, Pictures and Churches with which this place is stuffed and loaded. Our two excursions from Rome to Tivoli, and to Albano and Frascati charmed her, and the air and the woods and flowers were to me too a relief after so thorough a drenching with sightseeing as the ' necessities ' of travel here impose on one. Then too the finding Mr. and Mrs. Morgan here has been a great relief and pleasure, and they have been most kind to us.

Rome as a city of ruins is to me very disappointing, so few of these are in an intelligible state of preservation, and such as there are are representative of people and events separated by such a vast interval of time or inter-calcated one with the other so abruptly that they appear to have no more in common than the fossils of widely separated geological beds whose strata had been faulted and dislocated. Here you have something of the Caesars', at the next turn something of Constantine's, with perhaps a monument of the Kingly period opposite it, then a vestige of the Republic, then of the Volscians and the Goths—and so forth, to which you have to add the memorials innumerable of that compli-cated era when Heathendom and Christendom took it turn about to vex history. One heartily wishes that some one of those masters of Rome had destroyed every vestige of his predecessors' works, or else let them all alone ! As it is, what little History of Rome I had is reduced to the ruinous state in which I find the memorials of it ; and the objective vestiges of events not worth recording, take a firmer hold on the memory than the written records of world-disturbing events which have left no visible evidence of their identity or effects. I spent hours in the Palace of the Caesars, and as many in Hadrian's Villa, and I declare that if you were to change their positions and take me blindfold to either, I

should not know which I was at ! Each has a stadium and
its Palaestra, and a theatre and a library and so forth, and
in both cases these are separated by gigantic mounds of
ruins that have no distinguishing character of any sort.

I think however that I can picture Rome as it was in the
Republic better than before—a city of no imposing character
as a whole, formed of a vast multitude of low buildings with
slanting roofs belonging to the lower classes, with here and
there a magnificent vendua, and scattered public buildings of
still greater magnificence contrasting strangely with the hovels
around. In fact a sort of Benares with architecture of a
very different type. Judging from the frescoes, the ordinary
peasant's and townsman's house must have been very like
what we see now in and around Rome, and the house with
the atrium, peristyle, impluvium and so forth must have
been confined to the wealthy and few and far between.
Again it appears to me that the vast extent of the public
buildings and private ones of the upper classes, the prodigious
amount of material put into these, the vast amount of wrought
stones of incredible magnitude and hardness, and the lavish
decoration of mosaics that required much time or many hands
to produce them, all bespeak a prodigious disproportion of
a very poor working class whose labor was forced or paid
for by the smallest coin and coarsest food. The civilisation
that produced such buildings as we have in Italy, Egypt and
Assyria had I suppose this element in common of a prodigality
of forced or very cheap labor. These stupendous buildings
are the Stonehenge and the Monoliths of a barbarous people
in so far as means employed are concerned. It is true that
St. Peter's is as big a thing as any that the ancient Romans
produced, but it does not represent the ' brute strength '
that the Colosseum does, as an expenditure of human labor
or muscular power.

This last sage remark turns me to the Churches, an endless
theme. The first thing that must strike any traveller from
the North is the difference in the style of architecture from
the Northern Gothic, and the fact that one is adapted to
pictorial decoration, the other not, and this leads to the
enquiry how it was that architecture and painting, being
sister arts, should be so utterly divorced in the North ! Of
the Churches here the grandest in my view are the S. Maria

degli Angeli and the Pantheon ; these surprise you with the
vastness of their proportions on entering, which St. Peter's
does not. In fact, to appreciate the latter you must go into
the galleries of the dome and look down on human beings
like ants below, and up to the tier upon tier of gigantic
heads in mosaic that rear themselves aloft, curving inwards
to the navel of the dome. This last—the interior of the
dome—produces an overpowering effect, heightened perhaps
by the height at which you stand, and the slimness of the
rail that separates you from the gulf below and the infinite
space above crowded with grim faces glaring at you.

The dome itself, seen from outside, is very inferior in
proportions to St. Paul's, which, seen from Waterloo Bridge,
is exceedingly beautiful—as indeed is the whole building
which had the advantage of being the design of one man of
surpassing genius in many lines of thought and action.

As to the pictures and sculpture, it is dangerous to begin
upon them in a letter, and so I will draw this epistle to a
close.

By this time the plans for ultimate retreat from the cares
of Kew began to take practical shape. As he thanks Darwin
for his book on ' The Formation of Vegetable Mould through
the Action of Worms, with Observations on their Habits '
(published October 10, 1881) he tells of his purchase of a piece
of land at Sunningdale, to effect which he sold privately his
superfluous books and collections, for he could neither go on
accumulating collections at Kew, nor house what he already
possessed in any other residence.

I take shame to myself for not having earlier thanked
you for the Diet of Worms, which I have read through with
great interest. I must own I had always looked on worms
as amongst the most helpless and unintelligent members of
the creation ; and am amazed to find that they have a
domestic life and public duties ! I shall now respect them,
even in our Garden pots ; and regard them as something
better than food for fishes. I am interested in observing
how they shun some soils at Kew, apparently from want of
vegetable matter in them.

I have been very busy for the last six weeks owing to Dyer

and my daughter being on the Continent : they returned
last week. I have been busy, too, negotiating for the purchase
of a plot of land near Sunningdale whereon to build a ' Tusculum,' and am on the point of closing with an offer of
six acres of ' Bagshot sand,' including a hill of 300 feet commanding a superb view, and in a country of Scotch fir and
heather.   Another year I shall hope to be able to build
a cottage, an awful undertaking for me.   The situation,
1½ miles from the station, from which I can reach Kew
in 1-1½ hours, will be very convenient.

The building of the new house gave Hooker's old friend,
George Maw, the opportunity of making him a present, from
his own factory, of all the floor and grate tiles required for the
house.

Mrs. Asa Gray, knowing his partiality for such things, sent
some beautiful tiles for the study fireplace, which he afterwards
said jestingly kept her in warm remembrance.

The house was comfortable.   Hooker's first use of it in the
autumn of 1883 was to make it a winter home for his father-in-
law, Mr. Symonds, whose health had obliged him to resign his
living at Pendock.   Invalid though he was, and always in
suffering,

his spirits are as good as ever. I never knew such a man
for not knocking under ! He reads, writes poetry for Joe,
and keeps up his warm and generous interest in everything.
There is in him no trace of the ' selfishness of illness.'
(To Hodgson, May 28, 1884.)

The house was called ' The Camp.'

We took the name of the spot for it [he tells W. E.
Darwin, March 4, 1884] ; it is the site of a camp formed after
the battle of Culloden, the troops from which were such
scoundrels that they could not be kept in the town ! and
were camped in this spot, which was continued as a camp
during all the French wars.

These ' scoundrels ' were usefully employed in making the
Lake and Waterfall at Virginia Water, the large stones for the
latter being taken from the camp and its neighbourhood.   The

site of The Camp was again used as headquarters when the
volunteers were assembled in 1853.

The house agent, in a happy mean between ' villa ' and
' country house,' would probably describe this ' Tusculum ' as
' a charming country residence,' with ' every modern convenience.' The trees then planted have grown up high, so that
not even the chimneys, much less the red brick walls and Bath
stone mullions, are visible from the road, and the garden, rich
in rhododendrons, is a haven of peace.

The house itself fulfilled every expectation, but the levelling
and laying out of the ground in particular ran into greater
expense than he first had in mind, so that he began by ' furnish-
ing very scrimpily.' ' It is, however,' he adds, ' a great amusement, and as the value of land here is rapidly rising I can look
upon it as a tolerable investment.'

In the early eighties Time began to take a heavy toll of
Hooker's friends and contemporaries. Such loss made him
turn the more warmly to his best and closest intimates. To
Hodgson, now eighty, he writes on February 7, 1881 :

> My dear Brian,—I am haunted with the idea that
> to-morrow is your birthday, and writing as I am from London
> I have no means of verifying the supposition. Be this as
> it may, you will, my dear old friend, accept my most affec-
> tionate and most heartfelt greetings on the present occasion.
> I am rejoiced to hear so good an account of you as your wife
> sent to mine the other day. Now that dear old Colvile
> is gone, I cling more than ever to my only remaining Indian
> Chum ; and look more wistfully than ever to the hope of
> a permanent reunion in the unknown land. God bless you,
> my dear old friend, and Susie and all you love and hold
> dear.

Hodgson replied on the 9th :

> My dear Joe,—I hasten to express my heartfelt thanks
> for your affectionate recollection of my birthday. What
> you say about the trio of friends—Colvile, yourself and
> myself—has a significance you can hardly realise as regards
> me, in whom peculiar circumstances acting on a very sensitive
> nature have all my life gone to minimise the power and the

will to make friends ! Dear Colvile hit off exactly when he
called me ' the most unclubable of men.' . . . But however
to be accounted for, the fact remains, that I am the most
unclubable of men ; and, as a necessary consequence, I
cling to you now that Colvile is gone, with all my heart as
being almost the only thorough friend I have left. You
are younger than me who have entered my 81st year.
But neither of us can look to many years here below, and
therefore we naturally begin ' to look,' as you say, ' more
wistfully than ever to the hope of a permanent reunion
in the unknown land,' with the friends we have known on
earth.

A little later he exclaims to Darwin (June 18, 1881) :

We have lost no end of friends this year, and it is difficult
to resist the pessimist view of creation. When I look back,
however, my beloved friend, to the days I have spent in
intercourse with you and yours, that view takes wings to
itself and flies away ; it is a horrid world to be sure, but
it could have been worse.

Two months later died Darwin's elder brother, Erasmus, a
man whose intellectual gifts and great personal charm were,
owing to ill health, only known within the circle of his immediate friends and relations.

To Charles Darwin

August 29, 1881.

I have just seen the announcement of your brother's
death and must send you a few words of heartfelt sympathy.
I have somehow come to think those the happiest who,
like myself, lost an only brother when very young ; it seems
now as if they could then be best spared—a blunder no
doubt—but we know better what we lose after having lived
so long together as you and your brother have.

It was in your brother's house, near Park Lane, that I
first became acquainted with you—and shall never forget his kind face and kinder welcome—that was nearly
40 years ago !—I well remember thinking him then quite an
elderly man and yet I see he was then under forty.

But a heavier loss was soon to follow. On April 19, 1882, died Charles Darwin, the friend of forty years, in science the ally and inspirer, in personal affection and intimate sympathy the closest of his circle. Hooker's sorrow and weariness were broken in upon by the request for an obituary notice to appear in *Nature*. Happily he was spared this task to which he felt sadly unequal.

Kew : April 21, 1882.

DEAR HUXLEY,—Romanes, after asking me to write the notice of Darwin for *Nature*, now telegraphs that you had, unknown to him, been asked by the Sub-editor to undertake it, and had accepted.

I am right glad of it, as I am utterly unhinged and unfit for work and am not feeling well in my præcordia, and have not been for some time—pray say nothing of this, but I sometimes fear I shall have to seek rest if I would not that it were found for me. Nothing but the feeling that I was shrinking from duty induced me to assent to Romanes's request.

If I can help you with any notice of Darwin's early life I will come over to you on Sunday.

Up to the time of his going to Cambridge, though he had flirted a little with Nat. Hist., he had no notion of pursuing it, and had devoted himself to fox-hunting and partridges.

I did not feel our loss yesterday, but to-day I am depressed terribly, and a touching letter from Mrs. Darwin quite upset me.

I have heard nothing about the Abbey, though Spottiswoode promised to telegraph the answer to me. I have no fancy for the bitter taste of these ceremonials.—Ever, dear old boy, yours,

J. D. HOOKER.

Kew : April 24, 1882.

DEAR HUXLEY,—It is well indeed that I turned Darwin over to you—the only idea I had parallel to yours was a comparison with Faraday. I have sent your eloquent and most impressive éloge on to Keltie,[1] with a note to send proof to you.

---

[1] Dr. John Scott Keltie (1840) was for some years sub-editor of *Nature*, becoming in 1885 Librarian, and 1892–1915 Secretary of the Royal Geographical Society.

You are right ; it is too soon for any sort of biographical notice of life or works.

As for myself, I have had a ten days' bout of my Anginic pains, night and day, and am in a state of nervous worry, with Bentham failing fast (82) and pressing the Genera Plantarum on me, and no end of work in the Garden.

In short I have my warning note struck.

On the 26th Darwin was buried in Westminster Abbey. Hooker was one of the pall-bearers.

Bentham, the other friend of his youth and fellow-worker of his age, lived long enough to see the publication of the ' Genera Plantarum,' the monument of a quarter of a century's work. At eighty-three, with work still to do, he was alert and vigorous in mind, though growing frail in body, and sadly lonely, having no near kith or kin to look after him. But the interest of his great work once gone, he rapidly faded away, and died on September 10, 1884. ' Bentham's loss was and is a great loss to me,' he tells Berkeley, October 29 ; ' and I do not get over it.' In the obituary notice of him in *Nature* (October 2) (which a passing note tells us cost him two days' hard work), Hooker declared that he had ' no superior since the days of Linnæus and Robert Brown, and he has left no equal except Asa Gray.'

To Hooker he left the copyright of his British Flora and a mass of papers and family things. Busy as he was, he hardly knew how to deal with all this. He writes to Asa Gray (February 15, 1885) :

I am greatly troubled over Bentham's British Flora. He has left me the copyright and duty of bringing out a new Edition which Reeve is calling out for. I must not alter the character of the work, and yet how to do it justice without introducing a good deal of new matter is the question. It has been a very useful work, enticing many to take up Botany who otherwise would not have done so.

And on April 5 :

I am puzzled what to do with all the family things he left to me. I wish some Benthamophobist [?–philist] would offer me £1000 for the lot, portraits, MSS., swords, medals, autographs, and *hoc genus omne*. There is all Jeremy's correspondence I suppose—*piles* of letters, from all manner of people to him, apparently never opened since his death, —and bound volumes of Sir Samuel's correspondence &c., &c. I have no time to open them even, and no wish. I wish to goodness he had left them to his niece.

And yet again, on May 26 :

I am puzzled what to do with his autobiography, which he left to me without further instructions than left me unfettered what to do with it. It is contained in about 700 closely written pages of his small hand, 4to, and goes down to 1883 (?). It was begun to please his wife, and continued to please himself, finally broken off by his illness. It is full of curious matter, and, to the like of us, is most interesting, but I am in doubt how far any considerable public would be interested.

I have still a huge mass of his correspondence to deal with and a huger of his uncle's and father's, and am at my wits' end to know what to do with it all—being an undigested mass of papers.

As for the botanical papers, they must be for some botanical authority to sort. Not for Kew, however—our hands are full enough of unpaid work [not] to care for more, and keeping up the Icones without Bentham's head and pen will be no such easy task, worked as we all are ; for we must both digest and describe, as well as select materials.

Bentham's fatal illness and the impending duties of executorship for him and for Mr. Symonds, whose life for several years hung by a thread, prevented a long planned visit to the Asa Grays in America in the summer of 1884. Hooker had not meant to attend the meeting of the British Association in Canada, but being appointed an official Vice-President, proposed to take Toronto on the way.

This visit, to his great regret, never materialised, though, as appears from the following, it was planned again for 1886, for since their joint trip to Italy in 1881, he had had ' no holiday for a single week, but the 10 days in Paris last winter.' ' If we ever do visit you,' he writes, February 15, 1885, ' it will be a quiet one to Boston, which I should dearly enjoy and still look forward to, but until I retire I see no chance of it.'

But by that time circumstances had changed, though till then the hope was constantly in the foreground.

To Asa Gray

April 5, 1885.

I do indeed trust that California will set you both up, and that we shall find you both flourishing when we cross the Atlantic to visit you. This I assure you we have still in our minds, and only to-day, when my wife said, can we not go to Switzerland with the children this autumn? I sternly answered : No, we must lay by for a return trip to the Grays next year, if possible—and she meekly assented, or rather joyfully consented. . . .

Your account of the views in Mexico and the Cypresses makes me quite green with envy.

As older friends dropped out, the correspondence with Huxley, always active, gradually takes a leading place in the record of friendship. Huxley was eight years the younger, just as much his junior as Darwin had been his senior. That also was a lifelong friendship which lasted over forty years, unbroken by a shadow of discord, but constantly strengthened by fellowship in work and aims and proven trust in each other's character.

They had entered the same profession, for medicine was the one practicable channel to biological science ; they had the same after-career of scientific opportunity on an exploring ship tempered by the personal regime of naval discipline ; they had the same unceasing impulse to unriddle the palimpsest of Nature and find a new basis for speculative truth in place of the stifling infallibilities of the time ; they shared the struggle, the obloquy and the triumph that were symbolised by their friendship with Darwin. In the end, each in his turn was President of the Royal Society ; and in the roll of the Society their names stood next one another in the list of Copley and Darwin medallists.

As Huxley wrote of their friendship in 1888 :

It is very pleasant to have our niches in the Pantheon together. It is getting on for forty years since we were first ' acquent,' and considering with what a very considerable dose of tenacity, vivacity, and that glorious firmness (which the beasts who don't like us call obstinacy) we are both endowed, the fact that we have never had the shadow of a shade of a quarrel is more to our credit than being ex-Presidents and Copley medallists.

But we have had a masonic bond in both being well salted in early life. I have always felt I owed a great deal to my acquaintance with the realities of things gained in the old Rattlesnake.

Huxley's accession to the Presidency of the Royal Society in succession to William Spottiswoode, who died on June 27, 1883, was a matter of great concern to Hooker. In his view the office should be filled by a man of science, not merely a man of wealth or of high social standing ; a man who had shown good business capacity, but not simply a business man of scientific attainments or one who had made a fortune by turning scientific invention to practical account. The first step was to choose a temporary President from the Council till the next general meeting of the Society on November 30. Hooker, when called upon as ex-President to ' prick the list,' had no hesitation in his choice of one who to his other qualifications added that of admirable management as Secretary of the Society.

Thus he was very glad when his friend after some hesitation accepted the temporary office, though refusing on grounds of ill-health and overwork to be nominated for subsequent election to the permanent post. Nevertheless he continued his persuasion : ' You must not throw aside all possibility of the Presidency. I regard the Society's position as very critical,' and after enlarging on the various factors involved, remarks :

Lady Hooker won't hear of my pressing you to take the chair, because of your health ; on the other hand she won't hear of —— much as she likes him personally. What does Mrs. Huxley say—these women have a curious 6th sense not given to men. (July 2, 1883.)

Backed by pressure from many other friends, this persuasion succeeded, and Huxley was President till ill-health
compelled him to retire in 1885.

Being no longer on the Council, Hooker was not concerned
with the management of the Royal Society during this period ;
one letter, however, shows that he was consulted in regard to
the workers in meteorology which had been a special interest
of his own presidency.   Should the title 'Royal' be conferred
upon the Meteorological Society ?

To T. H. Huxley

August 23, 1883.

I believe that the Metl. Socy. is a very respectable hard-
working body, but know no more about it.

I think that the multiplication of 'Royal' Societies is
an evil—and that the fact of there being a Meteorological
Council of the R.S. renders it inadvisable to dub the Meteorological Royal.

The Scotch Met. Soc. will be the next claimant.   If any
of these bodies would give us decent weather I would consider their claims not only to Royalty but to Divine honours.

One or two letters of miscellaneous interest may be quoted
here for their personal note, whether in light or serious vein.

To Brian Hodgson

October 6, 1883.

I have not thanked you for your last kind letter.   I do
think that the vigour you show in your hand-writing is
marvellous.   My hand gets more cramped every year, and
I am about ashamed to send such scrawls even to ' the like
of you,' but your writing is as fresh as your affections, and
I can't say more.

Yes, Buddhist Literature is making enormous strides,
and this must be an immense gratification to you who are
the Grand Lama of the lore ; but what a curious discovery
is this now hinted at, that the early Chinese history is
borrowed from the Chaldean !

You will laugh to hear that I am about to be made a
Freeman of the City ; the 'Salters' Company,' which boasts

of a Royal Prince, an archbishop and a few swells of that stamp, have asked me to accept the honour of membership and freedom along with Huxley. It entails a dinner and a speech, both of which I detest. There is something rather funny in Huxley and me, both 'old Salts,' being picked out by the Salters' Company, though of this 'fittingness' of the compliment I suppose they never thought. I have turned up the Company in the Almanack and find that it is one of the oldest in the City. Hitherto the very few scientific men ever so honoured have been elected by the worshipful ' Spectacle makers.'

There is a humorous touch in the rule laid down for the interpretation of Providence according to one's own predilections.

To the Same

May 28, 1884.

[Hodgson was in Italy. Has he met a gouty M.P. of their acquaintance on his honeymoon ?]

Happily his gout was checked before the marriage day, for we expected to see him hobble up on crutches to the Altar as he did in the House to give his support to the Government. Then you see that Providence disapproved of the latter deed and approved of the former ! What a wonderful thing the finger of providence is—*if people would only understand it*—that's my philosophy ; and yours ?

On exploration without scientific observation he set moderate store.

To the Same

June 22, 1884.

I went to the Geographical Society the other night to hear Mr. Graham's account of his ascent of Kinchin ; he seems to have got up to 24,000 feet or very near it, but has not made an observation of any kind, sort or description. He was accompanied by a Swiss guide and is no doubt a bold mountaineer. Curiously enough he did not suffer from difficulty of breathing or discomfort of any sort ! and he coolly put all other descriptions of such suffering down to the imagination. I had to speak afterwards and could not help saying that he reminded me of a man, who being never sea-sick himself, would not believe that there was any such thing in others.

Sir R. Temple spoke absurdly in praise of the paper, talking most ignorantly as if no one had written on the Himalayan chain before. So I took the liberty of reminding him that such men as General Strachey, Hodgson, Thomson, and many more had told us all about the Himalaya that Mr. Graham had, though none had performed the same plucky exploits, or ascended nearly as high.

His youngest son was born in January 1885. To Asa Gray he writes playfully of his bearing the name of his ' Judicious ' ancestor, and the paternal inheritance of a skull sloping up high behind, a shape so fashionable among one of the western coast tribes of Redskins, that they press their infants' heads between boards to produce it in the extremest form.

To Asa Gray

                                                February 15, 1885.

We have returned to Kew, with the new child christened Richard,—my wife's whim,—I don't approve ; ten to one he won't be ' judicious ' at all,—all the more as his cranium is just like mine, i.e. like a Chinook's, tremendously long from chin to occiput (sketch) *facsimile.*

I wish when you go North, you or Mrs. Gray would get made for me a Chinook flattener to work the other (back) way, and I will ask my wife to clap it on for a year or two and see if I can't make him judicious.

Early in October 1885, soon after his sixty-eighth birthday, Hooker resolved to resign the Directorship of Kew, retiring from office at the end of November. He had held the post for twenty years, and for ten years before that had been his father's assistant, gradually taking more and more of the burden of administration upon his own shoulders. Now, though the enormously increased work, scientific and adminis-trative, was divided with his assistant, ' the world,' in his shape, ' was too much with him,' laying waste the hours he would fain have devoted to that lifelong work, the Indian Flora. Now, also, though full of vigour, and indeed continuing an ordinary man's share of labour for another quarter of a century, he would not wait to have it whispered either that he was growing feeble or that, being hale, his ideals belonged to a bygone stage of progress.

Accordingly he retired to his 'Tusculum' among the pinewoods of Sunningdale, whence he could easily reach Kew to work in the Herbarium at his Indian Flora and other books.

True to his principle that testimonialising paid official services is (except in very rare cases) inexpedient, and even mischievous—liable to great abuse, a burthen to the poor, and setting a very bad example (as he wrote to Huxley in March 1879), he got his intimate friends to choke off any proposal of the kind. He was pleased, however, by the warmth of the official expressions in regard to his own and his father's services.

To Asa Gray

December 2, 1885.

I am pretty busy—changing quarters, putting old wine into new bottles, stuffing the contents of a big house into a small one, making over charge of Garden duties, and excogitating plans for putting Dyer at his ease in the shape of providing an office, and such scientific assistance as I can get for him. I am deep in Indian Laurels (they are perfectly dreadful). I have just sent Bentham's Flora to press. I am on the Councils of the Royal and Geographical, and I have to find time for bed and meals—I forgot that I have the Bot. Mag. ever before me too.

My wife lives at the Camp and comes up and down after the furniture, books, and goods and chattels of all sorts. I am taking most of my books down and shelving two rooms at the Camp. I wanted to part with the birds and some Wedgwoods, but she will not, so the Camp resembles a Dry Goods Store. As for me, I shall be here till Xmas, except Saturday to Monday at Camp. It is ghastly sitting with empty shelves and no pictures, but then I am utterly quiet and get through a lot of work and correspondence, for all the world writes to condole or congratulate, with

me, or the public, or both, and I feel inclined to say to the
world, like the vain actor when applauded : ' Bless you
my people."

As for myself, I have nothing but vainglories to detail.
Lord Iddesleigh has written me really a beautiful private
note, regretting my resignation and adding that Kew will
be to me what St. Paul's is to Wren ! I have thanked in
my family's name (as including my Father).

The Secretaries of the Colonies and India have both
addressed the Treasury officially, deploring their loss of
me, and hoping that my services to them will procure me
a good pension !—(We shall see. My 'opes are not 'igh !).
I make out that they owe me £930 as pension ; perhaps
they won't see it. I feel very keenly the cutting adrift
from my official relations with so many Public Offices, but
I guess I have not seen the end of them, for visions of
Treasury Committees float before me. I shall cut London
Society generally, except perhaps that we may take lodgings
in Town for a month in the Season, that is when our friends
will be too busy to care for us ! Meanwhile I have taken a
little house at Kew for Willy, and shall keep two rooms for
ourselves.

Compliments, however, issue from public offices more
speedily than cash, and the long delay of the Treasury in
deciding the amount of his retiring pension was a serious
inconvenience. Four months after his retirement he writes to
Huxley (March 27, 1886) :

I have just had a very handsome acknowledgement of
service in a despatch from the Government of India—but
' fine words butter no parsnips,' and I am £—600 at my
Bankers and don't want to sell out if I can help it. I think
I must poke up the Treasury.

It took the full year, to November 1886, before the Treasury
informed him that his pension would be paid on a certain scale,
and Christmas before the proportions allotted for his various
services were finally decided. It is well for the public administration to practise economy, but to haggle over the retiring
allowance of distinguished public servants is not the happiest
exertion of this public virtue, nor the most encouraging to those

whose services have all along been remunerated on a scale accommodated to the prospective gilding of their declining years, if they survive so long.

His term of public service extended over forty-seven years, seventeen in the Naval service (1839–1855) and thirty in the Civil service, under the Board of Works, his name remaining on the Navy list till 1870. Under the terms of his appointment to Kew, ten years' nominal time was to be added as a ' special award ' for Kew in calculating his retiring pension.

On the latter point the Treasury's reluctance gave way to the explicit testimony of the Board of Works. On the naval question they were more stubborn.

In 1870 by Order in Council all Assistant Surgeons who had not served afloat during the five previous years were compulsorily retired on pay of £200. Hooker came under this order. Before 1870 there had been no retirement scheme for Assistant Surgeons. The continuance of his name on the Navy list, which brought him no higher pay, had been thought useful to Kew in various ways, and was especially so when he went to Syria with the Hydrographer. During the seventeen years from 1839, he was serving either afloat or ashore, always under orders in India or publishing the botany of all the Voyages in the South from Cook's onwards, and had no idle half-pay time. His last service afloat had been eight years before he came under the Board of Works as Assistant Director of Kew.

The Treasury now proposed to disregard this Order in Council, and to calculate his naval retired pay on the current scale of £54 15s., as if it had existed in 1855. To this proposal, put forward at first tentatively and in private form, Hooker objected strongly, even though it appeared possible by cutting down his rightful naval allowance to make his ' special award ' for Kew larger than what he was entitled to receive—an act of grace instead of simple equity.

Backed by the Board of Works and by counsel's opinion as to the inability of the Treasury by an administrative act to alter the naval pension awarded by Order in Council and confirmed by Parliament, he continued to press his point. As he expressed himself to Huxley (November 11, 1886) :

You may call me a mule for my pains, but I will have the whole matter thrashed out.

I do not like bothering others with my evils, but I know your nares will dilate at the scent of strife, even as do those of one parent of the above-named quadruped, and of which I may be the other.

In the end, his claim as to naval pay could not be resisted, but the Treasury had the last word, striking off the civil pension what was added to the naval.

CHAPTER XLI

RETIREMENT, TO 1897 : BOTANICAL WORK

HOOKER was now sixty-eight. With his ' Tusculum ' ready
to move into, it did not take long to settle down to the purely
scientific work the completion of which was the main object
of his retirement. This was an ordinary man's full work,
but in the first expansion of relief from official burdens he
writes to Huxley, then slowly recovering from a severe breakdown, as though such work were mere idleness : ' I am glad
to hear you are lazy ; it keeps me in countenance. My indolence is, as the Yankees say, phenomenal.' (May 3, 1886.)

For many years he journeyed to Kew three or four times
a week and spent the day working in the Herbarium. Old
age was slow to shackle his energies ; he was ninety-three
before he confessed, ' I am getting very lazy in my old age.'
But at eighty-two he was ' younger than ever ' ; ' I often
feel,' he exclaims, ' that I have no business to be so well as
I am,' and it was only after thirty years' familiar indifference
to the vagaries of an intermittent pulse, that at eighty-three
he yielded to his doctor's orders not to travel alone. But
this was of less concern to him than his slowly increasing
deafness and the fitful onsets of eczema which lamed him from
time to time. And at eighty-four he began to accuse himself
of loss of memory, more perceptible perhaps to himself than
to others, describing thus, for instance, inability to call up
the details of once familiar research and botanical study,
or, in other fields, to keep clearly apart the distinctions of the
different forms of Buddhism.

As an octogenarian, in 1900 influenza robbed him to a

great extent of smell and taste; but his sight remained; and save that in 1904 one eye troubled him, he could repeat contentedly at ninety as at eighty that working at the simple microscope, his eyes were as good, his hand as steady and agile, as when he had first begun such work. By that time, too, his patience, as he said, ought to be inexhaustible. And though in 1895 he admits to Mr. La Touche, ' What you say about anxiety pressing on the old is true indeed ; I feel very acutely that " the Grasshopper is a burden," ' he adds with a quietly sardonic touch, ' Then too I have so little to complain of and so much to be thankful for, that any little grievance takes Cyclopean form.'

These ' little grievances ' are no doubt the things he mentions in 1897 when sympathising with his friend's worries :

> Certainly the Clergy are, taken as a whole, the most long-suffering class of the community ; and then you, like me, suffer from what I regard as the worst evil of old age, the accumulation of petty duties, calls and inroads upon one's time, temper and pocket. The very calls for subscriptions mount up to something appalling, and most of them are reasonable calls (which is the worst of it !).

For himself, it was as he prophesied it would be with his friend George Maw (May 26, 1886) :

> I do hope you may feel your release from business trammels as keenly as I do that from official drudgery— though I fancy you should enjoy it more than I do, as you had not the many bright spots in the daily intercourse with people worth knowing that I had.
>
> It will, I suppose, be with you as with me, no cessation from work—that would be no pleasure ; but the feeling that you can break away at any time with ' none daring to make you afraid,' and best of all, without the feeling that you are neglecting duty, is a most enjoyable state of existence for me.

Now he had the work he desired, not the work he was constrained to do. He still had a year to serve on the R.S. Council ; he continued on that of the Geographical Society to which he owed much, for ' it has always upheld Kew, and it gave me its Gold Medal ' (1884). His Committee was actively engaged in devising means for the encouraging the study of Geography. And there and elsewhere he was ever ready to serve the one cause which he would not grudge as a distraction from the overriding claims of the Indian Flora, the cause of Antarctic exploration. Both here and at the British Association throughout the nineties he warmly supported those unwearying efforts of Sir Clements Markham which brought about Captain Scott's first expedition in the *Discovery*.

The definite break with his old activities is clearly illustrated by a letter to Sir John Lubbock (Lord Avebury), declining to be put forward as his successor in the Presidency of the Linnean Society.

The Camp, Sunningdale : March 7, 1886.

MY DEAR LUBBOCK,—When I resolved upon retirement from official life and work, it was from well considered principles from which I cannot depart. These included severance from active participation in the labors of Scientific Societies, as an absolutely essential condition for concluding some at least of the almost life-long tasks that I have in hand.

Could I see that Science would be more advantaged by the breaking than by the holding to my resolve, I should be justified in reconsidering it ; but in my view everything points the other way.

Were I to undertake the Presidentship of the Linnean, I should feel it to be my duty and my endeavor to show the same disinterested zeal in personally conducting the Society's affairs, that I hope I have shown in every other position of trust that I have held. To divert my thoughts from its requirements would be impossible. No Secretary could relieve me of that feeling of responsibility to Science and the Society which the Presidentship demands, nor of the craving to carry out measures for its improvement in every detail.

Even were I living at Kew, the work of reconstituting the Society as the headquarters of Botany (as the Geological, Chemical and Zoological are of their Sciences) would have been difficult for me as one of the old school ; and even had I the leisure for it, it would from my present residence be impossible.

Moreover, the President has a social as well as a scientific position to maintain, for he is a representative man, occupying a high position in other Scientific bodies. Such a position the Director of Kew has also to maintain, and the two would have worked admirably together. This latter position (social) I have sacrificed under the conviction that Botany will be better advanced by my passing the rest of my days in the study ; and I hope to be allowed to advance the interests of the Linnean to more purpose by contributing papers to it, than I should be occupying its Chair.

I need not repeat what has passed in conversation between us, about the fatigue to a septuagenarian of conducting duties in places so distant from each other as this, Kew, and London ; nor as to my conviction that anyone occupying the Chair as merely honorary President would not be to the Society's advantage. It is enough for me that no possible arrangements could relieve me of the feeling that I was not doing for the Society what I might, could and should do, or avert the regret with which I should retrospect a Presidentship of inactivity.

My wife and I would have been delighted to have done at Kew, that for the Linnean primarily, which we attempted to do for Scientific Society at large when I was President of the Royal and since that time ; and when indeed we included every Fellow of the Linnean that we personally knew or who held official positions there ; and that this opportunity is lost to us is one of the regrets entailed by our leaving Kew, and I assure you we both feel it in reference to this invitation to the Presidentship.

It remains to thank you most warmly for your kindness and consideration in this matter, and to express my earnest hope that a successor to the Chair may be found who will be as thoroughly appreciated during his term of office and as sincerely regretted when he retires as yourself.   Yours affectionately,

JOS. D. HOOKER.

For the next twelve years, then, his cardinal labour was to finish the Flora of British India, a life's work in itself, begun, it will be remembered, on a vaster scale by himself and T. Thomson, but abandoned for want of support, to lie for fifteen years in the limbo of the incomplete with only the dim regretful hope that the half used material might some day be worked up by younger hands.

The first part of the new issue had appeared in 1872; the last was to appear in 1897. Of these, the last three volumes, issued after his retirement, contained 2500 pages.

Some points from the correspondence connected with this work will be noted later. To continue the bare outline of his occupations, two of his old Kew interests were still in his hands, the 'Botanical Magazine' and the 'Icones Plantarum,' those illustrations of new and rare plants from the Kew Herbarium, for the continuation of which Bentham had left a considerable legacy (£7000). Two volumes of the 'Icones' (200 plates), from 1890 to 1894, were devoted to the Indian Orchideae, while in 1895 his 'Century of Indian Orchids' appeared in the 'Annals of the Calcutta Botanical Gardens.' Bentham had also left him the duty of revising his 'Handbook of the British Flora.' Of this he brought out four fresh editions, the fifth to the eighth, between 1887 and 1908.

In 1886 he published the third and final edition of his 'Primer of Botany,' 'the rashest and most profitable of my undertakings,' as he called it to Asa Gray, adding in particular a section on The Movements of Plants including the carnivorous ones; then a fourth edition of the 'Student's Flora' in 1897; while in 1896 he edited from a manuscript in the British Museum, Banks's Journal during Captain Cook's first voyage.

In 1888 he was commissioned by the Government of the Straits Settlements to publish a Flora of those colonies, in conjunction with Dr. King,[1] the head of the Calcutta Gardens,

---

[1] Sir George King, K.C.I.E., M.B., LL.D., F.R.S. (1840–1909); Indian Botanist, Superintendent of the Royal Botanic Garden, Calcutta, and of cinchona cultivation in Bengal 1871–1898; greatly increased production of quinine and established a method of distributing it at a low price; organised a Botanical Survey of India and became the first director 1891–98; founded *Annals of Royal Botanical Gardens, Calcutta,* and contributed monographs; published 'Materials' for a Malayan Flora. Sir Joseph's correspondence with Sir G. King was very extensive, but most unfortunately on the death of the latter his letters were all destroyed.

who was to be at Kew all the autumn, arranging all prelimi-
naries with him.[1]

Overshadowing these secondary occupations, however,
came a mass of Darwinian work ; correspondence about
the Life which Sir Francis Darwin was preparing (1887), and
the arrangement and entire revision of the colossal ' Index
Kewensis,' Darwin's invaluable legacy to botany, which,
set afoot in 1886, appeared from the press during 1893–95.
While he was toiling over the microscope at the Indian Grasses
his time was drawn upon for this long task : ' The huge
Darwinian " Index Kewensis " drags its slow length along,
at the rate of 2 sheets,—about 2500 names, authors and
native countries to revise in press per week. It will take
1½ years more to finish it.' (To La Touche, April 9,
1894.)

Mention may be made of the obituaries of De Candolle,
1893 ; and Dr. David Lyall, his fellow Antarcticker, 1895 ; of
the Eulogium on Robert Brown, 1888 ; while various reviews
and monographs arising out of his botanical work may be found
in the Bibliography, of which ' Pachytheca ' 1889 may have
special mention.

The new era opened enthusiastically.

To Asa Gray

Jan. 24, 1886.

We are very comfortably housed now, and I have just
got my books into place. I find working here and at the
Herbarium vastly different from in the study at Kew.
I can now concentrate my attention (I hope I will) and
write off the Magazine without interruptions of all sorts ;—
indeed I find the withdrawal from the Directorship a stupendous relief ; and, many regrets notwithstanding, it is sweet
to be independent, and to be free of that thirst for power
and position which alone enables one to carry on official
business with some ease and less friction.

---

[1] 'Happily the foundations of it are laid, in my *Flora of British India*, which
includes the Straits Flora, but which is far too bulky a work for the use of Colonists, and does not expressly deal with the timbers, drugs, dyes, textiles, oils,
waxes, and guttas ; and the nomenclature of which is in such sublime confusion,
that the Govt. has called for such a Flora as we shall undertake.'

As for Botany, I am working hard at ' Fl. Brit. Ind.,' but am not yet done with *Litsaea* (alias *Tetranthera*), of which I have utterly failed to make natural groups. I shall have, I suppose, 60 or 70 Indian species to be dissected. Without fruit they are very hard to delimit.

A temporary chagrin was the necessity of reprinting Part I. of the ' Genera Plantarum.' The demand had been larger than could reasonably have been expected, and the book was out of print. ' I never grudged any job more,' he continues. ' By the time I shall be recouped for that I must reprint Part II., a heavier job.'

To Asa Gray

Feb. 23, 1886.

I am still at *Laurineae*, but nearly done. The one-celled anthered *Litsaeaceae* I have followed Bentham in bringing under one genus, Lindera—though I feel satisfied that they should make 3 or 4 by habit and inflorescence. The latter is very curious and often difficult to make out. Bentham did not attempt to understand it—had he attempted it the Gen. Plant. would not now be finished ! Laurineae is one of those Orders that he would have wished me to do, not because I could do it better, but because I have more patience with that sort of analysis that is required and which is necessary, even if I do not make so good a use of it. I did however suppose that Bentham had done generically the S. American ones for Schomburgk and Spruce.

You never, I think, tried your hand at such a job of exotic plants as classifying any of these obscure, arborescent, tropical Orders. They are a great strain, but I prefer such work to such jerky work as Bot. Mag. But you have your full share of troubles in your own Compositae and other Orders.

I am now printing Indian Polygonums, 70 species !

To Asa Gray

June 20, 1886.

I have just completed Laurineae and am utterly dis-satisfied with the result. No doubt I have tripped up

Meissner[1] over and over again, but really my work is, though fatter by far, no better of its period than his was, and the whole has raised old Nees[2] a good deal in my estimation. It is at any rate much better than his Acanthaceae.

I am now at Euphorbiaceae, of which Boissier[3] has of course made too many species, chiefly by looking at scraps in isolated herbaria. How could he have expected to find undescribed plants of Hayne in Herb. Vienna, Petersburg, &c.

Of botanical news I have none. We are busy with more Icones, with what of Bentham's money has come in. His affairs are not wound up yet and I am sick of them. They were to have given no trouble to anybody! Much he knew! It makes me miss him all the more. I occupy his room at the Herbarium, where I am about three days a week. The more I see of Oliver, the more I wonder at his marvellous knowledge. He has far the greatest knowledge of Phaenogams of any *two* Botanists that ever lived. You cannot puzzle him.

To Asa Gray

Sept. 27, 1886.

I am more and more absorbed in Indian Botany, and have thrown aside all idea of making headway with—any desire to keep up with even—heads of Chemico-botany,

---

[1] Charles Frédéric Meissner (1800–74). His father, of Hanoverian origin, settled at Berne, where his son was born. He was educated at Yverdon and Vevey, and afterwards at Vienna, Paris, and Goettingen. He contributed monographs of various families to De Candolle's *Prodromus Linnæus, Botanische Zeitung*, Hooker's *Journal of Botany*, Warming's *Symbolae Botanicae*, Lehmann's *Plantae Preissianae* and the *Flora Brasiliensis*, and also published, 1836–43, his *Plantarum vascularium genera*. He formed an extensive herbarium, sold at his death to Columbia College, New York.

[2] Christian Gottfried Nees von Esenbeck (1776–1858), physician and botanist. Educated at Darmstadt and Jena 1796–9. He was Professor of Botany at Erlangen in 1817; and chosen President of the Imperial Leop. Carol. Akademie of Natural History the same year. Subsequently he held the Chair of Botany at Bonn 1819, and Breslau 1830, when he was also Director of the Botanic Gardens. Among his many botanical works is the *Handbuch der Botanik*, 1820–1. He also wrote on Entomology and Philosophy.

[3] Edmond Boissier (1810–85), botanist and traveller, born and educated at Geneva. Most of his work relates to the Mediterranean Region, Spain and the Orient. He published his *Voyage Botanique dans le Midi de l'Espagne* between 1839–45, and in 1842 travelled in Greece, Anatolia, Syria and Egypt. Between 1842 and 1854 appeared the first series of his *Diagnoses Plantarum Orientalium Novarum*, in 1848 his monograph of the Plumbagineae, and in 1862 his elaboration of the genus Euphorbia, and the *Icones Euphorbiarium* in 1866. His great work, the *Flora Orientalis* in five volumes, was published between 1867 and 1884.

and Micro-phytology. I may content myself with a casual
grin at young men calling themselves botanists, who know
nothing of plants, but the 'innards' of a score or so. The
pendulum will swing round, or rather back, one day.

I have no copy of Fl. B. I. here, but look forward with
amusement to seeing what I have written on page 893.
I often muddle in press what I have written clearly enough
on paper. I look with great dissatisfaction on *Piperaceae,
Laurineae* and *Myristaceae,* but think I may with different
feelings on *Polygonaceae* and the reforms in *Loranthaceae.*

. . . I enjoy my freedom from harness more than ever,
though so much poorer. They have given me a shabby
pension from Kew, reducing what would otherwise have
been given as a ' special award ' *because of my Naval Pension !*
though this was awarded for services rendered three years
before I became Asst. Director. . . . I shall have to econo-
mize for the next few years and work as I have done (and
liked). . . . However I am well, happy and contented,
and have been a most fortunate man in family, friends,
launch into life, and opportunities, and if my income is not
so good as my father's was, it is because I have had such
a family to educate and support—otherwise I should have
been *wealthy.*

As to the extremes of the New Botany, the ' casual grin '
recurs in his letter of the 22nd to Professor Oliver, whose son
Frank he hopes will decide to take up botany.

When settled he would not be disinclined I think to
take up a large Genus in a small Nat. Ord. I should like
him to attack one with minute flowers, like the Phyllanthi
I am at, the male flowers of which are often not 1/40 in.
diam.—they might delude him into the belief that his work
was histological! and nothing but the very minute seems
worthy of the attention of the modern school.

Similarly, a couple of years later, writing to a young botanist
about the possibility of standing for a professorship, he
remarks :

There is already a strong feeling apparent, that vegetable
physiology and anatomy alone do not supply the wants of
the public—and that some knowledge of plants in general,

their uses, physiognomies and distribution, should be
taught :—a knowledge of plants, in short, as well as of their
' innards ' and movements.

To F. Darwin

September 9, 1894.

I am glad you are going to teach the Medicos a little
practical Botany. It is lamentable to find that all this
botanical teaching of the greatest Universities in England
and Scotland does not turn out a single man who can turn
his botanical knowledge to any use whatever to his fellow
creatures. Where should we be if Medicine, Law, or any
other pursuit were taught after that fashion ?

All through 1887 the Indian Euphorbias were ' on his brain.'
He struggled with the confusion of species under one or more
names in the Wallich Herbarium. Bentham, who distributed
the plants after Wallich returned to India, ' had evidently not
as yet got his Botanical Optics.' The work was most distracting in the genera with microscopic flowers, one of which he had
on the table for two months. He tells Asa Gray (March 8)
that he has kept *Baccaurea* for a *bonne bouche*. ' The pile of
specimens is quite 5 ft. high ; it goes on growing by specimens thrown out of other genera. I suppose that no genus
of Phaenogams is so little known, not even Calamus ! ' Still
he confesses : ' The more I see of Bentham's work on the
Euphorbs. in Gen. Plant., the more I am lost in " wonder, love
and awe." ' (January 7.)

It was November before he could record that he was printing the last sheet of the Euphorbiaceae, just eighteen months'
work. But even then,

> The next part will hardly take all in. I am getting on
> with the Urticeae, and making a list of the Monocots, so
> as to see how I am to get all in. Two-thirds of my synonymy
> and citations are perfectly useless, but King, Thomson and
> others thought I should make a sweep up of all, so that
> Indian Departmental Floras need not do so.

From time to time during the progress of the work, he contributed descriptions of new and rare species to the Icones

Plantarum, particularly, as has been already noted, of the Indian Orchids from 1890 to 1894. When sending two such descriptions to Professor Oliver, he writes (September 22, 1886) :

> I must really get some more Indian things for the Icones. There are plenty, and having to describe them at any rate, it would be easier for me to contribute to that work than for any one. But I, like you, seem to hate anything that drags me out of the track of methodical work.

Some further illustrations of his labours on the Fl. B. I. may be drawn from the correspondence with Mr. Duthie [1] the Director of the Saharunpur Botanic Garden. This correspondence is typical of his interest in more than the immediate needs of his own task ; his desire of putting others in the best way of serving Botany as a whole and building up their own reputation. Indeed, when in 1875 Duthie was designated for Saharunpur, Hooker invited him to steep himself in Indian Botany at Kew. There were moreover many Indian Natural Orders which needed to be monographed : all the Kew material should be available if he undertook the work.

Thus (September 9, 1890) :

> It is good news that you are going to collect in the Central Provinces. I do not expect much novelty, but a considerable accession to dim knowledge of the Geographical distribution of Indian plants. You will find some Himal. species that you could not expect in that region, and probably some novelties consisting of plants of the Eastern Ghats, a country not explored since Roxburgh explored the Circars.
> What we want now is rather observation than specimens, especially of the smaller flowered plants, and above all of Orchideae.[2]

[1] John Firminger Duthie, botanist, B.A. and F.L.S., Director of the Botanical Department of N. W. India and Assistant for India at Kew in 1903. He sent collections of upwards of 1200 dried plants to Kew from Kashmir in 1893, besides 200 grasses in 1896. He published a *List of N. W. Indian Plants* in 1881, and has written a great many papers on grasses and on other botanical subjects.

[2] He insists again and again on this point, notably with regard to the Grasses and the ' unmanageable ' groups of Festuca ; of which Mr. Duthie had sent him a rich collection. ' Indeed what we want now from India is not more collections, but critical observations on the spot, with illustrative specimens.' (Feb. 16, 1896.)

May 24, 1893 :

When you are again up in the hills, it would be well to ascertain *exactly*, what are the *Aconites* whose roots are eaten with impunity. . . . The subject of the alkaloids in the roots of *Aconites* is now being studied with great care, and there are anomalies amongst the Indian ones (as named) which I can only explain by a confusion of nomenclature. The Pharmaceutical Society is publishing separate researches on the subject.

. . . It is impossible in the Herbarium to limit the Alpine varieties of several of the species, and to distinguish those of one species from those of another. A careful following of each species from a lower to a higher elevation, would be a great thing to do.

. . . The species of Iris of India are in a shocking mess, I wasted days over them. Pray observe them well, especially the ' orris rooted ' wild and cultivated.

January 7, 1896 :

I can quite sympathise with you in your disappointment at not being sent along with the Pauri Commission. . . . It may indeed be doubted whether the result would be worth the expenditure ; what is now wanted is, not more collections, but far more knowledge about what plants we have. The number of genera, of which the species want careful examination, is very great ; in the Orchideae, for instance, all the terrestrial species require revision ; Griffith set the example of what could be done by the co-ordination of observing with collecting. . . . The long and the short of it is, the N.W. Flora wants revision throughout, by careful study of each species, its range, variation, and obscure characters.

February 22, 1895 :

Have you ever yet taken up the Indian *Irises* systematically ? It can only be done by garden specimens and drawings. You have also a fine field in the terrestrial Orchids, which want analysis on living specimens.

February 16, 1896 :

Grasses are terribly difficult and full of snares ; but

what really is more wanting in India is that men like yourself
should carefully work up the floras of one or more definite
areas, not *too large* ; and also take up such cultivated plants
as *Pennisetum typhoideum, Andropogon Sorghum, A. Nardus,*
&c. &c., carefully note their habits and varieties over such
areas, and, as far as they travel, we can get specimens from
beyond it.   Just as I told you *Iris* wanted it.   Mr. Foster
[Prof. (afterwards Sir) Michael Foster, F.R.S.] has done
more for Indian *Irises* by getting people in India &c. to send
roots, than have all other Indian botanists put together.
The rage for collecting has cold-shouldered observations in
the field.   Old Roxburgh [1] stands alone.

The systematist's labours in this part of the Flora were no
less than in the preceding part, '1400 species rescued from
veritable chaos.'   His request for more observations on the
Orchids was called forth by his immediate difficulties.   He tells
Mr. Duthie (September 9, 1890) :

I have just done *Habenaria*, about 100 species, and you
have no idea of the difficulty I have experienced in deter-
mining the structure of the stigmatic surfaces and processes,
and above all of the rostellum.   Accurate drawings of this
latter organ would be invaluable.   In some, so called,
Peristyles it appears to cover the pollinia glands, and, if so,
what becomes of the character of *Orchis* ?   I have been off
and on *Habenaria* for now 5 years, and am very dissatisfied
with the result.   The longer I am at it, the more I feel that
Fl. Br. Ind. is little better than a sweeping up of materials,
with approximate references and synonymy.

The grasses were still more difficult to tackle.   Morpho-
logically he found them very interesting, and the types of all
the huge Order occur in India.   But the work was all microscopical ; the only method was to draw the floral organs of
every species and often many specimens of each.   Systematically, confusion was rampant.   The descriptions of the

[1] William Roxburgh (1751–1815), a botanist who, while in the Indian medical
service, not only collected plants, but set up an experimental station to improve
the cultivation of valuable plants.   He was Director of the Calcutta Botanic
Garden from 1793 to 1813.   His accurate descriptive work was made doubly
valuable by his copious drawings, which were copied for Kew at the expense
of Sir William Hooker.

old Indian authors were often impossible to recognise, thanks
to insufficient data and incomplete characters. Even in
Roxburgh's careful work, a good many species could not be
guessed at ; in many of his otherwise excellent drawings,
Hooker felt sure that the enlarged fruit figures did not belong
to the plant named.

Every detail therefore had to be worked over anew. And
when this ' laborious job ' was done, it would be so far unsatisfactory that no ' good sections '—i.e. clear subdivisions—could
be found in the genus.

As he puts it to Mr. La Touche (April 23, 1894) :

> [The Grasses] are dreadfully difficult and systematically
> a chaos of imperfect descriptions, erroneous identifications,
> confused synonymy and imbecile attempts. We have up-
> wards of a century of collections and not an attempt at a
> classification. Each Botanist in his own country has worked
> at his sweet will in ignorance of his predecessors' and
> contemporaries' work, with imperfect materials and often
> no books—' Hinc illae lachrymae.'

' I think,' he grimly remarks, ' my results will open the
eyes of Botanists.'   (To T. H. Huxley, August 16, 1894.)

A further complexity was introduced into the difficulties of
the subject by the fact that a set of the Indian *Andropogons*
had been sent to Professor Hackel [1] to be worked up for his
Monograph on the Order.

> I have been trying Hackel's Monograph [he writes
> on May 24, 1893] and find it most difficult to work with.
> For diagnostic characters he prefers the most obscure and
> difficult to detect, to the obvious and natural. I long to see
> another Kunth [2] in Germany.

Panicum is in a chaotic condition ; there is a pile of
materials five or six feet high from half a dozen collections,

---

[1] Eduard Hackel, an Austrian, one of the greatest authorities on grasses,
the author of various papers and publications on the grasses of Carpathia,
Portugal, Spain, and the Alpes Maritimes. He has published a monograph on
Andropogon and one upon Festuca. Professor in a gymnasium at St. Pölten.

[2] Carl Sigismund Kunth, botanist. He published his *Flora Berolinensis*
in 1813; revision of various Natural Orders as Bignoniaceae, Leguminosae,
Malvaceae, Tiliaceae, Gramineae, &c. His *Nova Genera and Species Plantarum*
appeared 1815-25.

topped by Mr. Duthie's last bundle from Kashmir, and his yet
more important collections from Central India, which linked
Northern and Southern India. For four months he does not
get through one species a day, identification being often a matter
of counting nerves in the glumes under a microscope in an
enormous number of specimens. The synonymy is 'frightful';
for example, all the species of three other genera have been put
into *Panicum* at one time or other. It was impossible even
to 'divide it into good groups; all characters inosculate hopelessly. The only way to get on is to make pencil sketches
of the spikelets and their parts on scores of sheets of specimens, often of one species only.' *P. sanguinale* alone, of
which he unearthed 86 synonyms, kept him for three weeks.
A large form of this, noted in the Himalayan Journals as
being grown for food in the Khasia, found mention nowhere
else, and could not be safely identified with herbarium speci-
latest wrongly labelled or described. Worse still, even the
mens investigator, Hackel, had made a new species of a hairy
state of one of its forms and "has already been floundering,
by taking up individual specimens for description." (March
15, 1894.)

To Mr. Duthie

March 15, 1894.

Occasionally I, for a change, take up the *Andropogs*,
which Hackel has worked out well; but tooth-drawing is
nothing to working with Hackel; you are in agony till you
have waded through his line after line of characters, many of
which apply to the whole genus—if not to all the vegetable
kingdom I was going to say. He has not found out that it is
a perfect matter of indifference to most grasses, whether they
wear hairs or not. Still I should not complain; if Hackel
had not done the Andropogs, I do not know who would;
and I feel sure no one would have done them more conscientiously.

I wonder what you will say to Clarke's [1] *Cyperaceae*, the
most important of all contributions to the Flora.

---

[1] Charles Baron Clarke (1832–1906), botanist and collector. Educated at
King's College School, London, and Cambridge, bracketed Third Wrangler 1856,
and elected Fellow of Queen's College 1857. He was called to the Bar in 1858,
and in 1866 joined the Educational Department in Bengal. He acted as Super-

To the Same

I thank you for sending me the report on your Kashmir tour, which has interested me very much, besides being of use in giving me exact information as to the localities where you made your collections, for many of which localities one may hunt for ever in the ordinary maps and gazetteers without finding them.

I am in the middle of your grasses, by far the finest collection ever made in Northern India. But oh how difficult ! You have many more species than Thomson obtained. The N. Western grasses differ more from the Sikkim than I could have supposed ; but then I have no good Sikkim materials. My June and July collections were almost destroyed during the rains, living as I was in a tent of two blankets, with no collector proper.

I have been for nearly three weeks at *Poa* and am utterly beaten. As to Nees's, Royle's, Munro's and Grisebach's [1] names, they are all wrong. It is impossible to name *Poas* on single specimens. Except *P. persica* (*var. soongarica*), which is Royle's *Festuca Amherstiana*, there is not a definable specimen in the genus, and I am at my wits' end what to do. I cannot even sort the specimens ; your specimens leave nothing to be desired as such. Indeed, but for their *magna copia* I could get on well enough ; it is easy to sort collections of single specimens, but yours show such series of forms and in such good state that they run all sorts of apparently distinct things into one. I shall take counsel with Stapf,[2]

intendent of the Calcutta Botanic Garden from 1869 to 1871, and in 1879 was placed on special duty as assistant for the Flora of British India at Kew. He returned to India in 1883 and served till his retirement in 1887,when he returned to the Kew Herbarium and worked as a volunteer till his death. He contributed largely to the Flora of British India, De Candolle's *Monogr. Phan.*, *Journal of Botany*, and the *Journal Linn. Soc.*, &c.

[1] August Heinrich Rudolph Grisebach (1813–79), botanist and Director of the Botanical Garden at Göttingen. He published his *Plantae Wrightianae e Cuba Orientali* 1860–2, his *West Indiens Geographische Verbreitung der Pflanze* 1865, *Vegetation der Erde* 1872, and *Pflanze lorentzianae* 1874.

[2] Dr. Otto Stapf, Ph.D., was his excellent coadjutor at Kew, to the value of whose hints in the revising of the clavis he specially refers, and furthermore adds (May 20, 1895) :

'Stapf has worked up the Indian *Poas*, and after carefully revising his work, I do not think it could be better done, but the want of definable, or constant characters in the genus will, I feel sure, render it impossible to name a single species by book alone, with any confidence.'

and cut the Gordian knot as best we can. When I have done the grasses I will send you a list of your species.

He complains that the immense variety among the specimens as their numbers increased made systematic results sadly uncertain ; the incessant revision necessary delayed progress with the proofs. ' I am perpetually on the scent of false reductions or false references of a doubtful or badly described species.' (January 7.)

Finally, June 26, 1896 :

I have put *Poa, Festuca,* and *Bromus* into Stapf's hands, as he is well up in the S. European and Persian species. In respect of *Poa* his work agrees well with Hackel's determinations of your plants, but as to *Festuca* they are wide apart. After two struggles with the latter genus I gave it up, and I shall be curious to know what future observers make of Stapf's determinations. I can testify to the extreme care with which he has worked, but my impression is that the whole of *ovina* and *rubra* and *duriuscula* groups are utterly unmanageable.

The fact that the same plants were being sent from India for description to Hooker and Hackel simultaneously might have led to embarrassment, especially as the former was, so to speak, in the Indian Government employ. The only clash, however, took the form of a happy coincidence.

To the Same
March 27, 1895.

I left Kew Herb. yesterday with my usual bundle of grasses for description at home, amongst them a new genus, which I examined some time ago and for which I intended the name *Duthiea,* when to my astonishment I found your letter on my table, and the very same grass for which Hackel had proposed the very same generic name !

Leaving Mr. Duthie to arrange the matter, he adds, for his own part :

I am only too glad to have Hackel's identifications in which I put the greatest faith, and I shall adopt his names

for all new species in the place of my own whenever we agree.
The latter [agreement] will I do hope be the rule, but considering the very different materials, in amount of specimens and
number of collections upon which we work, there must be
discrepancies in details.

To Mr. Duthie

Jan. 7, 1896.

I am not sorry that Hackel has cut me out of *Duthiea* ;
he could and has discoursed on it in a way I could not have.
My impression is that the plant is as good an *Avenacea* as
Festucacea—but I am not going to interfere with Hackel's
determination.   After all Hackel's paper shows that the
tribes and sub-tribes of *Poaceae* are a sad lot.

He was doubtful, however, as to the natural position near
Bromus assigned to *Duthiea* by Hackel.   A new *Duthiea* turned
up in a fresh consignment of Indian grasses,

which satisfies me (and Stapf) that I was right in placing the
genus next *Danthonia*.   When Hackel's paper appeared I,
of course, followed him, and in the clavis of genera (printed
off months ago) I put it next to *Bromus*.   I shall ' hale it
up ' in the body of the work to its right place.   It corrobo-
rates my view of the close affinity between *Bromus* and
Avenaceae.   (Feb. 16, 1896.)

Amid such complexities of this, ' the hardest work I ever
did,' the solution of which depended upon the critical observations on the spot that he so much desiderated, it was hardly
to be wondered at, however annoying, that errors crept in.
For these he was constantly on the look out, and on February 6,
1898 (the Flora was published in 1897 and he was then at work
on the Ceylon Grasses for the completion of Trimen's Flora of
Ceylon), he writes to Mr. Duthie :

You may be shocked to hear that my new genus *Ney-
raudia*, over which Stapf and I spent no little time (to make
sure !) turns out to be a known not uncommon *Triraphis* !
and is not even Arundineous.   The fact is that Gramineae
are still in a shocking state.   Stapf is making discoveries
constantly of mistaken affinities amongst well-known genera,

and hopes to arrive at a more natural grouping than has hitherto obtained, of the *Poaceae* especially.

And on August 16 he is staggered by ' a hideous blunder in the Fl. Brit. India which fell under my eyes accidentally two days ago, and was nearly followed by a stroke of paralysis.' He found that the plant listed as *Panicum latifolium* of Linnaeus had nothing to do with the *latifolium* of Kunth, and there was no *latifolium* in the Kew Herbarium ! It should be *zizanioides*.

But the blundering he could not trace turned out to be no blunder at all. ' Now I recollected ' (he writes on Oct. 21) ' that I took the name from Linn. Herb., so I went there the other day, and sure enough a fine specimen of *oryzoides* [*zizanioides*] is named in Linnaeus' own hand *P. latifolium*, so *oryzoides* is *latifolium* Linn. Herb. nov. Sp. Pl.'

One fresh observation incorporated in the ' Flora of British India ' came from the work of his old friend Bertram Mitford, later Lord Redesdale,[1] who, not content with growing many kinds of Bamboo in his beautiful garden at Batsford, published a botanical study of them in his ' Bamboo Garden.' Hooker, to whom the MS. was shown, was no little interested, and pronounced his description quite clear and good except in one or two small matters of wording, against which, to ward off cavillers, he pencilled suggestions, especially where the turn of the phrase appeared to convey, in strict botanical language, more than should be intended.

His friend demurred, though with diffidence as against such authority ; to which Hooker replied (Aug. 23, 1896) :

Never hesitate to challenge me. I have not seldom been convicted of error in the use of terms. I quite saw what you meant, but hold that putting *persistent* where you did, does in botanical language *assert* that the sheaths are persistent on the culm and branches respectively.

[1] (1837–1916.) He was created Baron Redesdale in 1902, having inherited the estates of his uncle, the Earl of Redesdale, in 1886. His literary classic, the *Tales of Old Japan*, was the fruit of his diplomatic career in Japan. Then he entered the Office of Works during the Ayrton affair, and lent Hooker much aid and sympathy. Later the famous gardens of Batsford enabled him to pursue his botanical tastes to the full.

But when, granting that the suggestion lamed the sentence, he suggested that it be rewritten, so as to draw the currently accepted distinction between these, the next letter revealed that his friend had made entirely new observations on the persistence of portions of the sheath (known as ' pseudophyll ' and ' limb ' respectively) in the branch as well as in the culm of the Bamboo, and Hooker responds (August 26) :

There you ' 'ave me on the 'ip,' as a Cockney friend once addressed me in a friendly dispute. I took it for granted, in my ignorance, that pseudophylls were restricted to culm sheaths. That they are not suggests a close examination with the view of ascertaining whether or no a transition may be found from the pseudophyll to the true leaf-blade. It would be satisfactory in either case to correlate them with some functional or morphological character of the plant. . . .

The matter of the pseudophyll is of great moment to me. I cannot find its special attributes described, or even alluded to in any other accounts of Bambuseae that I have as yet consulted than yours, and I must bring it in in a note prefatory to the account of the tribe in the ' Fl. Brit. Ind.' (by Gamble [1]), over which note I shall ask you kindly to cast your eye when I get to Bambuseae. Also I expect I shall have to introduce observations from the ' Bamboo Garden ' under the Indian species, as to which I must have your good offices. . . .

I have asked Stapf to look out for any other grasses with all the sheaths normally deciduous from the node by a clean, clear-cut line, as in Bambuseae. I think I can name cases where the true blade when sessile on the sheath disarticulates, but none where the true blade is petiolulate. All such reminiscences are, however, very untrustworthy. I only wish I had had the point in my mind when working up the Indian grasses. It is most interesting to find a field of research opened up by a study of the last tribe of the Order !

The Camp, Sunningdale: September 1, 1896.

DEAR MITFORD,—Since I wrote yesterday I have tackled a proof sheet in which *Bambusa* sees the light ; so I thought

---

[1] James Sykes Gamble, C.I.E., M.A. (Oxon.), F.R.S., F.L.S., late Conservator of Forests in India, and Director of the Imperial Forest School, Dehra Dun.

it best to spend the morning in attempts to compose a rider
on 'the Bamboo Garden,' and the matter of the blades.
I enclose it herewith, with tremor, you must kindly look
over it for me and give your candid opinion like the good
fellow you are, and let me have it back.

Ever, my beloved Censor, Yours,

J. D. HOOKER.

September 8, 1896.

A thousand thanks, my dear Mitford. The difference
between our says resolves itself into mine being the work
of an at second hand hasher up of material not his own,
and which he has not fully grasped—yours is the result
of long and careful autopsy.

The long and short of it is that what I would like and
crave is a brief statement of the dry facts from your pen
with leave to give it as such. In case of your agreeing, to
show what I want, I have extracted from your able exponent
what I should like to see. I have drawn a pencil round the
parts of your MSS. which I have not included to save you
the trouble of collation—the rest is verbatim. Kindly look
at it. Excuse haste—this is our afternoon for visitors.

Ever yours,

J. D. HOOKER.

Sept. 10. I sent off the paragraph with the emendations,
and with it a thankful heart.

In 1889 he published in the 'Annals of Botany' (vol. iii,
pp. 135–40) a paper on the curious organism *Pachytheca*.
For half a century this was a puzzle to botanists. It was
a seed-like body found in the Ludlow bone-bed, and first
described by Hooker in 1853 in a note to H. E. Strickland's [1]
paper on the fossils there found (*Quarterly Journal of the
Geological Society*).

These specimens did not allow of transparent sections
being cut for the microscope, but the observations that could

[1] Hugh Edwin Strickland (1811–1853), naturalist; accompanied William
John Hamilton in a geological tour through Asia Minor, and traversed Greece,
Constantinople, Italy, and Switzerland, 1835; drew up rules for zoological
nomenclature, ultimately with some modifications accepted as authoritative.
Among several important scientific writings were *Ornithological Synonyms*,
1855, and *The Dodo*, 1848.

be made justified Hooker in considering it to be a sporangium of a Lycopodium or allied plant.

In 1875 more specimens were discovered in the West Malvern limestones, and sent to Hooker through Mr. Symonds of Pendock.   Sections made convinced Hooker (as also various cryptogamists) of its structural connexion with the Algae, though there was much difference as to its existing affinities. A stumbling-block was the apparent discontinuity between the filaments of its inner cavity and the tissues of the surrounding walls, which even suggested the possibility that the former belonged to an intruded parasitic alga, or the mycelium of a parasite.

Similar bodies were constantly found in Great Britain and in America, and in 1882 Principal (Sir J. W.) Dawson of Toronto communicated a paper on the subject to the Geological Society.   In his view *Pachytheca* was a seed, rather than a spore case, belonging to a primitive form of Gymnosperm (*Prototaxites*).   But his arguments did not find much support ; others variously suggested that it was the float of a seaweed, or even of animal origin ; Professor Thiselton-Dyer, who believed he had detected the missing connexion between the contents of the cavity and the wall, traced a morphological affinity to *Codium* in both *Pachytheca* and *Prototaxites*, but saw no evidence of the former being a sporangium of the latter.

All through the eighties Hooker sought more light, and corresponded with various investigators.   Professor (Sir E.) Ray Lankester confirmed his general views ; the specimens were examined by the Rev. J. D. La Touche, himself a keen geologist, who was publishing the local geology for the Shropshire Natural History Society ; Mr. W. Phillips, the algologist of Shrewsbury, propounded a definite form of the parasitic view which on many grounds Hooker considered untenable.

Robert Etheridge,[1] President of the Geological Society in 1880–81, who had ' again and again examined these bodies, '

---

[1] Robert Etheridge contributed various papers to the *Memoirs of the Geological Survey of New South Wales*, 1888–94.

summed up, saying (July 6, 1888), ' Your own note in the
Quarterly Journal for 1853 contains after all the pith of the
matter and the best figure.'

He would not rest content, however, without having a
thorough microscopic examination of the material, old and new,
made by a young botanist of Christ's College, Mr. C. A. Barber,[1]
a capital artist, who in the course of drawing, made out a
new feature which Hooker suggested might be callus plates.
A doubtful specimen which he showed to Professor Bayley
Balfour [2] had yet more doubtful signs of attachment, as of a
marine alga, which, if substantiated, might suggest, as the latter
said, the possibility of *Pachytheca* being a *gemma* which had
been detached from a highly organised Alga. But there was
nothing to bear out this suggestion ; and other suggestions as
to its affinities based on exact but limited knowledge of some
one class were unacceptable to wider knowledge.

Pachytheca gets more and more inscrutable [Hooker
writes to Mr. La Touche on Aug. 3, 1888]. My impression
still is that it is in all probability a type of structure of
which there is no existing type—and of such structures there
must have been thousands of old. Uniformitarianism has
run mad in Geology and Palæontology and every dead thing
is screwed into the category of living things. I always ask
what proofs have we that what are now land shells were
not (in their primitive types) sea shells, and *vice versa* in a
more or less degree. But this strikes at the root of all
Geology as founded on Palæontology.

Accordingly, after tracing the history of Pachytheca and
the theories about it, he wound up his paper by declaring that
no certain conclusions as to its real nature and affinities were
as yet possible.

His judicial caution was justified.

[1] Mr. C. A. Barber, assistant to (Sir) F. Darwin at Cambridge, was appointed to the new Agricultural Department of the Leeward Islands in 1891,
and in 1898, after three years at Cooper's Hill College, became head of the
botanical department for the Presidency of Madras, at Ootacamund.

[2] Isaac Bayley Balfour (*b.* 1853) was Professor of Botany at Glasgow 1879,
at Oxford 1884, and at Edinburgh in 1887, the position so long held by his
father, J. H. Balfour.

To Rev. J. D. La Touche

June 9, 1895.

I have this morning received from Mr. Murray of Brit. Mus. a notice published in the ' Phycological Memoirs ' of *Pachytheca*, which is enough to turn your hair gray—if it were not so already !

Pachytheca is found to sit in a cup ! from the base of which cup rootlets issue ! (sketch). The cup has distinct traces of a stalk, and so must have been attached to some large body. *Pachytheca* itself of these specimens retains its internal structure—but no structure has been found in the cup. They are from S. Wales.

A more ghastly proof of the futility of speculations on the nature of imperfect specimens could not well be. ' Ex uno disce omnes ' !

A letter to Canon Ellacombe, the well-known horticulturist and botanist, illustrates Hooker's experience of the pitfalls of Fossil Botany.

Dec. 19, 1887.

Your question is a very difficult one to answer. Fossil Botany has made enormous strides in the matter of publication of names and drawings since 1852, but not, that I am aware, under the hands of competent botanists familiar with the existing flora of the globe or the infinitely varied forms of recent plants, and above all with the incessant repetition of identical forms of foliar organs in the most different natural families of plants. Of certain identifications there can I think be no reasonable doubt, as of the genera *Liriodendron* and *Salisburia* [1] and of a few others where the fruit has been found associated with the foliage, as I believe, of *Liquidambar* and *Platanus, Acer,* &c.

When you come however to coniferous fruits, where to ascertain even the tribe of the recent genera, you should know the position of the ovule, and where the form of the foliage and cone is so polymorphous, I must confess to the gravest doubts. Nor would I accept the evidence of *Athrotaxis*, except on the statement of a botanist well versed in recent Coniferae.

[1] Nevertheless no competent botanist would be surprised at receiving from New Guinea or China, plants of totally different Natural Orders showing the foliage of these genera.

It is an ugly fact that, tempting as is the study of Fossil Botany, every competent botanist with a large knowledge of existing floras, and that has tried his hand on it, has given it up, notably Brown, Brongniart, and Lindley, or these have subsequently confined themselves to specimens exhibiting *structure*, as fossil wood, &c.—whilst Oliver, Bentham, &c. have only shaken their heads when asked to identify a fossil plant. If you are ever at the Herbarium and will look at the multitudes of figures of leaves in Gardner, Lesquereux, and other works, the vagueness of the identifications will strike you at once. There is a standing joke at the Herbarium, if you have a plant the affinities of which puzzle you, ' fossilize it and send it to a palæontologist and he will give you the genus and species at once.'

To sum up, in a *general way* the work of Vegetable Palæontologists has thrown much light on the older floras of the globe, but far too much is made of the supposed facts, which in detail are wholesalely unreliable.

With regard to *Salisburia*, I was the first to show the strong resemblance of the Coal Measure fruit called *Trigono-carpon* to that of *Salisburia*, founded on a careful comparison of the *tissues* of the several layers of the fruits obtained by slicing silicified specimens of *Trigonocarpon* and fresh ones of *Salisburia* (published in ' Phil. Trans.') but I am quite prepared to find the same tissues in *Cycadeae*, in *Podocarpus*, in *Dacrydium, in Cephalotaxus* and other genera ! Had leaves of *Salisburia* occurred with *Trigonocarpon* the evidence would be good. As it is, possibly *Salisburia* fruits occur in the Coal ; and abundant *Salisburia* leaves occur (not in the Coal) but in tertiary rocks without *Trigonocarpon*.

I have myself tried my hand on the identification of fossil collections of plants by their leaves, &c. and what I find is this, you have, say 100 forms of leaves from one bed, evidently belonging to many genera and families. You identify one of the most peculiar with a plant of Japan we will say. Well, you have no difficulty in matching all the others with plants of Japan, and you conclude that you have an old Japanese Flora. But if you had started with a S. American identification, and had an equal knowledge of S. American plants, you would have been as successful— probably more so—for the simple reason that the *forms* of leaves are protean and repetitive and that in the fossil you have neither insertion, stipulation, surface, texture, vernation or colour to check you in your headlong course at identification. Lastly sweep up the floor of a Herbarium, after a good case of rejecting bad specimens from a heap of plants, and see what a fossil botanist would make of the disjecta membra !

He was much interested about this time in his friend Huxley's excursion into practical botany. The latter, seeking health in long summer visits to Arolla and the Engadine, began to study the structure, life-history, and affinities of the Alpine gentians, and went on to investigate the gentians of the whole world in order to make out their evolutionary affinities and distribution. Indeed his devotion to the gentians led the Bishop of Chichester, who was staying at Arolla, to declare that he sought the ' Urgentian ' as a kind of Holy Grail.

He seemed to find that the distribution of the gentians corresponded very closely with that of the Crayfishes, on which he had written a well-known monograph. But Hooker doubted any general identity between zoological and botanical regions.

Athenæum Club : Feb. 25, 1887.

DEAR HUXLEY,—I have finished the Gentians, and have been much instructed by the first part and interested in the last. The junction of your line with Mueller's and both running into the same terminus is the ' bright particular ' point in it.

As to the Taxonomic part, *Gentiana* is one of many old genera founded on a few heterogeneous materials, the result of which is, that for years it is not only not reformed but all sorts of things are added to it—it takes 7 devils worse than the first. I found Palms in the same condition when I worked up the genera for Gen. Plant. *Areca* was the equivalent of *Gentiana*.

I am very uneasy about your Zoological views : none previously prepared will do for plants—and you will not help us. I have left your MSS. at the Athenæum.

Sunningdale : March 14, 1887.

DEAR H.,—Awfully sorry;—my fingers got vilely cramped. The last allusion was to your Geographical distribution of animals, which won't run on all 4s with plants, for that New Zealand won't make a Botanical province. But the more I try, the more difficult I find it to limit Bot. provinces at all—there are only two—land and sea !

I said something about *Gentiana* being one of many genera established on a few heterogeneous materials, and which grew by the accession of new matter to the various salient points. A ' Gen. Plant.,' reviewing genera as wholes, should, if it could, have broken such up. I do not remember what else I may have said.

To the Same

Sept. 25, 1887.

The retroversion of the anthers in *Gentiana* was first described by myself in the ' Flora Antarctica,' as character-istic of the Southern species. It does not occur in all the genus, and it is now necessary to correct the anthers' charac-ters with the others of your groups. If I remember aright, some dehisce when still ' introrsum spectantes,' others not till after reversion—depending on the form of the Corolla.

As far as I can remember, all the gamopetalous Corollas are polypetalous (whether thalamifloral or calycifloral) in the earliest state—certainly very many are. I think Payer [1] so represents a great many in his ' Organogenie.'

Later, when his old friend had settled down in the bracing air of Eastbourne, to his regret that it was not in the pinewoods of Sunningdale near the Camp, he fostered this new taste for botany by sending him plants for his newly made garden, some in particular with the bantering remark that as they do well on any neglected dry rockwork, they should succeed under his tender cares !

[1] Jean Baptiste Payer (1818–1860), Membre de l'Institut (Acad. des Sciences); Prof. de Bot. à la Faculté des Sciences de Paris et à l'Ecole Normale Supérieure. Author of *Traité d'Organogénie Comparée de la Fleur*, published 1857.

CHAPTER XLII

DURING 1886 and 1887, as the Life of Charles Darwin was
advancing towards completion, Hooker had much correspondence with (Sir) Francis Darwin, reading the first proofs and
making various notes and suggestions out of his close knowledge of his old friend's work, and the scientific circles of the
time. One note is of interest for all biographers, and in this
direction, the Life, when published, left nothing to be desired.

I think you have rather a paucity of footnotes referring
to men's position, works, &c. Remember how little the
next generation will think of E. Forbes, Hancock [1] and many
great Guns of your father's lifetime. It added enormously
to the interest of the life of Lyell to be told in footnotes
who even the now second-class workers were of whom he
spoke, and who were luminaries in his day.

The question was raised as to Darwin's purpose in spending
eight years upon his monograph of the Cirripedes.

To F. Darwin

Dec. 31, 1885.

MY DEAR FRANK,—When I can get at the letters I may
find something that will throw light on the question you
raise—but I am helpless till my Library is shelved and
painted, when I shall bring down the letters, which, with
my books, are all in boxes at Kew, waiting.

[1] Albany Hancock (1806–73), zoologist. Received the Royal Society's
medal for his paper on ' The Organisation of Brachiopoda,' 1857 ; F.L.S. 1862 ;
collaborated also in works on Mollusca.

I do not understand that passage of Huxley's to imply, as you seem to think, that your father first went in for Barnacles deliberately thinking they would be good training; but that he took to monographing the Order under that impression, and in this Huxley is I know right.

Your father had Barnacles on the brain, from Chili onwards! He talked to me incessantly of beginning to work at his 'beloved Barnacles' (his favorite expression) long before he did so methodically. It is impossible to say at what stage of progress he realised the necessity of such a training as monographing the Order offered him; but that he did recognize it and act upon it as a training in systematic biological study, morphological, anatomical, geographical, taxonomic and descriptive, is very certain; he often alluded to it to me as a valued discipline and added that even the 'hateful' work of digging out synonyms and of describing, not only improved his methods, but opened his eyes to the difficulties and merits of the works of the dullest of cataloguers.

One result was that he would never allow a depreciatory remark to pass unchallenged on the poorest class of scientific workers, provided their work was honest and good of its kind. I always regarded this as one of the finest traits of his character—this generous appreciation of the hodmen of science and of their labors, and which culminated in the 'Steudel' [i.e. the Index Kewensis], and it was monographing the Barnacles that brought it about. The fact is that no one goes into such a piece of work as his Barnacles upon a cut and dried motive. When once begun various motives supervene or grow that direct the course adopted to this and that end. Your father recognized in conversation with me three stages in his career as biologist, the mere collector, in Cambridge &c.; the collector and observer, in the *Beagle* and for some years after; and the trained naturalist after, and only after, the Cirripede work. That he was a thinker all along is true enough, and there is a vast deal in his writings previous to the Cirripedes that a trained Naturalist could but emulate.

I have no more to say but that it would have been marvellous if your father had not felt the want of such a training as monographing the Cirripedes would give, and if he had not consciously taken advantage of it with his eyes open to its value in weighing all evidence pro and con evolution.

If you will let me make a suggestion, it is that you alter the expression ' could be considered well spent,' for eight years so spent by any other man would establish his reputation for all time, and whether as a discipline to your father, or for its results, I cannot conceive his spending it better, at that period of his career especially.

Probably it all came out in this wise—the original idea was to work out the problem of the complementary males : this he told me over and over again. When once begun he told me that he felt the want of training and discipline in every detail of work ; he applied to me (1844–6) for microscopes and lenses and for lessons in dissecting under it, for information as to the relative value of male and female organs in plants, of characters afforded by buds and flowers, fruits and seed, and no end of matters as to synonymy, priority, and the practical details of descriptive biology. We even dissected and drew together ; he all along calling himself a learner in these matters of research.

You are welcome to send this desultory scrawl to Huxley, or to make any other use of it. I have been interrupted over and over again, for I am writing all after page 1 on New Year's Day—of which I wish you and yours many returns and all good with them.

As he goes on with his notes on his correspondence with Darwin, he exclaims : ' I am staggered at the inordinate share of myself that your " Life " will contain, even if I am ever so brief.'

The next letters speak of the meeting at the Linnean Society when the joint Darwin-Wallace paper was read, and of the severe criticism passed on the *Quarterly Review* of July 1860 by Huxley in the chapter he contributed to the ' Life ' (ii. p. 182).

To F. Darwin

Oct. 22, 1886.

I was present with Lyell at the meeting. We both I think said something impressing the necessity of profound attention (on the part of Naturalists) to the papers and their bearing on the future of Nat. Hist. &c., &c., &c., but there was no semblance of discussion.

The interest excited was intense, but the subject too novel and too ominous for the old School to enter the lists before armouring. It was talked over after the meeting, 'with bated breath.' Lyell's approval, and perhaps in a small way mine, as his Lieutenant in the affair, rather overawed those Fellows who would otherwise have flown out against the doctrine, and this because we had the vantage ground of being familiar with the authors and their themes.

Bell, the President, in the Chair, was, though a personal friend of your father's, hostile to the end of his life. Busk, who was present as Secretary, said nothing, nor did Bennett, the Bot. Sec. Bentham was also there, and silent.

I do not remember Huxley being present, you might ask him.

Huxley has sent me the proof of his contribution to the 'Life.' I do not think it too severe. The *Quarterly* then held the highest place amongst the first class Reviews and was most bound to be fair and judicious, but proved unjust and malicious and ignorant. It went indefinitely beyond ' severity ' and into scurrility, and for all Huxley says he cites abundant proof. It is not for us, who repeat *ad nauseam* our contempt for the persecutors of Galileo and the sneerers at Franklin, to conceal the fact that our own great discoverers met the same fate at the hands of the highest in the land of Literature and Science, as represented by its most exalted organ, the *Q.R.*

I talked to X. about it in as strong terms as I could, when he turned round to me and asked if I really believed the doctrine, and on my response he pointed to the poker, and with fatuous solemnity said, ' Dr. Hooker ! I would as soon believe that that poker bred rabbits.' It amused me to think, that if the Apocalypse had said that pokers bred rabbits he would have believed it devoutly, and thought your father wicked to disbelieve.

To return to Huxley, I suggested his replacing the word ' person ' by ' reviewer,' in the bottom of the first slip, and to omit ' tricks of ' in alluding to Owen's style, because it weakened the force of the passage. As for the rest, if

not vigorous it would not be Huxley's, and if we ask a man
for a specimen of himself, we must let him appear in his
own colours.

When he made these suggestions to Huxley the day before,
he wrote :

These are the only points as to which your articles could
be hypercriticised in the matter of taste. For the rest I
would not alter anything. . . . The *Quarterly* does not
get one iota more than it deserves, or than the public should
see it gets.

To F. Darwin

The Camp, Sunningdale : Oct. 23, 1886.

DEAR FRANK,—I was not aware that the Bishop had
acknowledged the Review in his Essays. It certainly does
appear that this should be stated, but I would like to see
the passage acknowledging it, before considering how
much or how little of it should come in. No one who was
present at the Bishop's attack on Huxley at the Oxford
Meeting could wonder at Huxley's delighting in ' paying
him off.'

I believe that the *Q.R.* has just treated Gosse as badly
almost as it did the ' Origin,' and if so Huxley's dressing is
not inopportune. It is abominable that a Review of such
standing should seek out ignorant and incompetent and
even prejudiced and hostile reviewers to write in such
cases.

I quite feel with you that it is a pity that the ' Life ' of one
so far above all fierceness of disposition should have to treat
of matters requiring such stern and hot handling. But the
Q.R. was, from its influence and position, the head and front
of the offending, and if the history of Evolution has to be
dealt with, it must be brought to the front to be pilloried ;
and given Huxley as executioner, the rest follows ! Nothing
short of recasting the whole of his contribution as regards
the *Quarterly* would meet the case. Were Owen or the
Bishop in Huxley's place, and the tables turned, you would
have a contribution of malignant sneers and innuendoes.
It is the old story, ' the greater the truth, the greater the
libel.'

To F. Darwin

1887.

Unfortunately we have not got the best point in Huxley's answer to the Bishop. It was to this effect—' The Bishop asks how would I like it if my *Mother* had been an ape. I answer, putting aside the bad taste of the allusion to a relationship which I have in my own case regarded as calling up the tenderest memories, and having regard to that argument from the ' Godlike gift ' of language which the Bishop has put forward as paramount against Mr. Darwin's theory, that I would rather I had a parent wanting that " Godlike gift," than a parent who devoted that great and godlike gift to the perversion of truth or to diverting the minds of an audience from the facts that support a great scientific hypothesis by ridicule,' &c.

This was the sense of it—the words I cannot recall. The telling point was that Huxley showed how keenly wounded he was by an allusion to a relationship which he regarded so tenderly [having] been driven home in so indelicate a manner.

I shall see Huxley to-morrow and if I can get him to attend to me will endeavour to obtain from him a more definite account, for use or not, and will let you know.

Hooker's own letter to Darwin on the Oxford meeting (see i. 525) did not appear in the ' Life '; he feared it was ' far too much of a braggart epistle.' But he added :

Have you any account of the Oxford meeting ? If not, I will, if you like, see what I can do towards vivifying it (and vivisecting the Bishop) for you. I had utterly forgotten that letter of mine, and am amused to find that it recalls the scene so clearly. (Oct. 30, 1886.)

His account was printed in the ' Life,' vol. ii. pp. 320–323, November 21 :

Here is my screed. I do not like it altogether, but can do no better. I should like Huxley to see it if you put it in print. Pray Anglicize it where necessary. . . . I have been driven wild formulating it from memory.

Later the question was raised whether the Bishop's taunt to Huxley dragged in the name of his mother or his grandmother, as Hooker's memory had it ; and on March 10, 1887, being twice disappointed of meeting Huxley, whom he wished to consult on the point, he writes :

> I find, however, on enquiry of others, that they did not understand the Bishop to allude to Huxley's *Mother*, but his Grandmother, so pray make no alteration in what you have written as to the Oxford meeting except Huxley approves. It is impossible to be sure of what one heard, or of impressions formed, after nearly 30 years of active life.[1]

The following letters refer to the Darwin Obituary (*Proc. Roy. Soc.*, 1888), afterwards republished in Huxley's Collected Essays, vol. ii. His memory of what had happened thirty and forty years before was rarely at fault, despite his depreciation of it ; yet looking back, it was in such a far away vista of the past that he was moved to exclaim ' Darwinism is all a dream to me now.' (November 3, 1890.)

To T. H. Huxley

March 25, 1888.

I have not seen Dana's obit. notice of Gray. I suppose it is in Silliman—I will send for it and tell you what I think.

I never attached much importance to Gray's philosophy of Darwinism. He ' illuminated ' the text, but did not advance the subject in a scientific point of view ; only in a general and popular one.

Darwin has nowhere that I can think of dealt with the causes of variation. My impression is that he regarded them as inscrutable, and I doubt his assenting to the view that they were in any scientific sense limited or directed by external conditions—except in so far as that conditions which kill an organism limit its powers of variation !

Organisms vary from whatever you please to call type, under no known fashion ; and this whether the conditions are favorable or unfavorable to life : if they are favorable so much the better for them.

I very much hope that you will carry out your Primer idea. I feel myself ever apt to go astray on the subject,

---

[1] See the unveiling of the Darwin statue at Oxford, 1899, p. 432.

and hark back on pre-Darwinian ideas that were not necessarily anti-Darwinian, but which should not be confounded
with these.

The Camp, Sunningdale : March 27, 1888.

DEAR HUXLEY,—Dana's Gray arrived yesterday and I
turned to pp. 19, 20. I see nothing anti-Darwinian in the
passages, and I do not gather from [them ?] that Gray did.

I did not follow Gray into his later comments on Darwinism, and I never read his ' Darwiniana.' My recollection of his
attitude after acceptance of the Doctrine, and during the
first few years of his active promulgation of it, is, that he
understood it clearly, but sought to harmonize it with his
prepossession—without disturbing its physical principles in
any way. He certainly showed far more knowledge and appreciation of the contents of the Origin than any of the reviewers, and than any of the commentators, yourself excepted.
He almost alone had faithfully studied it from beginning to
end, and for this both Darwin and I have given him credit.

Latterly he got deeper and deeper into theological and
metaphysical wanderings, and finally formulated his ideas
in an illogical fashion. Before this I had given up reading
him, and as I have said, I have not read his ' Darwiniana.'
Such workings of the mind have no more attraction for me
than those of Maurice [1] and J. Martineau.[2]

Be all this as it may, Dana seems to be in a muddle on
p. 20, and quite a self-sought one.

It remains for you to put the whole matter clear in the

---

[1] Frederick Denison Maurice (1805–72), the saintly divine who fled from
the narrowness of a dissenting ministry to law and literature ; then took
Orders at Oxford (1830-4), hoping to realise a Christian unity based on a
spiritual fact instead of dogmatic opinions, and linking with this his Christian
Socialism and the foundation of the Working Men's and Queen's Colleges. His
sharp opposition to ' parties ' and his maze of metaphysical subtleties led to
lifelong controversies. Hunted out of his Professorship at King's College for
heresy, he was supported by more liberal opinion when appointed to St. Peter's,
Vere Street (1860) and to the Chair of Moral Philosophy at Cambridge (1866).

[2] James Martineau (1805–1900), Unitarian divine, born at Norwich, studied
at the Manchester (New) College (1822-7), in which he became Professor of
Mental and Moral Philosophy, 1840, migrating with it to London, 1857, and
becoming Principal 1869. From the first, as pastor in Dublin and Liverpool,
he made his mark in preaching and controversy; but study at Göttingen (1848)
turned him from the old deterministic unitarianism to freewill philosophy.
Hooker probably had in mind not the ' Types of Ethical Theory,' 1885, nor the
' Study of Religion,' 1888, but the earlier Essays and the controversy over
Tyndall's Belfast address.

matter of expression by words,—and this no one can do so well as you, or better than by a Primer.

Spencer has alone I think tried to account for variation on scientific grounds. I thought at the time his ways were good but not all in the right direction—but I forget what they were !

The Camp, Sunningdale : May 2, 1888.

DEAR HUXLEY,—The evolution of Darwin is excellent, it makes quite a Natural Order of him.

You will find an X on page 1 in reference to Darwin's father. I understood from D. that his father had not only scientific proclivities, but ambition, and that he presented to the R.S. a communication on some optical subject, which, being rejected, disgusted him, and led to his stifling his own early scientific tendencies and scoffing at those of others. If worth following this up Frank might confirm or refute my memory.

Your treatment of two subjects, Medicine and the Church, is capital—the latter (quoad Paley and Pearson) is very diverting from the cynical way it is put.

Nothing could be better than your history of the Evolution of the doctrine of that ilk, and the showing up of the erroneous, imperfect conceptions of some of its favorers.

Your treatment of the main subject, under heredity, variation and multiplication as the factors, is very happy, and will instruct as well as elucidate.

Would you not say where I have put an X on p. 10, that among the non-adaptive concomitant characters some may be even disadvantageous so long as they are not predominant ?

I well remember the worry which that subject of tendency to divergence caused him. I believe I first pointed the defect out to him, at least I insisted from the first on his entertaining a crude idea which I held, that variation was a centrifugal force,whether it resulted in species or not. P. 7. Is ' malice '[1] the best term to apply ? I am sure you can find a better— it was a case of ' one for his nob '—(irony ? banter ?) it was a ' taunt.'

The whole has interested me warmly, and I am particularly glad to see references to his tastes and feelings that

---

[1] The word was used more in the sense it bears in French than English.

traverse an article on Darwin in the *Atlantic Monthly*. That disproves his want of culture.

Now dear old boy do take care of yourself, and do not rashly follow the prescriptions of any doctor on the faculty, but use your own judgment, and follow nostrums tentatively.

In this Epistle I have sacrificed lucidity of expression to penmanship, so I hope you can read it. The sheets go by this post. Keep a copy for me. Ever yours,

J. D. HOOKER.

In 1887 he was awarded the Copley Medal, the highest award of the Royal Society for scientific discoveries or the advancement of science. 'The Copley quite took my breath away,' he writes to Huxley, November 7. ' Much as I have had to do with that award, I never once thought of myself as within the pale of it.' And to Asa Gray, November 15 :

Dyer tells me how kind and generous, I fear too generous, you were about it. The secret was well kept, for I heard not a whisper till the award was made. It is an honour which I never expected, often as I have had to award it as P.R.S., and oftener to take a part in awarding it.

I never for a moment put myself into a thought of it— and am not now clear that it is a ' Statutory ' award, being intended for *bona fide* discovery. As, however, I am informed that my name was brought forward *last*, and all preceding ones at once withdrawn, I must concede that there are some good grounds for the departure from precedent. I do feel it to be a tremendous honour.

Quotation has already been made from his speech returning thanks for the medallists at the Anniversary dinner in which he compared the state of botanical teaching in his youth with that of the present day, and told the story of his earliest essays in botany and the hereditary impulse which he followed.

My father and my grandfather, Dawson Turner, were distinguished botanists, and both were Fellows, at a comparatively early age, of the Royal Society ; so that when I was startled by the intelligence that I had been awarded the Copley Medal, my first thoughts were that I had arrived at that distinction by a process of Evolution ; that I was, in short, the puppet of Natural Selection. I was, however, soon confronted with the truth, that I was by no means 'the Survivor of the fittest!' I have said that my father and grandfather were both Botanists, and singularly enough they both began their studies with the Mosses, quite independently of one another ; and my friend Mr. Galton, whom I am glad to see here, may be interested to know that I am a born Muscologist.

He referred, as he always gratefully referred, to his father's influence in launching him on his career ; then added :

I have one more advantage to record, and it is the greatest of all,—it was the friendship and encouragement, for forty years of my life, of a man to whom I looked up, as the Pole Star and Lode Stone of my scientific life. The name of that man is uppermost in the thoughts of every one here. It is Charles Darwin.

The conclusion of his speech ran as follows :

Mr. President, I have exceeded all bounds already, but if I may be allowed a few minutes longer, I would, taking advantage of the patriarchal age which your treasurer has assigned to me, say a few words for the encouragement of the younger scientific men here present. A septuagenarian may indulge in retrospection ; indeed it comes natural to him to do so ; and when I heard of the award of the Copley Medal to me, I could not but ask myself to what quality or exceptional condition of mind I could attribute it that I had attained to so unique an honour. Heredity, early training, advantages, opportunities, experiences, and even research itself, are fruitless, if there is not some inward motive power to compel us to exercise our faculties, and some inward heat, some fervour, to ripen the fruits of our labours— I can truly say that I am conscious of no genius, exceptional powers or talent ; but I have a talent, and it is one that is possessed by every one in this room, and by many I hope in greater degree than I possess it. It is not talent in the modern meaning of the word, but in the old French meaning of wish or will, and I cannot better express the sense in which I possess it, and you all possess it, than in the words of a very modest motto adopted for his rule in life, by a very great man, who died four hundred years ago—Prince Henry of Portugal,[1] the Father of Navigation and Patron of Navigators, who chose for his motto ' Talent de bien faire,' ' the wish to do well.' To such as have this wish, and will use it with all their might, even a Copley Medal is attainable.

To W. E. Darwin

Dec. 7, 1887.

How I did crave for your father's sympathy in the matter of preparing my speech of thanks at the Royal dinner for the Medal, and how he would have sympathised and encouraged me. It was an effort for such an unaccustomed orator (and faulty one) as I am, but it went off so well that I have been asked to print it !—This eclipses the Copley in my opinion !

And in similar vein to Huxley, December 5 :

The success of my after dinner homily at the R.S. is to me far more wonderful than getting the Copley. You who are one of the few who know how morbidly nervous I am— can guess my condition of two days' nausea before the dinner, and 2 days of illness after it. I am not speaking figuratively. It is mere nervous upset.

When the Copley was awarded to Huxley the following year he wrote (November 4, 1888) :

I am rejoiced at the news of the Copley being awarded to you : and that our names will stand next one another in the glorious hierarchy of the R.S. is a real pleasure to me (whether as past Presidents or Medallists). Ask Mrs. Huxley to accept my most cordial congratulations.

In 1888 the Linnean Society celebrated the centenary of its foundation by (Sir) J. E. Smith, the friend of Banks and fortunate purchaser of Linnæus' collections.

[1] Prince Henry of Portugal, surnamed the Navigator (1394–1460), first distinguished himself at the conquest of Ceuta 1415, after the death of his father João I. He lived at Sagres, near Cape St. Vincent, and while at war with the Moors, his ship reached unknown and unvisited parts of the ocean. He formed a school for instruction in the science of navigation, and his pupils discovered Madeira in 1418, and one of his mariners sailed round Cape Nun as far south as Cape Bojado in 1433, and in 1440 to Cape Blanco. Cape Verd was also discovered in 1446 ; three of the Azores in 1448. A national celebration of his memory took place in 1894 in Portugal.

To [Sir] F. Darwin

I am in for an éloge of Robert Brown for the Linnean
Anniversary and am re-reading his miscellaneous writings
with increased admiration. I know no botanical writings
at all comparable to those on Morphology, taxonomy and
classification, for sagacity, profundity, range of knowledge,
scrupulous accuracy and clearness. He took in the whole
phænogamic kingdom. Every young botanist should go
through a course of reading these miscellaneous works.
They are as much above all others as Wellington's despatches
are to those of subsequent warriors.

However, he now found the dinners which accompanied
these functions oppressive, and refused every such invitation
that it was possible to refuse.

To the Same

May 31, 1888.

I really cannot martyr myself any more to dinners of the
kind—they completely knock me up. I was ill the whole
day following the Newton function, and again now after
the Linnean. Not a week now passes without my having
pressing invitations of this kind—another came by the same
post with your letter.

By way of exception, he attended the Ipswich meeting of
the British Association in 1895 in honour of the inauguration
of Botany as a separate section, and in June 1896 the Kelvin
Jubilee at Glasgow ; otherwise meetings, like public dinners,
were avoided, especially the distant conference of botanists
at Berlin in 1893. Paris being more accessible, he would
have attended, as a 'very old Associate,' the centenary of
the Académie des Sciences in October 1895, and actually
began to 'grind away at a conversation book' to rub up
his 'wretched French,' the chief drawback to such a visit,
but circumstances counselled prudence, so that he wrote to
his friend La Touche (October 22) :

I was not sorry to give up the Paris trip. I like seeing
festivities but not taking part in them ; I can hardly speak
French intelligibly—cannot converse at all—and I dreaded

the weather breaking up as it seemed disposed to do. It is now raining here with a N.E. wind which would be deplorable in Paris. Add to this I am deep in work with my Indian Flora at home and in the Herbarium at Kew ; and that all my friends say I was wise! not to go, at this season, to Paris.

Occasionally he regretted his absences, as when at the Oxford meeting of the British Association in 1894, Huxley proposed the vote of thanks to Lord Salisbury for his presidential address, and in the city which had rung with the first loud fight over Darwinism, gracefully countersigned his somewhat hesitating acceptance of the doctrine of evolution. And again, later in the autumn when he missed two speeches by his friend, the one on receiving the Darwin Medal from the Royal Society on November 30, the other at a dinner given by Messrs. Macmillan to Sir Norman Lockyer, who had edited *Nature* since its foundation twenty-five years before.

To T. H. Huxley

Aug. 16, 1894.

I was very glad to see your ' hand of write,' as the Scotch say, again. I saw with surprise that you were exposing yourself to the Saturnalia of the British Association. I was much tempted to go, but am so bothered with deafness and eczema auricul*arum*, that I funked it. So I accepted an invitation for a fortnight to Glenfinart on the Clyde, where I enjoyed visiting old haunts of sea, loch and mountain.

I was much struck with the first part of Lord Salisbury's address, but had hardly patience to read the last, which is silly—really I thought he had more gumption. I am much more disposed to believe that inability to grasp the subject, i.e. to conceive of the operation—than a surrender is at the bottom of his hesitation. The fact is, that like many other physicists, he with difficulty entertains any but mathematical reasoning.

I hope that your ' Discourses ' are not exhausted. I call them my ' Pick me ups,' they are such refreshers.

To the Same

Dec. 2, 1894.

MY DEAR OLD BOY,—The award of the Copley and Darwin medals[1] gave me complete satisfaction and supreme pleasure ; and as for your speech at the dinner, it made me glow all over. I should like to have been there, but bronchitis dogs my footsteps in the night air, and between ourselves I fear I am in for it for the winter, I hope in a very mild form.

Dyer tells me that your address at the *Nature* banquet was exceedingly good in substance and manner, and ought to be printed for its worth as a warning voice. Do think of it.

The ' Journal of Sir Joseph Banks ' was published in the autumn of 1896. A curious history attached to the book, especially as the preservation of the text was primarily due to Hooker's grandfather, Dawson Turner. When Hawkesworth [2] first edited ' Cook's Voyages,' he made a mélange of Banks' and Cook's journals, which were put at his disposal, interspersed with reflections of his own. There was nothing to distinguish his sources, and he only made such selections from Banks as would interest the general public. Cook's own journal had recently been published by Admiral Wharton ; it was time to make clear the real and very distinguished part played in the great adventure by Banks.

Now while the bulk of Banks' property was left to the Hugessens (his wife's family) his library and herbarium were left to Robert Brown, the botanist, who was his librarian, with the proviso that on his death they were to go to the British Museum.

Robert Brown being unable to write the Life of Banks, suggested that it should be undertaken by another friend of them both, Dawson Turner, to whom the papers were

[1] To Frankland and Huxley respectively.

[2] John Hawkesworth (1715?–73), author, said to have succeeded Johnson as compiler of Parliamentary debates for the *Gentleman's Magazine*, 1744 ; edited Swift's Works, 1755 ; LL.D. Lambeth 1756. His *Edgar and Emmeline* was produced at Drury Lane 1761. He published an *Account of the Voyages undertaken by the Order of His present Majesty for making Discoveries in the Southern Hemisphere* in three volumes, 1773.

handed over. Dawson Turner had the whole carefully transcribed by his two daughters. Of this Hooker writes in his Preface to the book :

It was when on a visit to my grandfather in 1833 that I first saw the original Journal in Banks' handwriting. It was then being copied, and I was employed to verify the copies of the earlier part by comparison with the original. I well remember being as a boy fascinated with the Journal, and I never ceased to hope that it might one day be published.

But Dawson Turner did no more, and original and copies were returned to Mr. Knatchbull-Hugessen [1] (afterwards Lord Brabourne), and for many years rested in the Manuscript Department of the British Museum, to become the property of the Trustees after the death of Lady Knatchbull. During one of the periodical attempts to have the Life of Banks written, the transcripts were transferred for examination to the Botanical Department, and were thus saved when in the middle eighties Lord Brabourne refused to accept the view of the Museum authorities as to the ultimate property in the MSS., and carried off the box containing the originals. Later he offered these for sale to the Museum, but being dissatisfied with the price offered, had them sold by auction at Sotheby's, lock, stock and barrel. The result was pitiful. The 207 lots into which the Journal and correspondence were broken up, realised but £182 19s., and a collection, the peculiar value of which lay in its being preserved entire, was scattered to the winds. Letters with well-known signatures were resold as autographs, the rest destroyed. One large portion of the Journal was afterwards traced to Sydney, an appropriate resting-place. But the full material on which Hooker worked was the salvage of his own family's labour, taken from the very papers upon which he himself had worked as a boy, sixty-three years before.

In a Journal such as this, posted up from day to day, there

---

[1] Edward Knatchbull-Hugessen, first Baron Brabourne (1829–93), M.A. 1854, who took the surname of Hugessen 1849, was M.P. for Sandwich 1857 ; Lord of the Treasury 1859–66, etc. ; privy councillor 1873 ; and raised to the peerage 1880. He was the author of a charming volume of children's stories.

was much work to be done in reduction and verbal correction ;
but the longest task was the identification of the plants mentioned under native names.

To Mrs. Darwin

Feb. 27, 1896.

I am busier than ever—getting through the last Vol.
of the 'Flora of British India,' the Grasses, the hardest
work I ever had ; and editing Sir J. Banks' narrative of
Cook's first voyage. This interests me much, for it nowhere
appears that Banks ever worked as a Naturalist proper
so to speak, whereas his Journal shows that he employed
himself with extraordinary zeal and industry to collecting
and observing throughout the voyage—that he is in fact the
earliest of those Voyager Naturalists of which Darwin is
the greatest. The Journal is admirably kept ; even at sea
he never let a day pass without an observation : and the
accounts given in ' Cook's Voyage ' of the manners and
languages &c., &c. of the people are all from Banks' Journals.
It is of course well known that Hawkesworth was permitted
to draw what he pleased from Banks' Journal when publishing
Cook's and that the result is a composite work ; but this
result gives no idea of what Banks really did. Reggie [1] is
aiding me most efficiently.

His less specialised scientific interests are particularly reflected in the correspondence with the Rev. J. D. La Touche.

Mr. La Touche, under whom at various times Hooker placed
three of his sons for coaching, was a man of wide interests and
stimulating influence, a Shropshire clergyman whose broad
liberalism was barely contained within the strict pale of the
Church. He was a keen geologist, and fostered local activity
in science. Hooker gives him practical advice as to running
a field club ; appreciates his son's geological work in the
Himalayas. But the bulk of the correspondence—Hooker
wrote him seventy-six letters between 1886 and 1898—abounds
in references to the books they read, their current interests,

---

[1] His fourth son. Sometime Secretary to the Royal Statistical Society ;
afterwards Head of Branch under Board of Agriculture and Fisheries. Author
of many papers published in the *Journal of the Royal Statistical Society.*

popular education, politics, wherein they had a friendly
difference, La Touche being a Radical, Hooker a philosophic
Conservative, a strong Unionist, but not a Tory.

To Rev. J. D. La Touche

May 10, 1886.

(There was a question of attending a local scientific meeting
at Shrewsbury, to which after all Hooker did not go.)

It is a great pity that such gatherings have degenerated
into a mere ' picnic with flirtations '—it was not so with
one, the meeting I attended at Ludlow with you some ten
years ago.   But really I do not wonder at it, when I see how
rare is the taste for Science.   It is only when a contingent
of the Scientific men of a great town like Liverpool, or
Birmingham, or Newcastle, can be counted on for attendance
that there is any chance of keeping the Scientific character
of local assemblies.   Well I shall hope to visit you some
day at your own house, and that is what I should like better.

To the Same

Aug. 22, 1888.

I quite sympathise with your views as to the future of
the Caradoc Field Club—but am as you know very sceptical
as to the influence of outsiders in such matters.   The force
must come *from within.*   I well remember telling the Club
this at a Ludlow (?) meeting ages ago.   Men who have it
in them want no talking to as an incentive to work.   All
over the country these Clubs have been urged to take definite subjects up, but what is every member's business is
no member's practice.   All were formed with the professed
view of investigating the Nat. Hist. &c. of their several
areas—how many have elucidated one single branch of it ?
Few of such local Faunas and Floras as have been published
trace their origin to the Club, but many to the energy of
individuals fond of the subject and irrepressible.   On the
other hand, the Clubs do a vast amount of good in keeping
up an esteem for Science in the country, and by their meetings
they afford opportunities for discussing scientific matters
and giving object lessons in every branch of science ; and
what one wants to see at these meetings is more devotion

to discourses and object lessons, peripatetic lectures and the like, and less mere amusement.

Be all that as it may, there can be no question of the advantage of getting those of the members who are competent to devote themselves to the description of the county —but to carry this out I should go to individuals, not to the Club, nor to those at the Club. I should get their adhesion first and report it to the Club.

It is the same with a Scientific Society, it is of no use that the President calls on the Fellows to do this or that—all applaud and no one acts.—What you do is to get up a small meeting of interested men, draw up the plan of work, and when complete announce it to the Society *together with the names of the future workers*. All this takes time and much thought, but it is the only practical plan. My experience is, that talking to a Club is talking to the wind if you cannot yourself by subsequent personal intercourse, and at great time and trouble, both educe zeal and direct its course.

You have I take it two objects in view, giving a better tone to the at present aimless rambles, and the elucidation of the Natural History of the County.

The same method applies to the rambles as to the solid works. I should engage competent members to be scientific leaders, who should conduct excursions with definite objects at each meeting.

Let one leader take the plants, another the geology, and so forth, and further, require that every member joining the meetings should attach himself for the day, or half the day, to some one or other of the leaders and stay with him. I would allow two hours in midday in some convenient (meal) time for ' promiscuous intercourse '; but that over, each goes to one of the appointed leaders for the rest of the excursion.

If the day is divided into a morning and afternoon ramble, you might allow a change—the botanical ramblers of the forenoon might join the geological of the afternoon—but there must be no chopping and changing at other times ; and this rule must be rigidly enforced. Of course any one is at liberty to go away altogether, or even absent himself temporarily, but not to join any other peripatetic Philosopher.

But I should not announce or allude even to this or any

other plan of procedure till *the whole* is *thoroughly* worked out, even to the supplying of substitutes in case of the failure of attendance of any of the leaders.

You may or may not have time before the 25th to organise such a plan of proceedings for that day—if you can so much the better, and a circular would announce it to all the members. But, whatever you do, have no half measures, let it be clearly known that the rule of sticking to one leader for the appointed time will be *rigidly enforced*. Members might be asked (not forced) to say beforehand which leader they will follow. It will be the leader's duty so to conduct his party that all temptation to stray into another party is avoided. If the ramble is to—say the top of a hill, they should arrange that no two parties meet there, for the moment you allow of ' promiscuous intercourse ' it is all up and the thing degenerates into an agapemone. ' Them's my notions,' if you think they are available, and see your way to the leaders, pray come here for a night and we will talk over rules.

P.S. Arrangements for a rainy day are most desirable, and the best thing I can think of is, that the leaders (or others) should give talks or lecturets to the members. It would depend on the remuneration whether these should be delivered to the whole Club at a time, in which case the lectures would follow one another ; or if several apartments were available, lectures on different subjects might go on at the same time— but here again no wandering from the lecture once commenced should be permitted.

It will be the same with the lecturets as with the rambles —a thoroughly matured plan is necessary and at least six resolute members to carry it out. Each Lecturer or leader should have his second in command.

Upshot—if very much work is to be done, either in the way of making the rambles instructive, or the labors of members more systematic, much personal supervision, forethought and time is necessary. The reason why the Clubs have done so little is, the want of organisation and supervision to the end of making them *instructive*.

I do believe that if such a plan of the rambles as I suggest were once established, you would get a far better class of attendants. People would know that definite information

as to the Natural History &c. of their own country was to be
had for the listening, and the leaders would be *marked* men.

J. H.

To the Same

Oct. 26, 1888.

The great thing will be to lead the Club back to definite
work with a definite object, and if you succeed in this it will
be an achievement. But even then a difficulty remains,
and that is to induce the workers to put up their work into
shape for publication ; however it is a case of ' nothing
venture, nothing win.'

A scheme of reform was proposed, and later bore good
fruit.. It was a curious coincidence that by almost the same
post Hooker received a similar communication from New
Zealand as to maintaining a Natural History Club.

In 1892 he was asked, with Lady Hooker, to preside at a
meeting of the Club, to give a short address and see for himself
the betterment of local effort. But though he went as a guest
and greatly enjoyed the excursion, he refused speech and
presidency, not only because with the Flora of British India to
finish, he had abjured all lectures and addresses, but because,
in his opinion,

the plan of asking outsiders of position to take the
leadership of such exceptional provincial scientific meetings
is a mistake. These meetings afford the opportunities for the
members of bringing forward their own provincial scientific
magnates, whom they should then show they honour, by
putting them forward.

The same principle is insisted on in the following letters.
Shrewsbury was erecting a memorial statue to Charles Darwin.

The Camp, Sunningdale : March 4, 1894.

My DEAR LA TOUCHE,—Anent the Darwin Memorial for
Shrewsbury, I feel sure that the best plan by far is to have
a copy of the statue in the Nat. Hist. Museum, and it ought
to be at the expense of the Salopians themselves. The
general public contributed most liberally, the scientific men

of London especially, to that in the S. K. Museum, and to call upon it for a second and a local one, appears to me to be unreasonable. It seems to me absurd that a great town like Shrewsbury should have to go round the world with the hat to secure its own credit! in fact I do not think it creditable, and I should think would not be well responded to.

A scholarship is rather a thing for a University to honor one of its Alumni with, and for a local object a statue would get more support than a scholarship. A school for promoting science would be a first rate thing, but it involves great labor and paid supervision. To be effective it would cost thousands to establish and much to maintain. I cannot see the objection to a copy, provided the original is good. The noble statue of Jas. Watt, put up by the Glasgovians, has been copied elsewhere.

The only objection to a statue is that it looks so miserable in the rain and grimy air of a British town. It should be in bronze, but that would be enormously expensive. The alternative would be a half size in marble, to be placed in the Museum, or in an alcove built for it (if there is no room). With good photographs of the one in S. K. Museum, a careful young sculptor should have no difficulty in reproducing the original in ' petto.' These are my ideas.

I quite agree with you as to the futility of enlarging W. Abbey for the mortal remains of illustrious men—but I have always favored a Campo Santo for memorial tablets, busts, &c., &c., which should have inscribed on them the burial place of the deceased, as well as his birth, death and deeds.

Ever, my dear La Touche, most sincerely yours,

J. D. HOOKER.

Huxley's ' Hume ' has just come. I have written to ask Mr. Darwin if he wants it before it goes to you.

To the Same

Feb. 18, 1897.

I am very pleased to get so good an account of the Darwin statue. There will be time to think of inviting some one to unveil it when it is got into marble—a long job. I differ from you as to this matter of unveiling. I do not like to see this delegated to outsiders. Why should every town

that wants a function of that sort performed send to London
for an operator ? To me the fittest thing appears to be that
the Lord Lieut., or Sheriff of the County, or its M.P. should
perform the office, inviting all the Listers and Huxleys and
Lubbocks to attend at the ceremony, and, if they will kindly
do so, say a few words on the occasion. It is all very well
if you want to get up a subscription, to get a Prince, Duke,
or P.R.S. to come and tickle the pockets of the town. In
the case of unveiling, it should be a source of pride to the
townsman or countyman, whereas it is a huge bore to the
outsider. 'Them's my sentiments.'

The Committee approving of the idea, he added (March 8) :

I am sure it is right. The looking to London for all out-
comes is a mistake and emasculates the country's energies.

The sculptor at first proposed to add a symbolic ornamentation on the back of the chair in which the figure is seated ;
but thinking this out of keeping with the perfect simplicity of
the rest, Hooker advised successfully that this be abandoned.

For my part I think that the Committee are most fortunate in the matter of the execution of the work, in respect
of both pose and likeness ; and were it mine to deal with,
I would let well alone and not 'try to make the elephant
dance,' as they say in India.

The unveiling took place in the beginning of September
1897, and Hooker attended the ceremony.

Knowing his friend's perennial interest in the Himalayas,
Mr. La Touche used to send on his son's letters from India.

To the Same

Dec. 28, 1892.

I should have thought that there were abundant traces
of the Glacial period in the Himalaya.

I was reading a review the other day of some one's book
on the Earth's crust, who thought that it was subject to
pressure from mountain masses ; if so the bases of the mountain masses must be a mile deep, and it struck me that this
might account for the curious phenomenon of the strata

of the foot-hills dipping away from the axis of the chain, so commonly observed. The intervening valley would then be a line of fracture parallel to the range.

I do not see the necessity for a shift of the Earth's axis to account for a glacial period in the Himalaya ; all you want is greater deposition of snow, either owing to a wetter climate or to a greater elevation of the chain.

When the N.E. of Asia was a sea (inclusive of Caspian, Aral, Black, and probably E. Russia), the Himal. must have had a much wetter climate. I never thought it necessary to suppose that the Ice of the Glacial period extended to one and the same latitude in a given longitude during the whole of the Epoch. Why should we suppose that it extended to mid. N. America at the same time that it extended to N. Italy ? The two Continents, then as now, differed wholly in climatic conditions, and as wholly, if not more so, during the glacial than the present period. Be all this as it may, we have no conditions existing in the globe to be compared with the Himal., when the Aralo-Caspian sea extended over N.W. Asia.

I should like to see a map of the Himalayas with the perpetual snow indicated as now, and as if the snow-line were 1000 ft. lower, and another with it 2000 ft. lower. I suspect that such an accession, covering so many now un-snowed Mts. and valleys, would bring about a pretty state of matters, and a sea to the North might bring about this.

To the Same

Aug. 11, 1893.

Thanks for Mr. Middlemiss'[1] paper. I am so awfully busy that I could do no more than glance through it, but that was enough to frighten me from doing more, for the very technical terms are new to me. The complication of the Strata is enough to make one giddy, and I recall a rhyme I made in the Himalaya and sent it to Darwin :

> Stratification is vexation,
> Foliation's twice as bad ;
> Where faults there be
> They puzzle me,
> And Cleavage drives me mad.

[1] Charles Stewart Middlemiss. In 1883 he was appointed Deputy Superintendent of the Geological Survey of India.

The following refers to a pamphlet by Sir W. L. Buller.[1]

To the Same

June 25, 1895.

I am sending you by post a pamphlet on Darwinism and other matters by an old friend of mine and capital Naturalist, who is author of a splendid work on New Zealand Birds.

The pamphlet has charmed me, reminding me of White's Selborne. I have only one criticism upon it—which is as to the necessary correlation, see p. 102, of the huge forms of Palæozoic life with a tropical vegetation, and the assumption that this is so now ; mentioning the Siberian Mammoth, the Irish Elk, &c., as having had tropical climates, whereas the food found in the teeth of the Mammoth, and stomach I believe, is of the leaves of the Birch and Willow now found in the Tundras, and there is no reason to suppose that the vegetation of the bogs in which the Irish Elk is found differed from that now prevailing—and Elephants abounded at the Cape of Good Hope when first discovered, or rather when first occupied. Also the Greenland Whale feeds on the Arctic Beröe, a very minute sea jelly-fish.

[1] Sir Walter Lawry Buller, K.C.M.G., F.R.S., of Canterbury, N.Z. ; Resident Magistrate and Native Commissioner in various districts of New Zealand ; served in the Maori war (medal and mention in despatches) ; a distinguished ornithologist ; Governor of the Imperial Institute ; author of *The Birds of New Zealand* and other scientific publications.

CHAPTER XLIII

FROM the wide variety of topics touched upon in the more personal part of the correspondence of this time, a few may be chosen to illustrate the writer's enduring freshness of mind. His letters differ in complexion according to the correspondent. With the botanists, they deal in botanical science, pure and applied, in travel and research, and scientific news. With men of science who were old intimates, affectionate freedom of personal topics enriches the tale of current activities. With friends whose common ground was mainly in the general interests of life, easy talk on books, art, politics, education public and private, social questions, are added to the personalia, and with the surviving friends of his youth the old time atmosphere envelops the evening of life with the unfading reflections of its happiest morning colours.

As to politics, suffice it to say that while at no time he took an active part in them, his native caution, his love of continuity and prudent discipline, led him to a moderate conservative standpoint. The ideal democracy was non-existent, and of democracy on the largest scale as he saw it, he wrote to his friend Ayerst Hooker [1] (December 29, 1891) :

> I have just received from America a new life of Thomas Hooker, short, but most instructive. I have given it to Reggie to read and send on to you. It attributes to said Thomas the source of the democratic Institutions of America, but in no wise reconciles me to it. A democracy sounds very well for a uniformly educated people ; but when the

---

[1] A member of another branch of the Hooker family descended from John Vowell, *alias* Hoker, Chamberlain of the City of Exeter, 1554 (uncle and patron of Richard Hooker), who was also the ancestor of Sir J. D. Hooker.

masses are not only ignorant, but wrong-headed, it could
only be a curse to all ; and Thomas Hooker arguing from
the point of view of a small body united in sympathy and—
well—superstition, is no sign of originality, and still less of
sound reasoning.

He disbelieved in the egalitarian tendencies of reforming
Radicalism as emphatically as in the true Tory's impene-
trability to ideas. In his official relations he had suffered
more from the activity of the one party than from the passivity
of the other, though neither offered special sympathy. Indeed
he writes to W. E. Darwin (June 9, 1895) :

My experience of the Public Departments is, that they
are riddled with small defects, and much idleness—and
that in this respect many want overhauling, but not all.
But that, as regards great reforms *of methods*, the chances are
that the new arrangements, not founded on experience, do
as much evil as good and are *certain* to cost more. There
must always be a tremendous friction between a spending
and a conserving Dept., and I do not see how a change
in the relations of the chiefs is to prevent that, or even to
mitigate it much. We are apt to forget that if Red-tape
has prejudice on its side, it also has experience.

During his long experience he found that with rare exceptions the official and governing classes cared little and understood less what science ultimately meant to the nation's
life. To men of their upbringing and circumstances science
was something alien, intrusive, disturbing to the established
order of education, thought, and action ; in its highest flights,
a thing abstract and dusty, typified by museums and professors ; and when it descended to earth, a useful familiar
labouring in the forges of civilisation. There was no question
of a national organisation of science for future national development ; aid was limited to the obvious by conscientious
administrators to whom economy only meant the cutting
down of expenditure. But when the question arose of reforming public departments from above, common sense forbade
him to expect a magical transformation.

As to foreign relations, suffice it to quote a few passages which have a curious bearing on the clash of arms to-day. One dates from November 1893, when the omens were unpeaceful and pointed to a coming struggle for the dominion of the Mediterranean :

To W. E. Darwin

Nov. 9, 1893.

I am dreamer enough to look for a time when America will forbid a European war ! What a splendid role this would be for a nation to undertake—to send us all to our tents and tell us that we may snarl at one another in the length and breadth of Europe as much as we please, but nothing more, and that if we go further she will intervene.

To Ayerst Hooker

April 27, 1890.

[Apropos of the Stanley reception at the Albert Hall. The impending partition of Africa into rival Colonies and spheres of influence boded no good to peace.]

I see much humiliation in store for us in Africa, and with I fear far more evil results than the Soudan or Transvaal.

Bismark was not the only man of blood and iron in Germany ; the whole of Prussia breathes arrogance, and this chirruppy young Emperor will soon be the tool of faction. He evidently has no conception of the difficulties and dangers of his position.

June 30, 1895.

What did you think of the pacific character of the Kiel gathering ? Seventeen nations, each with a loaded and cocked 6-shooter, calling themselves joint ' harbingers of peace ! ! ! '—the irony of the situation is lovely.

To W. E. Darwin

Jan. 8, 1896.

[The Venezuelan affair and the Armenian massacres threatened warlike complications.]

What an imbroglio we have been forced into ; and what an ass the German Emperor is.—We should lose less and

Germany more by a war than would any other nation in Europe.

By conviction of travel in our great dependencies he was somewhat of an Imperialist before the word became a political shibboleth ; he felt strongly the duties, and sometimes the compensations, we owed to these dependencies, and to the very end of his career was keenly concerned in the aid which scientific botany could lend to their comfort, health, and prosperity. His own active share in this was not small. It was very much due to Kew in his time that Jamaica was recovered from bankruptcy, and afterwards he kept in close touch with the work of the West Indian Governors and their botanical advisers, and was especially rejoiced at the issue of the Sugar Commission by Mr. Joseph Chamberlain when Colonial Secretary.

Education, again, was no party matter with him. Ignorance and the indiscipline of ignorance are the greatest enemies of the State. State education [1] was necessary, but let it be appropriate. The farm-hand must not be trained merely as a clerk for town life. Neither should boys of the wealthier classes be restricted to a literary education, for which at least half of them were unsuited. These and similar points, such as the course taken by the higher education of women, he discusses freely in his letters to Mr. La Touche, who as a school manager in his Shropshire village was much exercised over the working of popular education.

In the case of his elder sons he had tried to avoid, so far as was possible in the existing arrangements of the public schools, the extreme dose of classical instruction ; but the modern side was not then organised as it is now, and the newer schools which tried to meet the new demand did no good because they taught the new subjects in the old bookish way. It is only a minority of boys who are naturally adapted to reap the full benefit of this highly specialised form of literary education, and he

---

[1] As to the State's attitude towards religious teaching, ' I have always,' he writes, ' been for a purely secular State Education, affording at the same time encouragement to religious teaching by private effort.' (1896.)

realised that his own boys, like so many others, had not carried
off any enduring and precious intellectual harvest from their
inevitable share in the classical routine. Their later bent was
towards practice, whether scientific or administrative.

Hooker had always endeavoured to awaken his children's
interests and direct their minds by his own well-furnished mind,
for ' Nature he loved, and after Nature, Art.' Now that his
time was freer, he devoted daily hours to the early teaching
of his younger sons. As soon as they were established at
The Camp, he tells how the eight year old ' reads " Robinson
Crusoe " and the " History of England " every morning with
me for one and a half hours, enthusiastically.' Again we see
him reading Mrs. Markham's History with one of them before
breakfast and some simple Roman History after supper ; now
it is Ball's Astronomy for children—Star-land—or Huxley's
' Physiography,' a book of which, he tells the author who had
just sent him the latest edition, ' I have had a copy for each
of my four elder boys—but they disappear seriatim with the
youths themselves when they leave the paternal house. This
comes in the nick of time for Joe.' (December 27, 1887.)
He started them also, and very successfully, with colloquial
Latin from an Ollendorffian French handbook, in pursuance
of his belief that only a language spoken is a language learnt.
Having found companions for Joe in the grandsons of Colonel
Hannay,[1] he carefully watched their progress under a private
tutor, and when the elder boys went to school the same system
was continued for the youngest.

His lighter reading, ' Novels, Histories, Lives, an old man's
proclivities,' comes in for frequent mention in the letters, in
the absence of any friend to talk science or philosophy with,
or to contradict ! Novels run the gamut from W. D. Howells
and Mrs. Humphry Ward by way of the ' Briar Bush ' to ' Peter

---

[1] Colonel Hannay had much correspondence with Sir William Hooker on
botanical matters ; he evinced much interest in various departments of agriculture in India, but more especially in fibre-yielding plants and cotton, respecting which several of his papers were published in the *Journal of the Agricultural
and Horticultural Society of India.* He also brought a small experimental garden
of the China tea plant to a high state of perfection, and thus demonstrated what
was deficient in the adventure of the Assam Tea Company.

Simple,' ' a great relief between spells of work ' ; the Lives are
sometimes of his contemporaries, Leonard Horner, G. H.
Romanes, or Lord Roberts, or studies of the past like Menerval
on Napoleon or Mahan's Nelson ; and Boswell he read anew
with enthusiasm. Bryce's ' Holy Roman Empire ' matches
Green's ' Short History' ; but while Sir L. Mallet's Free Exchange
is illuminating, he cannot make head or tail of the Bimetallism
craze. The East and its philosophies are an unfailing lure ;
whether ancestor worship as described by Lafcadio Hearne in
his ' Korokoro,' or the relations between the greatest religions
of the East and the West as set forth in Lillie's ' Buddhism
and Christianity.' And not least from 1892–94 he enjoyed his
' pick-me-ups,' the successive volumes of the ' Collected Essays '
sent to him by his friend Huxley (see p. 311). This was a pleasure
not to be kept to himself, and he used to send on the volumes
to appreciative friends such as W. E. Darwin and Mr. La
Touche. To the latter he remarks (April 9, 1894) :

> I am glad that Huxley's works have interested you so
> much. His versatility is indeed wonderful, but not so wonder-
> ful as his power of rapid assimilation of mental food, in all
> shapes, and of all kinds. He is about the last of the ' thick
> and thin ' companions of my younger days.

To the Rev. J. D. La Touche
Dec. 28, 1892.

I am exceedingly sorry to hear of your annoyance with
the School board ; but must confess to having some sympathy
with the ratepayers. Education is a very good thing in its
way, but it is driving the rural population into the towns,
raising the price of labor beyond the limits of capital, and the
taxes of the poor themselves perhaps out of proportion to
the good *they* derive from it.

As it is, the well-to-do are spending tens of thousands on
the education of their sons, who are after all no better than
fit for the plough ! and the sons of the poor are getting the
same for nothing at the expense of the well-to-do. I know
that I am a blasphemer, and will only add that I am for
compulsory reading, writing, arithmetic and cooking, and
for a *limited* extension of higher education, to which the

Govt. should contribute.   But why on earth I am to be taxed to have every stupid child taught the niceties of English grammar, and of English history three-fourths of which is better forgotten, and the geography of every place under the sun—I do not see ; and all the more when it comes to educating the children whose parents could and ought to afford it themselves.

May 2, 1893.

I am a Radical in such matters—but your radicalism will not tolerate mine.

To the Same

May 24, 1893.

You must not think that I oppose education of the laboring classes, but I should like it conducted towards the future life of the average, and not to the high education of the few who can profit by the complex education of the Board Schools.   Mind you I am just as much against the higher school and College education of the *masses* of the upper classes ! surely it would be far better if much of their teaching were devoted to making them more useful members of society. . . .

To return to technical education, my notion of it is, that it should be begun early, at the expense of some of the Board's literature, classical English, &c.—and be accompanied throughout by semi-scientific teaching—i.e. the cobbler should be taught what tanning is, what bristles are, and how developed, and so forth.   If any Board-school child shows a genius for the higher education, push him on by all means to school and college ; but it is no use trying to ' make silk purses out of sows' ears.'

To the Same

April 22, 1898.

I get quite giddy over the letters and articles in the papers about Education.   It is, I think, beginning to dawn on some scholastic minds that you ' can't make a silk purse out of a sow's ear,' and that four-fifths of ordinary children or youths cannot be got even to see the beauties of language or poetry, Greek, English, or Latin, &c., &c., that our educated teachers (who do) are trying to cram into their

heads.   Tyndall used to say that you had only to show boys
scientific experiments to make them love science.   Bence
Jones [1] was nearer the mark when he said, ' all the boys care
for and call Chemistry is a blaze, a bang, and a stink.'   My
view is, to teach the mass the 3 Rs, and *something technical*,
picking out for a higher education the very few who have
shown talent or taste for higher things, and educating them
on as high as you can.   As it is we are throwing away millions
of money, and racking the brains of thousands of admirable
teachers, male and female, in attempting the impossible.
I do pity a Board teacher.

To the Same

Christmas, 1894.

I have just received as a present, from his sister (Mrs.
Hodgson), Townshend's ' Agricola and Germania ' of Tacitus.
I had never seen the little work before, and am very much
interested in it.   It is a great pity that boys were not made
to read such translations.   They can get no notion of the
subject or author by grinding at the original as in school
exercises of grammar.   It always amuses me to hear the
' tall talk ' of schoolmasters about the value of teaching
the Classics because it instructs boys and men in the *genius*
of the Greek and Roman authors and their languages and
literature.   The fact being that 19 boys out of 20 do not get
even an idea of the subject they translate by halting efforts
to put it into English—and as to the genius of the language,
they do not know what that means, and no one at school
cares to tell them.   Joe has been learning Mechanics.   I
asked him what was meant by Mechanics—his answer was
that it meant Hydrostatics, &c., &c.—but what it was he
had not an idea !   Turning to ' Chambers ' I find the
definition ' The Geometry of motion.'   A very little explanation would make this clear to a boy.

I have just opened B. Stewart's [2] Primer of Physics.

[1] Henry Bence Jones, M.D. (1814–73), was an accomplished physician and
student of chemistry, making many researches into the relation of chemistry
to pathology and medicine.   He became F.R.S. 1846, and from 1860 onwards
Secretary to the Royal Institution.

[2] Balfour Stewart (1828–87), physicist and meteorologist, was Director of
the Kew Observatory 1859–71, and Professor of Natural Philosophy in Owens
College, Manchester, 1870, till his death.   His most important researches were
in connexion with radiant heat.   In collaboration with Professor P. G. Tait he

WOMEN'S EDUCATION 331

The first paragraph is headed ' Definition of Physics '—
two pages follow with instances of physical facts—but no
definition of Physics is ever attempted. So too with the
Chemistry Primer.

To the Same

June 5, 1897.

I entirely agree with all you say in respect of the women
themselves ;—that the function of women in the scheme of
creation is not fully understood, by themselves that is, or
rather greatly misunderstood. What is wanted is, to raise
their standard of education throughout, without inter-
fering with their special function. I doubt, however, if
they have what you call ' a great future before them '
(I do not know, however, what you mean by this), i.e. in
a strictly intellectual sense. No education will give them
originality, and scarcely intellectual individuality. The
Epicene man is a poor woman, but the Epicene woman is a
bad man, or nothing. (What a miserable parody !) There
is no shirking the great fact, that the woman's function
is to be wife and mother, and that a degree certifying high
intellectual powers or qualifications is not that recommendation to the choice of a wife that it is to the choice of a husband,
who must live and maintain wife and children by his wits,
however manual his craft may be. Now the drift of all
this *very high* education of woman is to lead her to ignore
her true place in the scheme of creation—and most unfortunately it is pushed on at the *most critical period of
woman's life.* The excuse is, that so many women must
either live by their wits, or, if they have money and do not
marry, must have some intellectual food to keep them out
of idleness and possible mischief. This is all very true and
points to a good reason for giving to those women who must
live by their wits a substantial degree of some sort (such as
they have at the L.U.). Hence I rejoice in recognising a
public stamp of intellectual merit in such cases ; but do
not in seeing a rich man's daughter having any title of the

wrote *The Unseen Universe*, 1875, and was co-editor with Huxley and Roscoe
of Macmillan's series of Science Primers, himself writing the *Primer of Physics*
(1872), besides writing several successful text-books. He was one of the
founders of the Society for Psychical Research.

sort. But what do we see? the poor hardworking girls
alone take the L.U. degrees—the ladies aim at the incomparably inferior ones of Oxford and Cambridge! Practically
it is a question of rich and poor after all, which degree is
chosen; and this vitiates the desire for C. and O. degrees
as objects of ambitious desire. In the broadest aspect,
male and female of the human species are anatomically,
physiologically, and intellectually different (I will not allow
inferiority), least so in the last respect, but still manifestly
so. They therefore require different treatment in all these
respects as they grow up, and this treatment must be different
throughout life. There is no abrupt intellectual stop at
the period when woman's higher education begins, whereas
at that period the feminine functions are exacerbated, and
the mind more or less disturbed (as that of the man is not)
in consequence of the physiological peculiarities of the
sex.

To the Same

March 2, 1898.

Certainly the engineering trade still shows that the
mills of public education grind very slowly, and not finely
at all.

I am no enemy of public education, but what I would
have is, the giving it a more practical turn. Men cannot
live on intellectual culture, and a mighty small percentage
can make use of it if they have acquired it—fewer still care
to use it.

I am aghast at the folly of much of our intellectual
teaching of boys. It is in the hands, as a rule, of highly
intellectual men, who assume that the average boy will see
the beauties of Milton and Shakespeare, and Euripides and
Aristophanes, *if only taught to see them.* Now this is a profound error. Take [the mass of public school boys who] have
had 6–8 years of classical education—not one of them can
now translate a simple paper of Latin or Greek, or will look
into a classical author, or listen to the talk about one. I
do not say that this branch of education has done them no
good, but I do say that it is nothing as compared to the
time and expense, and that all talk of their having imbibed
the spirit or matter of the language is pure bosh. Perhaps
at the end of their education they could translate Livy

and quote Horace ; but they could not say bo to a goose
in Latin or Greek, or ask for food or drink even ! and
yet we say they have acquired a knowledge of the Latin
tongue ! and have had their intellect raised thereby.   There
are of course boys and men who will respond to all the
master teaches, and learn the colloquial Latin that the
master cannot teach or speak, but these are the rare exceptions, and I would give them every opportunity of acquiring
all that can be taught or learnt.   So it is with the English
Classics.   How many of all that are taught (or told) to
admire Milton and Shakespeare ever take up either as
a pleasure, or taking them up, have the divine gift of
readily appreciating their beauties.   But we do not give
the boys 6–8 years of our English Classics—why not ?
I would have every child whose parents could afford it
taught some Classics, both Latin, Greek and English,
and encourage those that had a gift for them to go on ;
but to continue rubbing the noses of 19 boys out of 20
in these subjects for 6–8 years is in my opinion utter
folly.'

The history of the East and its philosophies were a constant interest, especially when they touched upon the bases of
Western creeds.   There was a special fascination in the subject,
because he had ' " assisted," as the French say, at worship
in Buddhist Churches.'

To the Same

Jan. 23, 1891.

I heard a most curious thing last night, from Mr. Maunde
Thompson, principal Librarian B.M.—that the oldest
Chaldean inscriptions in the B.M. were dug up in London !
on the banks of the river, dating 4000 B.C.   Luckily some
Dutch tiles were found with them, which gave a clue to
the discovery that the position was occupied by a Dutch
merchant who traded with the Persian Gulf.   This was
before the Fire of London.   Probably the stones, which
are like those used for gate posts, came home as ballast,—
possibly were used as ballast by a boat down the Euphrates
or Tigris before getting on board the British ships that
brought them to England.'

To the Same

Dec. 18, 1893.

Lillie's Buddhist books are most curious, but very badly written—full of valuable suggestions. I cannot understand the Church ignoring the teaching of the Buddhists, and overlooking the startling fact, that it has taken over all the Buddhist worship, vestments, appointments, litany, orders of priesthood, nunneries and convents, &c., &c. *en bloc.* The history of all this was no doubt burnt at Alexandria. Some of it may turn up in Papyri yet. The dark ages were indeed dark when so pregnant a fact was disregarded or burked. Did the Christians absorb all of Buddhism but Buddha and his miracles, or did the Buddhists turn Christians *en bloc*, retaining their ceremonial? and when did it happen? it could only be where many of both were massed together—Alexandria? Rome? Constantinople? I can understand the conversion of a Buddhist Lamasery and Seminary to Christianity, for the Buddhists were not bigots, and I can understand the Christians (who everywhere allowed and winked at a great deal of idolatrous practice amongst their early converts) being dazzled by the pomps of the Buddhist church and adopting them. Both were ascetics, and both preached holy lives and good will to all men, &c., &c., &c. Both had hazy inchoate ideas of a future state and a Godhead—had community of goods and so forth, but whilst one had a ceremonial, &c., the other had none. It would have taken ages for the Christians to have evolved and established the ritual, &c., which the Buddhists had used for 600 years or so.

A very similar subject recurs a little later.

To the Same

April 23, 1894.

I should very much like to see the notice to which you allude of the Jesus in Tibet. I never heard of it. Theoretically it is likely enough, for I have often asked myself what were the antecedents of Jesus' late life, and why so short a time was spent in his preaching and teaching. Was he not supposed to be ætat. 30 when he was baptized? On the other hand, it is difficult to suppose that had he been a traveller there should be no allusion in his sayings to his foreign experiences. It is more probable that he worked as a carpenter and met with Buddhists whose doctrines he in part embraced. It is also difficult to suppose that one who had spent much time among the Buddhists should advocate the Jewish law to the extent of saying that ' not one jot or tittle of it should pass away,' &c.

To the Same

The Camp, Sunningdale: March 23, 1895.

I have read the Buddhist Life of Jesus, and am not edified. It is a lame story. The author, a very ignorant young Russian traveller, was travelling in India, and visited Tibet, where he broke his leg, and was domiciled for some weeks or months in a large Lamasery near Ladak. Having heard of ' The Life ' he asked about it and had it read to him by the Head Lama, or rather read to his interpreter, for he could not speak a word of Tibetan. He took desultory notes, and has pieced them apparently from memory. The History professes to give the life of Jesus as a missionary who travelled throughout the length and breadth of India preaching his Gospel, especially at Juggernath! He returned to Palestine, where he was put to death with some of the events of the Bible narrative more or less distorted. The Lamas told the author that copies of this narrative are in many Lamaseries in Tibet ;—and that is likely enough, as the Monks have little to do but write fables and histories, &c., and their libraries are innumerable and very extensive, and swarm with copies.

There is nothing impossible in the Indian part of the narrative, but all very unlikely, and to accept it from Mr. Novitesky's (or some such name) version would be absurd. On the other hand, nothing is more probable than that the Monks have written accounts of Jesus' death,—brought by their travelling brothers from Syria &c.

To the Same

May 10, 1894.

I should have returned this long ago, but I have been pretty busy, and wanted to think the thing over. It is a matter in which one can proceed only by guess work ; and

my guess is, that Issa was an Essene who carried the early
Christian teaching with its myths into Tibet, and by a very
natural process became identified with the founder whose
system he expounded and whose life he narrated. For
my part I cannot conceive Jesus having spent the best
years of his life as a traveller and yet no allusion to
foreign peoples, places, religions (other than Buddhist),
or things peeping out in his teaching. If he had so passed
his prime of life, it must have been well known to his neighbours—but except as ' the Carpenter's son,' no allusion that
I can recollect is made to his early life—except as a child,
and I cannot conceive the episodes of his childhood having
been remembered for 30 years by Orientals !—they were
imagined or made up with no intent to deceive afterwards
I suspect by his followers. My idea is that Jesus passed his
life (like hundreds of others, for Palestine is drilled with their
caves in places) as an Essene recluse, in meditation, and
came forth at last as a prophet and preacher. The tide of
events taught him the speedy destruction of the temple and
Jerusalem—and Buddhist pilgrims, or monks, taught him
Buddhist doctrines, proverbs and parables, rules of life,
duty, &c., &c., &c. It is impossible to guess how much is
true of what is attributed to him by his followers, or rather
his successors, who believed in his mission, but not in his
Godhead I suppose.

We naturally regard Jesus in respect of his family and
surroundings as we do ourselves ; forgetting that there is
no such family life in the East as we enjoy.

To the Same

Dec. 24, 1893.

What you say of A. B. and C. does not at all surprise me.
They are ' ne plus ultra ' mathematicians, have not a conception of biological science, and in fact are only *half intellects*
(I suppose I deserve to be burned), but so it is, that I have
often found such men to be impervious to reasoning out
of their own circle, in matters of natural science. With
biologists, who have to found everything, beyond pure
observation, on circumstantial evidence, the case is quite
different. For hundreds of biologists who are good mathematicians, you will not find ten vice versâ.

How many medical men do you suppose believe in the doctrine of the Incarnation ? A medical man's faith in a doctrine that contradicts his daily experience of obstetric practice, must be strong indeed ; a thousand times stronger than that of non-medicos.

It has often struck me that had the biological sciences preceded or run abreast with the mathematics and classics, we should long ago have had a religion of pure reason, such as Huxley has sketched at the end of one of his Essays—I forget which. As it is, biological science is hardly a century old, and just see what havoc it is making in doctrinal religion.

To the Same

Dec. 29, 1893.

I have just finished Huxley's last volume. The Essay on the ' Evolution of Religion ' is most remarkable and gives an astonishing idea of his grasp of mind, powerful reasoning, and admirable style. Certainly no one, theologian or other, has brought the subject before the ordinary reader in anything like the persuasive manner and rhetorical power he displays. It goes to Darwin to-day.

To the Same

February 18, 1897.

Your letter has interested me much, if only by the contrast it affords to our readings. I have been going through a long course of Boswell's Johnson, and of Boswelliana. I had already long ago read the Tour in the Hebrides, and Madame Piozzi, so I am pretty well up in the old Hero, whom one cannot help admiring (and disliking rather). But he had great nobility of character, and I much like the prayers and invocations he addresses to his Maker like a man, with all humility and earnestness, and yet in the language of one who felt it his duty to do so in his best style—neither whining nor pompous, not studied nor stilted, but as one deeply affected by the awful presence of his Maker. He has published his prayers and meditations somewhere, and I will try to get them.

As to poor, half crazy, clever, kindly, vain Bozzy, it was a shock to me, after having derived much pleasure from his writings, to find that he died at 55, the victim of drink! That he had bursts of this, as of other vices, he tells us (and Johnson) candidly, but I did not know that he succumbed to it, and at so early an age.

I quite agree with you, that the Board schools give quite as much religious instruction as is good for the children of the school, and that anything further should be supplied from other sources. But whatever I may think of the dogmatic religious teaching given in the voluntary schools, I do not suppose it has a particle of bad effect—it is all washed away. There is, my dear La Touche, an undercurrent of *jealousy* in the attitude of the Radicals, to both Landlords and Church, which blinds them to the good that these do and have done. My idea is that the proportion of good Landlords to bad = that of good Workmen to bad, if not greater. Changes for the better must come slowly, and in the process you must help the lame dogs over the stiles. I do hate doctrinaire Politics.

CHAPTER XLIV

THE current of events ran tranquilly during the fruitful autumn of Hooker's long life. The successive letters which follow serve to show the plain threads that were woven year by year into the fabric of daily life.

The year 1886 opens with a greeting to Hodgson:

> This reciprocates your affectionate good wishes and would if possible bear interest, but that I know is not possible, for you are my most affectionate friend in the world, and are ever in my thoughts.
>
> I go on laying out my grounds, chiefly with Rhododendrons and ornamental bushes and trees. I have no word yet about my pension, which is awkward, as I must now determine my style of living, which will depend upon it, and I have still outhouses and other expensive items to meet.

Next year the date of Hodgson's birthday (February 1) was missed, but not the greeting:

> Again your birthday has passed without the greetings from me that should have arrived on that very morning. Pray pardon me—it was not forgetfulness of you, for we were talking of you at the very time. . . . Your birthday is down in my Diary, but I do not look at it as often as I should, and so my own children's birthdays are as often overshot as not.

Miss Lyell, to whom the next letter is addressed, was the sister of Sir Charles, the geologist, and a lifelong friend of the Hookers, as her parents had been of his. Sharing both families' love of flowers, she sent him for his peaty garden some plants of the Star-flower (*Trientalis*), a rare and charming reminder of their Scottish hillsides. His other Highland favourite, the exquisite little Twin-flower (*Linnaea*), another of

> these mountain flowers,
> More virginal and fresh than ours,

of which he had not seen a living bloom for sixty years, had already been sent by her sister-in-law, Mrs. Henry Lyell, who was herself a lifelong friend of the family, being the daughter of Leonard Horner, Sir William Hooker's friend. She was Joseph Hooker's senior by just three weeks, and in their old age used laughingly to claim great precedence on that score. She outlived him three years, bright and alert to the last.

The patch of *Linnaea*, Mrs. Lyell's gift, was the best of mementoes. It spread and bloomed profusely under its familiar pines, by 1888 making a carpet 9 feet across each way literally clothed with flowers after Midsummer ; later three dry springs spoiled the original patch, and its glories were sustained by two daughter patches in shadier positions. Twenty years after and more he continued, whenever possible, to send Mrs. Lyell a spray of its blossom on her birthday in the first week of June.

Its immediate success in the wild garden is insisted on again in 1888, when two kinds of Wintergreen, another rare wildling, were sent to him, the one by Mrs. Lyell from Kirriemuir, the other by her daughter from the Engadine,

> the floweriest land [he assures her] I ever visited except Australia. Nowhere else have I seen such attractive plant and insect-life, and nowhere more beautiful and at the same time accessible mountain scenery. [Adding a postscript], If you see Mr. Huxley pray give him my brotherly slap on the shoulder, and ask him if the Engadine is not ' all right.'

To Miss Lyell

July 16, 1887.

We are all well and have been entertaining our American friends, Dr. and Mrs. Asa Gray, whom we accompanied to

Oxford and Cambridge—at both which Universities Dr.
Gray received an Honorary degree. We thus missed the
Jubilee demonstrations in London and I cannot say I regretted this, for I am too old to like a crowd, and illumina-
tions have no great attraction for me. I did, however, go
to the Queen's Garden party, which was a very pretty sight.
The Queen and all the Princes, Princelings and personages,
European and exotic, walked in a very long scattered procession through an avenue formed by the visitors, bowing
and shaking hands here and there, and ranging right and
left without formal order of march or other ceremony, so
everyone saw everybody without crowding or inconvenience.
I never before saw the Queen looking so happy and pleased,
and this she here looked, smiling and bowing and chatting,
leaning on an ebony cane—but what a little bit of a thing
she is, amongst a crowd of tall and stately Britons !

To the Rev. John Gunn (his uncle by marriage)

Sept. 19, 1887.

We returned only a fortnight ago from a short tour in
Normandy, where I took Mr. Turner's Tour. It was very
interesting to look at the original of Cotman's Etchings and
my Grandmother's drawings, and I can pronounce them all
to be admirable. We were charmed with the picturesqueness of the country and people. . . .

I saw a lovely Vincent at Christie's rooms the other day,
but dared not buy.

On this trip they spent ' two weeks with the Grays amongst
the churches of Rouen, Caen, Bayeux, St. Lo, Coutances,
Avranches, and Mt. St. Michel, and a fortnight at a desolate
watering place south of Granville.' The reference to Mr.
Dawson Turner's tour is explained in a letter of September 26,
1897, to Mrs. Henry Lyell. (See also pp. 197, 203.)

My father and mother went on a tour in Normandy
with Mr. Turner, Mr. Lyell, and some of my aunts. All
I believe sketched ; I think this was in 1815 : my Grandfather went in 1815, 1818, 1819, and was accompanied on
one or more occasions by Mr. Cohen (afterwards Sir F.
Palgrave) and Mr. Cotman, by whom many of the illustra-

tions in Mr. Turner's ' Tour in Normandy ' were made and
subsequently etched. No doubt a copy of the 'Tour' is
at Kinnordy. My father pencilled beautifully. I have
many of his sketches.

On September 15 died his father-in-law, Mr. Symonds,
who had spent so much of his last years at The Camp. ' He
is a great loss. Scientifically and intellectually he was the
life of a large surrounding.'

To Asa Gray

Nov. 15, 1887.

Read Bonney's [1] article in ' Nature ' on Huxley and the
Duke of Argyll in reference to Darwin's and Murray's Coral
reef theories; it is wonderfully good. The Duke's article
in the XIX Century (I think) was a very stupid one,—
but what struck me most was, the Duke's not seeing that
Darwin's theory was, whether right or wrong, a stroke of
genius, unaided by that knowledge we now possess of land,
sea, sea-bottom, chemistry and corals; whereas Murray's is
a conclusion arrived at through the labour of a staff of most
eminent fellow workers on the Ocean, and a knowledge of
all the facts and data that they were collecting around him
during the Challenger Voyage. As you say, ' the greater
truth, the greater libel '—so we may say of Darwin's theory,
' the greater error, the greater genius.' But I expect the
truth will lie between them, and that there will prove to be
two, perhaps more, ways of making Coral Islands.

To T. H. Huxley

March 4, 1888:

I went to Cambridge to hear one of Strachey's Geography
Lectures; the matter was excellent but very dry—as the
Frenchman said of English meat, which he bought from the
dog's-meat man.

In February and March 1889 his portrait was painted by
Herkomer, Kitcat size, for the Linnean Society, and looked
' a very old man indeed.'

[1] Canon Thomas George Bonney, D.Sc. (1833), alpinist and geologist,
Emeritus Professor of Geology at University College, London; was President
of the Geological Society 1884-8; Vice-President of the Royal Society 1899;
President of the British Association, 1910-11.

In June he spent a week in Cornwall, the guest of his
old friend General Sir J. H. Lefroy.[1] The sight of many
Himalayan rhododendrons acclimatised there filled him with
enthusiasm, and he writes to Mrs. Hodgson on the
30th :

Tell Brian with my love that I saw, in Cornwall, *many,
many* plants of the *Rhod. Hodgsoni* in the open air, 6 feet
across and more, and with leaves a foot long—they were past
flower unfortunately. They were planted in the woods and
throve luxuriantly. There were also noble plants of *Falconeri,
Aucklandii, argenteum, barbatum* and others—together with
Hodgsoni forming regular shrubberies, as if natives of the
soil.

' The Club ' brought about meetings with many interesting
people. Mr. Gladstone, though the Ayrton trouble had taken
place under his administration, was always on the best of terms
with Hooker, and shared moreover in his passion for Wedgwoods. The following impressions of a conversation with
him, and the inferences to be drawn from it, are of curious
interest.

[1] General Sir John Henry Lefroy, R.A., C.B., K.C.M.G., F.R.S., &c. (1817–
90); entered the Royal Artillery 1835; in 1839 was appointed to the Observatory for Magnetic Research in St. Helena and made the voyage in the *Terror*,
then starting in company with the *Erebus* for the Antarctic, a four months'
passage, during which was formed a lasting friendship between him and J. D.
Hooker.

In 1842 he was transferred from St. Helena to the Observatory at Toronto,
from whence in 1843 he made a journey with only one companion to Lachine
and Hudson's Bay, 5475 miles, by canoes and on snow-shoes, which established
his reputation as a geographer and was productive of most valuable results in
magnetic observations.

He returned to England in 1853 and took command of his battery, and in
1854 became Secretary of the Royal Artillery Institution, founded by himself in
1838; in 1854 he was appointed Scientific Adviser on Subjects of Artillery
and Inventions; in 1857 Inspector-General of Army Schools; in 1868
Director-General of Ordnance at the War Office.

From 1871-7 he was Governor and Commander-in-Chief of the Bermudas;
Governor of Tasmania 1880-2.

Among his publications are the *Handbook of Field Artillery for the Use of
Officers*, used in the Crimean War; *Memorials of the Discovery and Early Settlement of the Bermudas or Somers Islands* (1515-1685), 2 vols., 1879; *Diary of a
Magnetic Survey of the Dominion of Canada, &c.* 1883; also various scientific
papers.

An obituary notice of him and his work was written by Sir Joseph Hooker
and published in the *Proceedings of the Royal Geographical Society*, 1891.

To Ayerst Hooker

May 18, 1890.

I had the G.O.M. on my left at the dinner of ' The Club ' last Tuesday. He is marvellous ! as full of interest in everything in the shape of literature, Art, &c., as ever, and as pleasant and vivacious, and really modest and unassuming in conversation. I never saw a man wear so well at 80 ; looking at his hands, they are not like those of a man of his age ; and he enjoyed his dinner temperately. I soon found out that he was not allowed to see the *Times*, for he was deeply interested in the question of the mode of beheading of Charles 1st, but did not know of its now being discussed ! Nor did he know of Caprivi's speech [1] or that France had 20 millions of public debt.[2] I could not wonder at his living in an atmosphere of illusions.[3]

In August he went to Scotland ; but the tour was cut short by an alarm of diphtheria at home. The changes noted by him in Edinburgh are interesting.

To W. E. Darwin

August 7, 1890.

I did know it well 40 years ago, and now find the old features as good and grand as ever, and miles of new added, in good taste and with fine effects.

The old town is almost replaced by new houses, which however are not so conspicuously unlike the old as utterly to destroy associations and sentiment ; and when one remembers the intolerable filth, squalor, and stench of the old town, one cannot regret the change.

Nothing is more striking than the contrast in point of good clothing for heads and feet, that distinguishes the lowest orders of this and earlier days—though indeed bare filthy feet and legs, and towzly heads of capless hair are still too common. Above all I must mention that I saw but one

---

[1] His maiden speech as Chancellor : *The Times*, April 16.

[2] Apparently a floating debt of 28 millions to be met by a loan.

[3] He had known Mr. Gladstone for many years, admiring his powers though not following his politics. After meeting Mr. Gladstone at Sir Harry Verney's on March 8, 1878, he tells Mrs. Hodgson : ' I had, as usual, a long talk with him. He was enthusiastic about America and the Californian trees, and the methods of felling them and so forth. His memory is wonderful : he remembered passages in books on Western America that he had read 40 years ago ! '

drunken man (and no drunken woman !) between Holyrood and the Castle Hill.'

Mrs. Lyell had just printed for private circulation the 'Memoirs of Leonard Horner,' her father.

The Camp, Sunningdale : October 22, 1890.

I find it delightful in spirit, matter and style. The account of Holland after its Napoleonic barbarities is not only instructive, but quite new to me. I do not know where else a reader could get in so few pages such a living and moving impression of the state of the country and its causes.

Nothing can bear higher testimony to your father's calm temperament and noble soul, than that he too, after all he saw, should not have felt some spark of that fire of revenge that he so well describes as coursing in the breast of others, and as so inhuman and unwise : though, alas, so natural, and perhaps in some degree useful.

Indeed I had every reason to love and venerate Mr. Horner and Mrs. Horner, and I do so most truly. After Lyell's life, none could interest me more, personally linking as they do my early with my later life in bond of the kindest, warmest friendship, and most intellectual intercourse.

Of the two intimates whose death is next recorded, Gifford Palgrave was his first cousin,[1] his mother being a daughter of Dawson Turner ; and Mrs. Busk was the widow of George Busk the anatomist, a member of the inner circle of the *x* Club. She was an accomplished and gifted woman, with great social charm.

[1] William Gifford Palgrave (1826–88), whose mother was a daughter of Dawson Turner; scholar, writer and diplomatist. He early felt the call of the East; entered the Indian army, and left it to take up missionary work as a Jesuit in South India, and in Syria, where he barely escaped with his life from the Damascus massacre of June 1861. Then in 1862–3, partly with a view to missionary enterprise, partly with a semi-political commission from Napoleon III, he traversed Arabia, disguised as a Syrian Christian doctor, visiting places to which no European could penetrate. His *Narrative of a Year's Journey through Central and Eastern Arabia* is well known. Later he withdrew from the Jesuit Order, and entered the British Diplomatic service, where his astounding facility in acquiring languages stood him in good stead. He was entrusted with a mission to King Theodore, as a last resort before the Abyssinian war, to demand the release of the English captives. Afterwards he was Consul in Asia Minor, in the West Indies and Manilla, in Bulgaria, Bangkok, and finally as minister-resident in Uruguay, where he died.

To T. H. Huxley

The Camp, Sunningdale : Nov. 3, 1890.

I should have gone to Mrs. Busk's funeral, but I get so bronchitic this weather, that I am 'defended.' To-morrow I ought to go to Gifford Palgrave's, on every account. . . .

The bringing Palgrave's body from Monte Video is a curious episode in his history—after all his vagaries he died in the arms of Mother Church ! and to bury a full blown Ambassador of that creed in Buenos Ayres would cost an enormous sum—so large indeed that it could not be afforded.

Mrs. Busk's death is a great shock to us—a truer and better friend never lived ; but I am getting almost case-hardened to deaths. One feels them awfully, on wakening every morning especially ; I suppose jecur [liver] has something to do with morning melancholia.

To Francis Palgrave

November 9, 1890.

Poor Giffy ! one can only think now of his noble qualities. I was indeed gratified by finding I could attend at his interment, though it was accompanied with a flood of memories— some painful, and some peaceful. It brought back Hampstead days most vividly, and all that was grateful. How kind your dear mother was to me !

In 1892 correspondence with Mrs. Lyell resuscitates the picturesque but disreputable figure of Jorgen Jorgensen. She had been reading a book about him, published in 1891. Enquiry of Sir Joseph as to the journey to Iceland and the real part played by Jorgensen led to the following :

To Mrs. Lyell

April 11, 1892.

Oddly enough only the other day Miss Cracroft sent me a letter of Jorgen Jorgensen's addressed to Sir J. Franklin [her uncle] when Governor of Tasmania. My father, Mrs. Fry, and Sir Joseph Banks stood his friends when sentenced to death for robbing with violence, if I recollect aright, in England after his return from Iceland, and had his punishment commuted to transportation for life.

No doubt you know the history of his taking possession of the Island of Iceland, whither, though at the time a prisoner of war in England, on parole, he went, with a *letter of marque* ship, to get a cargo of tallow for a London firm, but which cargo he could not obtain because Denmark was at war with England. My father tells the story of his being taken prisoner by a British ship of war and brought home to England ; but not of his subsequent rascality that brought him at last to the scaffold.

The British Museum has all his correspondence with my father when in Newgate, written on gilt-edged paper ! and the sermon he preached to his condemned brethren ; and his life : (all were in the British Museum). The enclosed letter is half romance ; he did not give the dress (there was only one) to my father, who bought it, nor did my father give it to the Miss Smiths. It is now in S. Kensington Museum.

When I went to Tasmania in 1840, J. J. called on me in a half tipsy state and in rags, and begged for half a crown. On my return I found that he was dead, having been picked up in a ditch a few weeks before.

Kindly let me know in what book you are reading about him.[1]

---

[1] The above-mentioned letter of Jorgensen's is dated Watchorn Street (Hobarton), Oct. 26, 1839. The occasion he seized for addressing Franklin was ' the utmost satisfaction ' with which ' I learned the other day that your Excellency's merit has attracted the attention of the Royal Antiquarian Society of Copenhagen.' Thereupon he discourses of the character of Iceland, the discovery of America by Scandinavian adventurers, and of the visits to Iceland first of Sir Joseph Banks, and then of Sir W. Hooker and himself.

On this occasion he continues :

' I also purchased two uncommonly costly Female dresses, the only two which were left in Iceland of the ancient fashion. Maids wear distinct dresses from married women, and when a female is wedded her maiden dress is put by for the use of her first daughter when arrived at the age of maturity. The dresses are made of the finest cloth trimmed with Gold and Silver, and a large massive chain is placed round the neck with a precious medal appended thereto.

' The two dresses, the price upwards of One hundred pounds, I gave to Professor Hooker (Sir William), who again presented them to the two daughters of Mr. William Smith, then M.P. for Norwich. [My father purchased them himself ; one he kept, and it is now in the S. Kensington Museum. J. D. Hooker.] The two young ladies [no, Mrs. Smith's Frederick aged 14 and Octavius 16. J. D. Hooker.] attired themselves in the dresses one evening when going to Vauxhall, and every one believed that the Lady of the Icelandic Ambassador was in the Garden. The newspapers echoed this next morning. [True : my father went with them to Vauxhall, and informed me that they were discovered and mobbed. J. D. H.]

To the Same

April 15, 1892.

I shall get the ' Uncrowned King.' [1] You say he does not seem to be anything worse than a gambler—but did you read my letter ? My father had scores of letters from him *from the condemned cell in Newgate*, and I found him a *convict* in Tasmania. I have often heard my father tell the tale of his iniquities. He was the most plausible rascal he ever knew, and narrowly escaped the gallows. R. Brown was much interested in him from his having been with him in Flinders' voyage.

In the very letter I sent you he tells a lie—that he *gave* my father the Iceland dress. My father bought it. I suppose he thought himself safe in telling this to Sir J. Franklin at the Antipodes.

Ever, dear Mrs. Lyell, affectionately yours,

Jos. D. Hooker.

To T. H. Huxley

The Camp, Sunningdale : April 16, 1893.

My dear Huxley,—I am all alone and in the place of Hope—but hoping for what, beyond the completion of the Flora Indica, is hard to say. Well I am down to Grasses, which you may remember is at the bottom of all in the accepted classification of Phænogs., but there are 800 of them, from the top of the Himal. to Malacca and Ceylon, and no one has hitherto digested them. I hope to, for "it is dogged that does it"—words in which you rightly summed up my qualities.

I am very concerned to hear of your influenza and enfeeblement. Do take care of yourself.

I had a note from Spencer the other day asking information about Garden plants—he is still floundering on at acquired habits, &c. He makes no progress. In my apprehension, if it were a truth Nature would not be so d—d sensitive about it.

I am bothered with bronchitis, and eschew night hours, but go to Kew thrice a week despite Madam's objurgations —fair weather or foul.

---

[1] Presumably *The Convict King*, by J. T. Hogan, published in 1891.

Huxley had been asked by Sir Richard Owen's grandson and biographer to contribute an appreciation of his character and scientific work. He restricted himself to the latter.

To T. H. Huxley

October 8, 1893.

Though certainly I should never have expected it, I am not surprised at your undertaking the estimate of R. O.'s scientific work. No one is so well qualified, and I am sure that it will be done in a spirit of perfect good taste, and with judicial fairness—tending to the merciful. . . . Under any circumstances I will gladly look over it, and give you my opinion.

Yes, Jowett is gone—a phœnix. Dissolution is ever jogging our elbows; on Friday I was at my cousin Elizabeth's [1] funeral. I first saw her in 1820, and we have been fast friends for many a long year. On the following day I had another funeral to attend, one here, of the Mother-in-law of Flower's son, Mrs. King Chambers. I am kept well alive to the fact that people come here to die.

I have finished Vol. I. [of Huxley's Collected Essays] with keen interest. I like Descartes best, he seems to me the most subtle in analysis, and wonderfully lucid. I well remember looking at his statue in Touraine, reading the inscription, and turning away saying to myself that the converse, ' I am, therefore I think,' is quite as logical. I told you ages ago that I hated Metaphysics—but did not add, that my Metaphysics was six months of the ' Moral Philosophy ' class of a Scotch divine !

In December 1893 the death of his old friend John Tyndall by a tragic misadventure came as a great shock. ' Another of *us* has gone,' he exclaimed to Huxley. ' What a tragedy it all is—it seems to take a bite out of one's life. He, you and I took to one another in 1857.' And to another friend : ' He was quite the purest, brightest creature I ever knew to be a philosopher,' a man of whom he wrote to his cousin Francis Palgrave, five-and-twenty years before, when Tyndall had

---

[1] Elizabeth, Lady Eastlake, *née* Rigby; widow of Sir Charles Lock Eastlake, President of the Royal Academy.

over-hastily come into official conflict with him, ' He is so sterling and amiable, and his faults are so pointedly heart-affections, that I never can bear to see him hurt, and got the better of, without the strongest sympathy and wish to resent.'

Each old friend warned the other not to expose himself to the chills of a winter journey into the Surrey hills, but both went to the funeral of the old comrade whose love for those two had been beyond any others whatever.

Once home, his warmest concern was to learn how his friend had borne what he himself had found such a hard trial. For himself he had had a sound sleep in the train returning ; ' what a contrast to the previous agony, for it was all I could do to restrain my emotion.'

Huxley wrote a little memorial article on Tyndall for the January number of the *Nineteenth Century*, and asked for Hooker's reminiscence of their own first meeting.

To T. H. Huxley

It was at the Ipswich Association Meeting in 1851 ; but as I was courting at the time, I do not remember much else about it, than the presence of my beloved—yes I do,—I tossed off a wine glass of ink at the Red Lion dinner, which was handed me by a waiter, without any intimation that I was to receive *after it* a pen, and so be equipped for writing my name in the book.

To the Same

The Camp, Sunningdale : New Year's Day, 1894.

Together with all the best wishes of the day, I offer my congratulations on the ' Tyndall.' I think it is as judicious and as good as it could well be. Some of the lighter touches are delightful—notably the ' droll effect ' of some of his sayings. The Carlylean antitheses are very instructive. I have often watched the effect of such ' habits of thought ' in you both. The affecting episode of his last hours comes in perfectly naturally and well. I am glad that you introduced it, though I fancy it cost you a struggle to do so. The x Club comes in very appropriately—it is its Swan's song, I fear.

At our age we shall never know his equal for ' pure and high aims,' though such no doubt exist, nor ever hear of one who could hold his unique position.

I have never thanked you enough for the Essays, of which I have read all but the Darwinian, for they come out rather too fast for the reader, and so I lent the Darwinian whilst reading the earlier. The two which I should choose as being the first to re-read are ' Descartes ' and the ' Evolution of Religion '—I like both so much. The Educational were pleasant memories ; and here is Botany still holding its sway in the Medical Curriculum ! I do pity the poor devils of students—just to glance at the vocabulary of hard words which they are expected to attach some meaning to in some thirty lectures is appalling.

I do like you lecturing me on the preservation of my health,[1]—like it for its affection, and for the bitter irony of it—Satan rebuking sin. I do wrap up well and smoke a cigar in the train to keep off the possibility of cold draughts getting into my bronchial tubes, which are all right so far.

To the Same

Jan. 28, 1894.

No news is good news here. I hope you can repeat it after me ; long may it last to both of us.

I have a splendid fur coat, a historical one : it was Lyell's, who left it to Symonds, and he to me. I well remember hoisting Lyell into it on the last occasion on which he dined at the Phil. Club. With this and the last Vol. of the ' Essays ' (for which many thanks) and a cigar, I travel to and from Kew on three days a week—weather permitting.

In 1893 and 1894 Mr. La Touche was for a time threatened with blindness. In this connexion, two letters to him may be quoted.

September 29, 1893.

MY DEAR LA TOUCHE,—I am indeed glad to know that your Medical man has been able to pronounce so favorable

---

[1] Huxley had written (Dec. 30) : ' Now, my dear old friend, take care of yourself in the coming year '94. I'll stand by you as long as the fates will let me, and you must be equally " Johnnie." '

an opinion of your ' other ' eye.   You may now rest satisfied, and relegate any fancied obscurities of its sight to their psychical origin.   As to the use of the microscope, the very low powers I use do not strain the sight.   My use of it consists in making what are, with my half century's experience, really (for me) rough dissections under low powers.   When one wonders what Malpighi [1] and the older microscopists discovered with the miserable instruments that they possessed, one forgets how far practice and foreknowledge of the results to be sought obviate the need of high powers.   As to the stooping, that affects my vertebræ, not my eyesight.   A fortnight ago, at Kew, I most foolishly spent two consecutive hours dissecting grasses under the microscope, sitting *on a low stool, at a low table* ; on rising I was as stiff as a board, and I could not straighten my back for three days ! nor then without pain.   My eyes were unaffected.

April 9, 1894.

I do indeed most sincerely sympathise with you in your great affliction, and can only implore you not to let it prey on your mind, and to hold fast to the stores of information you have laid up to refresh your mind with in evil times. It is easy to say all this and more, but it cannot be easy to fall back upon such considerations at first—that will come, and you may rest assured that the anticipation of a failure of sight is far less endurable than the failure itself. If the empty headed blind are notoriously at peace, how much more should those blind be who have useful lives to look back upon, and well-stored minds to draw consolation from.   No doubt in your case the money matters are a seriously disturbing element, but you may depend upon that being removed in some shape or other.   So, my dear friend, do not be down-hearted.

Among his Wedgwood collection, Hooker possessed complete breakfast and dessert services in the rare Water-lily pattern.   William Darwin, who was lending specimens for a Wedgwood exhibition, wished to learn the origin of these.

---

[1] Marcello Malpighi (1628-94).   Besides his anatomical researches he published two volumes of botanical works in 1686, and his *Anatomes plantarum Idea* appeared in 1675-79.   A statue was put up in his honour in 1897 at Bologna.

To W. E. Darwin

April 2, 1894.

I have hunted through Jewitt [1] and Meteyard for any notice of the Nelumbium pattern—in vain—which is curious, as it is certainly noteworthy from its direct reference to the ' Botanic Garden,' and to one of the most striking passages of that work (if I remember aright).

I cannot guess who designed it, but I see no reason to doubt the current view, that it was in compliment to old Erasmus, and had reference to some marriage in the family. Curiously enough Mrs. Horace was here on Friday, and seeing a plate on the wall, at once recognised it, as having been made in reference to some family wedding. I did not take much notice of what she said, as I had not then received your note.

The pattern includes all three kinds of the Water-lily mentioned in the ' Botanic Garden,' with their several leaves, flowers and fruit. It might be worth while to accompany the exhibited piece with a copy of the passage in the ' Bot. Garden ' that has reference to the pattern.

I commiserate with you on your having to lend, for I do not see how you can safeguard against accidents. The three superb cases of Antarctic Birds that I lent to the Naval Exhibition were returned not only smashed—*all of them*— but some of the birds had to be reset up ! They paid for the damage—a goodly sum to the man in Piccadilly.

Shall I send you a copy of the passage in the ' Bot. Garden ' ? No doubt ' Erasmus ' was consulted if he did not originate the pattern. I know that Josiah consulted Erasmus as to his Etruscan and Egyptian patterns.

P.S. I find in Chambers' Biographical Dictionary that Erasmus married his second wife in 1781, and that the ' Bot. Garden ' was published in the same year. So it is very possible that the service was a wedding present, and if so a very appropriate one.

This, if tenable, hypothesis fixes its date of manufacture.

---

[1] Llewellynn F. W. Jewitt (1816–86), antiquarian. He learnt wood-engraving before he was twenty-one, and executed nearly all the drawings for *London Interiors,* and also contributed with pen and pencil to the *Pictorial Times* and the *Illustrated London News.* He had for a time the management of the illustrations for *Punch.* He was chief librarian at the Plymouth Public Library 1849–63, and the editor of the *Derby Telegraph* 1853–68. He established the *Reliquary,* and was the author of many books, including *The Life of Eliza Meteyard, The Wedgwoods, Ceramic Art of Great Britain,* and several handbooks on coins, antiquities, &c.

A request for the reference in Erasmus Darwin's 'Botanic Garden' found him away from home at Norwich, but he promised to look it up on his return. 'If I am not dreaming, my father used to quote it in his lecture when treating on Nymphaeaceae.'

Sunningdale : May 1894.

MY DEAR DARWIN,—I have found the passage. It refers to the *Nelumbium* alone, which is the central figure in the plates. It is in the 'Loves of the Plants,' Canto iv, line 345,—but is so pompous and vapid that I doubt your using it. My father used to interest his students vastly by quoting passages applicable to the plants he lectured on from the Poets, and I well remember the first few lines of that on the Nelumbium. I thought that Darwin had also portrayed the common Waterlily, but I cannot find the passage. Ever affectionately yours,

J. D. HOOKER.

Next year this china appropriately returned to the possession of the Darwin family.

April 18, 1895.

MY DEAR DARWIN,—I have met with a disaster in a terrible fire on the Common having got into my grounds, and done a deal of damage, including the destruction of my fine holly hedge, several hundreds of yards long. To meet the expense of renewing this and repairs of all sorts, I am going to sell some of my Wedgwoods, and should you care to have your grandfather's breakfast and dessert services, you might find it worth while to take Rathbone's valuation, or make an offer yourself.

. . . 22nd. I should not like them to go anywhere but back to the family. I am very fond of them, but I have no room to display them, and the Dessert service is too gorgeous to be quite shut up as it is now.

The purchase made, 'it is quite a relief to me,' he tells his friend, 'to feel that the crockery is going back to where it should have gone by rights.' It went, but some of it soon returned, for Mrs. Darwin made Lady Hooker a graceful present of several pieces as a souvenir.

In 1894 he was again in the West Highlands, where Mr. La Touche wrote to him about the ' parallel roads ' of Glen Roy.

Glenfinart House, Ardentinny,
via Greenock, N.B.: August 2, 1894.

MY DEAR LA TOUCHE,—I visited the Parallel Roads some 20 years ago, with Smith of Jordan Hill, since which much light has been thrown on their history, principally by Tyndall's observation that the great Ice Dam must have been exactly opposite what is now the Invereck side of Ben Nevis ; and then by Prof. Dickie [1] of Aberdeen, who examining the diatoms, still to be found under the stones, showed them to be fresh water species ! so I think that their sub-*aerial* formation must be admitted, and the *marine* be dismissed. You do not say what your friend's theory was. I do not see how he could escape the marine or the sub-aerial (i.e. glacial).

We are on a visit to a neighbour of ours at The Camp [Mr. Pigé Leschallas], who has bought a good many thousand acres on the banks of Loch Long, on the Clyde, with a first rate house thereon—otherwise a valueless property of mountain and moor, not grassy enough for sheep, or heathy enough for grouse. The scenery is however lovely of its kind, something like the upper end of Loch Lomond. It is all very familiar to me, for it borders on a few acres (2, I think) of property called Invereck that my father bought some 60 years ago, with a pretty cottage on it, to which we used to resort in summer from Glasgow. The site of that cottage is now occupied by a fine house, built by some wealthy Glasgow merchant, but the lovely scenery remains, and is of melancholy interest to me, as there I spent so many happy days with relatives all but one now gone. There I fished, sketched, and practised rough surveying, preparatory to my ardent aspirations for a traveller's life in unknown regions. Unluckily the climate is typical

---

[1] George Dickie (1812–82), botanist ; M.A. Marischal College, Aberdeen, 1830 ; Professor of Natural History at Belfast 1849–60 ; M.D., Professor of Botany at Aberdeen, 1860–77. He specialised on Algae and published works on the Flora of the East of Scotland and Ulster.

of the N.W. Highlands, and I have had three days of an abominable cold. Nevertheless it is all very enjoyable, for our host and hostess are all hospitality and with horses and carriages enable us to see the country with ease and more than comfort.

I am very much interested in the changes I see around me pervading everything : miles of villages along the coast where all was uninhabited in my young days ; the Clyde cumbered with huge steam vessels and clouds of smoke ; Railways for miles where I had to trudge ; Glasgow a splendid city in contrast to its former squalor, and the Clyde for miles below it, when I left nothing but grass and trees, now occupied by literally hundreds of building yards, all full of iron steam-ships building, many of stupendous size. The great strike of ship-craftsmen on the Thames some 20 years ago drove all the ship building there to the already prosperous Clyde, where the proximity to coal and iron mines furthered the industry marvellously.

But what strikes me most is the change in the Kirk ; the minister last Sunday *read* his sermon ! a thing that would have been the means of hooting him from the pulpit 50 years ago. The hideous, barn-like Kirk itself, now a neat and not unadorned building ; the long earnest prayers are cut short, and the sermon reduced to half an hour— the time made up with a double allowance of Psalms, Hymns, and the old sung Scotch ' Paraphrases ' ; the congregation all with shoes and stockings ! and no dogs admitted ! Gaelic is a thing of the past.

Of curious coincidences I may mention two : My host's bailiff at Highams (his place near The Camp) tells him that my father bought the little place he had near this (Invereck) from his family, who still have the papers. Now my host picked up this man in London !

The other is, that coming down here in the steamer from Glasgow, I fell into conversation with a Glasgow gentleman, who in course of conversation told me that he had a summer residence at Helensburgh, and that the place was formerly held by Dr. Hooker, who planted it. I asked him the name of it, and he said Burnside at Helensburgh, which is a place which my father rented for several years before he had Invereck and did lay out ! Of course he had not a conception who I was till I told him, and then he knew nothing of me.

The year 1895 was marked by another heavy loss. His old friend Huxley had been attacked by influenza in the last week of February. This was followed by pleurisy and other complications. When he rallied early in the summer, Hooker ventured to write to him, a cheery letter, on June 7, but his reply on the 26th, though still buoyed up by the strong will that kept his very senses alert, left no hope in his friend's mind.

To T. H. Huxley

June 7, 1895.

MY DEAR OLD FELLOW,—I have been wearying to write to you for weeks, but dared not till now that I hear from Foster that you are really better; and have indeed been out in the garden.

Foster tells me that he will go and see you shortly. He will tell you all the R.S. news.

We have a scare, my youngest having found diphtheria somewhere—at Dawlish or on the way home. There was plenty of the white film and pain of deglutition, but no complication. We have a sharp young doctor here, but I think he funked Antitoxin. However he got the little chap round all right, and the infant prodigy congratulates himself on knowing diphtheria by experience! He is quite ready to undertake a case.

I have been all right all winter—I am stone-ware you know, and shall have to be buried alive if at all, I suppose.

I was at Torquay the other day, seeing my invalid sister, and went on to Plymouth and over a huge iron-clad, the *Endymion*. It nearly turned my brain. For complexity of structure and function in a given space there is nothing to compare with it in the inorganic world, and Man only in the organic. The discomfort at sea must be extreme, you cannot walk straight for ten yards anywhere— nor can two walk anywhere abreast that I could see. You are ever knocking yourself about and breaking your shins or toes against brass, wood or iron.

The officers' quarters are good, but for the rest, rather

give me wooden walls, with wooden beef, weevilly biscuits, and bilge water.

But I vowed when I began this that I would not yarn to you till you are right well.

With love from us both to you both, and bottom hopes for your speedy recoveries.

Ever, dear Huxley, your devoted J. D. H.

I am sending you the ugliest and most insignificant rock plant in my garden. I sent it home from Hermite Island, off Cape Horn, in 1842 ! It is a Composite, *Cotula reptans*, with a puny head of flowers, like a daisy without the ray. It is a capital rock plant, growing on dry or damp earth—between, not on stones, and forms a lively carpet even on very arid rockery, green all the winter, and very aromatic. It likes water of course, but gets none here. Along of my blessing it has lived 52 years at Kew.

To Ayerst Hooker

June 23, 1895.

It was indeed a relief to hear that you had reassuring news of Ayerst [his son], for I need not say that we both felt deeply anxious. As one gets older one thinks more of young lives than of old in the way of Hope ; for as to us old ones, it is no use the expecting fortune to take our side against nature. Huxley's state distresses me much. I had a very brief note of affectionate regard from him the other day ; the last I suppose he will ever write, for outside the cover his wife has written, saying, that the effort had brought on sea-sickness, and she begged me not to let any of his friends know that he had written to me. Poor Mrs. Huxley is I believe as ill as can well be ; happily their children are all that one could wish, and so with you my dear friend I am solaced to know.

Yes, Huxley's attitude to the Atom is—well ! just like himself. It was not given out as an axiom of his belief, but as a smart rejoinder. At one of our little monthly coteries (of Huxley, Tyndall, Lubbock, Busk, Hirst, Spencer, Frankland, Spottiswoode and self), who dined together once a month for twenty years, and admitted nobody but an occasional foreigner ; the discussion about Atoms waxed hot between Frankland and Tyndall, Huxley quietly giving occasional dagger thrusts to both, and especially to Tyndall, who held that he saw atoms visually in his mind's eye, and who on his inability to describe to Huxley what he saw, was met by H. with the rejoinder, ' Ah, now I see myself ; in the beginning was the Atom, and the atom *is* without form and void, and darkness *sits* on the face of the Atom ! ' No doubt if Huxley had been asked for his view of the Atom, this would have been the substance of his answer, if not the wording.[1]

To the Same

June 30, 1895.

This has been a chequered birthday for me, as the day of a great sorrow and a great joy.

Huxley died yesterday ; we had been fast friends for 42 years, and our brotherly love was shared by none in its depth and strength. I heard from him on the 27th ! no doubt the last letter he ever wrote, in a sad hand, still clinging to hope, but giving an account of himself that precluded my accepting it. We began life as Assistant Surgeons, R.N., he a few years after me, and we forgathered like steel and magnet on his and my return to England in 1851.

Per contra,—Reggie has come out at the top of the candidates for the post of Assistant to the Director of the Intelligence Department of the Board of Agriculture.

The same letter to Ayerst Hooker tells of other interests during the preceding days.

We have just returned from three days in Staffordshire, to an old friend Godfrey Wedgwood, who has retired from business, and built himself a very pretty house at 700 ft. above the sea, some 8 miles from the potteries, out of the smoke, and quite in the country. · We went over Etruria and saw all the processes, and much to be admired. The Americans have developed a taste for the ornamental ware which the Wedgwoods have vastly improved.

The chief motive of our visit was, to be present at the opening of the Exhibition of old Wedgwood ware, in the

[1] Cp. p. 112 and i. 543.

Wedgwood Institute, at Burslem ; which was opened by
Mundella,[1] with an oration, in which he missed what I think
the great merit of Wedgwood ware—the ' Adaptation to
purpose.' I added a few words on this point—as that
Wedgwood plates always had a sunk border for salt and
mustard ; whereas in ordinary plates, these condiments
shoot into the gravy. And a better example may be drawn
from the modern earthenware teapot ; in lifting this, the
first thing is that you scald your knuckles against the body
of the pot ; secondly, the lid shoots off—example, you scald
the finger of your other hand, by pressing on it ; thirdly,
the tea shoots out and spurts out and splashes over the
teacup ; fourthly, the spout dribbles when you set the pot
down. Now in a Wedgwood teapot, first there is room in
the handle so as not to scald the knuckles ; secondly, the lid
won't fall off till the pot is held actually vertically ; thirdly,
the handle is so placed, that by a turn of the wrist, the tea
leaves the spout gently, and without your having to lift
your elbow at all ; fourthly, there is no after dribble from the
spout. I have tested these points in scores of the Wedgwood
pots, and the same care *in adapting to purpose* is displayed
in every pot, jug, plate, or other article that he made. You
may pile plate upon plate of old Wedgwood from floor to
ceiling, and the whole forms a rigid column ; and you never
can spin one of his plates in another.

Then too, Mundella made no allusion to the Medallions
of eminent contemporaries, of which Wedgwood executed
hundreds from the best medals, and sold them cheap. Nor
to his pyrometer and crucibles, the fountain inkstand, his
glazes, colors, and use of barytes, and lots of other ingre-
dients which he introduced into the art of pottery. In fact
Mundella confined his laudation to the ornamental feature
of Wedgwood wares, which are admirable adaptations of
classical ornamentation to his wares ; these are perfect in
their way, but at best only clever adaptations especially
of Greek models, and rather proofs of the skill and genius
of the Italian workmen whom he employed, than of his own
taste ; still this utilisation of his workmen amounted to

---

[1] Anthony John Mundella (1825-97), the social and educational reformer,
was Vice-President of the Council 1880-5, and President of the Board of Trade
1886-7 and 1892-5.

genius, on his part. To conclude, Wedgwood's life has yet to be written, but it must be with the aid of a competent potter, as well as an artist.

The Holloway College, near Egham, was but a few miles from the Camp, and for many years Hooker used to come over in the summer time and deliver to the students a peripatetic lecture on the trees in the spacious College grounds. On Commemoration days also he was a frequent visitor, that of 1895 in particular being now chronicled.

Some big-wig minister was to have spoken, but was prevented by the *resignation* ; and Lady Frederick Cavendish, almost on the spur of the moment, undertook the duty, and in all my life I never heard anything so apropos, and so charmingly delivered—it was the perfection of unstudied eloquence, with feeling galore. The words dropped from her mouth like pearls from a crystal goblet, every one to the purpose, of friendly advice and encouragement to the girls.

Through these years interest in the unsolved problems of the Antarctic was slowly growing, both in Australia and, thanks especially to Sir Clements Markham, at home. From time to time an appeal was made to Hooker's experience. Thus he writes :

To Asa Gray

November 15, 1887.

I am kept busy about the Australian Antarctic Expedition, for which the Colonials have asked our Government to subscribe £5000. The Government has applied to the R.S. to know its opinion as to the scientific results to be obtained, and I have at its request drawn up a Memo. embodying my views, which are, that before any attempt is made to explore beyond the pack, a pioneer ship should circumnavigate the globe between 60° and 70° South and determine the position of the ice in every longitude, for that, in my view, the pack moves in enormous masses, leaving open sea here and there for uncertain periods. I am convinced that if Wilkes had pushed on a few miles to the East he would have

got down to where we did—and that if we had tried a little
to the West of where we did we might have got South without entering the pack at all.   Again in the longitude where
Weddell sailed South to 74° 80′ and returned (owing to late-
ness of season) from a clear open sea *without* seeing even the
pack, Ross in that same longitude met in 63° 15′ a pack of
old ice so heavy that he could not even enter it.   So my
advice is to subscribe to the Colonists sending a ship to
look out for the soft places, previous to sending properly
equipped exploring vessels to do battle with the ice.   The
whole circumnavigation could easily be accomplished in one
season, and one Naturalist to use the tow-net and to dredge
at moderate depths would be able to bottle up a splendid
harvest of pelagic life.   But I do not envy the voyagers.   A
more desolate, boisterous, dangerous sea does not exist,
harassed throughout summer by gales, fogs, and snowstorms.
I find that we were once six weeks without getting an observation of the sun !   I suggest that our Government should
subscribe the £5000 in aid of scientific objects and make it
clearly understood that it accepts no responsibility.

To *Ayerst Hooker*

Oct. 4, 1895.

We went to the Brit. Assoc. Meeting at Ipswich for three
days, the first I had been at since 1882, the occasion being
the inauguration of a special section for Botany, of which
Dyer was President.

The only other thing that interested me was the further-
ance of another Antarctic Expedition, upon which I had to
speak, as the only surviving officer of that of 1839–43 !
I was in fact the 'Ancient Mariner,' and as such introduced
myself, reminding the audience that, like my prototype,
I was 'an old croak.'   I quite expect that we shall get an
Expedition out of Lord Salisbury's Govt.

I was introduced to Flinders Petrie, an interesting man,
who read a queer paper on the undesirability of interference
by civilisation (missionaries, &c.) with many of the habits
of natives.   It was difficult to see where, and at what, he
would draw the line.

Some one also read a paper on the cannibal tribes of the
Congo region, who *habitually* feed on human flesh, and sell

it openly in markets. No one remembered that is a very old story, and that Huxley somewhere (in ' Man's Place in Nature,' I think) gives a copy of a woodcut out of some old book, of the shambles with joints for sale on the table. The result of the practice was the finest race of people in the world, with ne'er an old, halt, blind or sick member of the community ! I shall wait for verification.

In 1896, when he received Mr. Douglas Freshfield's ' Caucasus ' (see p. 382 note), he had just entered his eightieth year ; but his old love of the mountains and mountain botany broke out as unquenchably as ever.

The Camp, Sunningdale : July 31, 1896.

DEAR MR. FRESHFIELD,—Now that I have made progress with your ' Caucasus ' I must express my admiration of the work, both as to matter and style. What I knew before I owe to you, but time had blunted my memory, and to what of it remains these noble volumes add a hundredfold in instruction and pleasure.

The lushness of the vegetation bordering on the Forest zones reminds me of the Himalayas, but I think that the special types such as Umbelliferae and Compositae that abound in both mountain systems attain a greater stature in the Caucasus. On the other hand the Alpine flora of the Himalaya in variety of type I think beats both the Caucasus and the European Alps. I should like to institute the comparison by a visit to the scenes of your ' head and heels ' labors ; but what can an Octogenarian do ? You may yet set foot in the Himalayas, and I hope you will. With your powers of grouping big features of rock and forestry a brief sojourn would not fail to bring out points of difference and likeness that would be of high interest and value.

Again thanking you for your truly magnificent gift, Believe me, sincerely yours,

J. D. HOOKER.

Lord Kelvin's jubilee in June 1896 as Professor at Glasgow was one of the exceptional occasions which he felt in duty bound to attend, whatever its fatigues, for he remarked to Mr. La Touche :

I am a Glasgow M.D., my father and Thomson's father were fellow Professors, and I sat in Thomson's father's class (Mathematics) with himself some sixty-five years ago ![1]   He was the youngest and cleverest in the class ; and [we] have been friends ever since.

On June 25, he continues :

(We were) invited to stay with Dr. Story, Professor of Church History, at the College, where we greatly enjoyed our thirty-six hours' visit.   Kelvin's speech was admirable, but amusing in respect of his description of his delightful conference with his Students ; for he is a very bad teacher, and the Students are only puzzled by applying to him after Lecture.   Nevertheless they are devoted to him as a man— he puzzles them with so much kindness, believing that he is making night day to them ! The Exhibition of his inventions was marvellous.

His eightieth birthday was an anniversary happily kept. It was not forgotten by Mrs. Lyell, who had just celebrated hers, and to whom he replied :

July 2, 1897.

MY DEAR MRS. LYELL,—I thank you most sincerely for your kind thoughts of me, your congratulations, and the useful birthday present which you so kindly sent, and which is already installed on my writing table for insertion of notes and observations on my daily work.

I quite enjoy the description of your birthday's jubilee, and am truly glad that you could have so many around you. Indeed mine was a very happy day, too happy I fear to expect a repetition of it.   For I had Willy, Charlie and his wife, Grace, Reggie, Joe, Dick, and Charlie's little boy arrived.

Hyacinth has I think told you of what we saw of the jubilee sights ; of them the illuminated Fleet far surpassed all others in striking character, in fact I never saw but two that beat it for over-powering effects.   One was the view of the glacier-clothed and Berg-imprisoned mountain chain of

---

[1] As he told Mrs. Paisley, December 18, 1907, 'he at the top of prizemen, I at the bottom.'

South Victoria Land, with Mount Erebus blazing in front. The other the first view of the Himalaya, as seen from Darjeeling, covering perhaps 100° of one of the horizons with perpetual snow, with Kinchinjunga, 28,000 feet, towering over all.

Thanks too for your kind words on my G.C.S.I., which took me quite by surprise, and all the more from my not knowing a single member of the India Council (except a bowing acquaintance with Lord G. Hamilton) even by sight! and indeed I only now know it by the *Times*! I had a few weeks previously received the rare honour of an official letter from the Governor-General and Council of India, reinforced by the Secretary of State for India, thanking me for my services to India. This I thought a superabundant reward for anything I had done. As it is, had I ever thought it within reach, there is no honour of [that] kind that I would have coveted more than that so unexpectedly bestowed upon me.

To Mrs. Lyell
September 9, 1897.

We have been very quiet all the summer, making very short visits, to F. Darwin at Cambridge; to the unveiling of the capital Darwin statue at Shrewsbury; to the Farrers for a couple of days; to my sister at Torquay for a week, and to Osborne to receive my decorations from the Queen. They are gorgeous, star and badge ablaze with diamonds, with the consequence that I do not like keeping such property in the house for fear of burglars! We have still a few more visits to pay, and then I hope to be quiet for the winter.

CHAPTER XLV

THE 'LION' LETTERS

HOOKER, it has already been shown, was a notable instance of inherited faculties, both general and specialised. Heredity, so signally illustrated in both sides of his family, was always a subject of vivid interest to him. It tinged his literary criticism, as when returning 'Mary Barton' and 'North and South' to Darwin, he writes (August 17, 1867) :

> The whole of the vraisemblable of the latter falls before the Darwinian Gospel—how could such imbecile parents have such a child as Margaret ?

Long afterwards, when Sir Francis Galton was collecting statistics on the subject, Hooker noted an important omission in his scheme. He writes to W. E. Darwin, July 19, 1904 :

> I have just filled up Galton's table, and am puzzled by his omission of the wife's father. In my case both Henslow and Symonds are worth mention, and had a notable influence on my children's minds. My Reginald got his mathematical ability from the former.

By a curious freak of heredity, however, none of Hooker's children inherited in their turn his strong bent towards natural science and especially botany, although of scientific and indeed botanical pedigree on both sides. Their talents found scope in many directions ; in business work leading to the Civil Service ; in doctoring, engineering, statistics ; in soldiering and colonial administration, but not in pure science. What his father had been to him, that he wished to be

to his own sons, here teaching, there supervising their
school-work; now turning discussion on scientific and
serious subjects, now trying to evoke the scientific spirit by
encouraging collections of plants or insects. But the over-
mastering impulse towards science, which alone makes the
scientific man, was not theirs. The nearest approach to the
desired consummation was in the case of the elder son of
his second marriage (*b.* Dec. 14, 1877), who, as a boy, growing up in close touch with the father now free from official
ties, was moved by a sympathetic ambition to follow in his footsteps and become a botanist. His early education was shaped
with this end in view; however, his ambition gradually faded
as he realised that it was rooted in sympathy rather than the
inborn scientific impulse; and at eighteen he definitely abandoned this in favour of the idea of electrical engineering, and
finally, after reading Science at Cambridge, the Army.

All the letters written to this son from his schooldays on
have fortunately been preserved, bound carefully into two
volumes under the title of 'The Lion Letters.' For in one
of the nursery games he used to play with the child, Hooker,
his beard representing a shaggy mane, enacted the part of a
lion, whence their pet names for one another, the Old Lion
and the Little Lion, regularly used in the letters. But the
'little' lion speedily shot up to a most inappropriate height,
and 'Little Lion' had to be abandoned for 'young lion'
or 'Cub,' the latter finally winning the day. Last touch,
and a charming one—when the 'young lion' married, his
wife was adopted as 'lioness.'

The note of the letters is their sane simplicity, full of the
affection that would keep complete touch between home and
school, while guiding the boy's mental growth by dwelling
on the things which involve observation, co-ordination of
thought, and accurate, attentive concentration—that 'intending of the mind,' as Newton called his own chief faculty—
without which the quickest intelligence is ineffective.

They record the home details which enshrine boyish interests,
but steadily add something that opens a wider vista. The
dog and the pony are not forgotten; but the historical associa-

tions of places visited are recalled and linked with some reading
in Scott's novels or English history. If a journey is described,
let it be followed on the map. The capture of rare insects
for the home collection is noted ; due report is made of the
caterpillars kept in captivity and of the ants in the formicary,
with all the sad history of the accidents to its population and
the end of a queen who seemed unable to survive her subjects.
And later, when the keeping of a rain gauge had been added
to the home interests, special notes on this subject appear.

Natural history plays a large part, for a young botanist
was in the making ; he encourages careful botanical collecting,
identifies and discusses plants which the boy has found, and
is pleased when the names have already been made out.
' Always,' is his advice, ' get the names of the Natural Orders
when you can as it is the greatest point to help to knowing
plants.' Drawing also is made much of. At eleven ' you should
always be trying to draw whatever comes in your way ; by
that means alone can you acquire facility and accuracy.'
And three years later : ' I am glad you are drawing plants ;
be very careful as to the setting on of the leaves and flowers.
The greatest advantage of drawing is that it teaches accuracy—
or ought to do so.'

Unceasing, too, is the desire to know how the boy's own work
is getting on ; what mark he receives ; what subjects he likes
best ; what books he is reading ; whether he is getting on with
his riding and his swimming. A characteristic remark is,
' You should be able by this time (13) to swim fairly well if
you give your mind to it.' By all this he could manage to
visualise the boy's actual life and understand where encouragement could best be given, during those absences when he
' missed him so much ' and would not willingly let a week
go by without sending a letter ' from the old den.'

As with the boy, so with the young officer serving in the
Boer War or in India. The cinema in London and photographs
sent home from South African towns help in the realisation
of his surroundings, but still,

> your account of your daily life and experience interests
> me more than anything else. . . . You cannot think how

anxious we are to know of your surroundings wherever you go—who are your friends, who you talk to, play with, ride with, work with—all such matters even down to your servants, dogs and cats brings you home to us, and all about your Regimental duties instructs and pleases me. . . . Have you time for reading? or educating yourself for the higher grades of your profession?

A secondary object of the correspondence was to train the boy to write a good letter, with something in it. Like almost everything else, this he believed to be in essence a matter of attention—the questions asked broached subjects which in any case it was well for the boy to reflect on, and might suggest others of his own finding. A letter worth the name must be more than the empty screed such as too often boy-nature is content with. One indeed is branded as ' such a very empty one that it did not deserve an answer,' though it did get a long reply ; but the note of praise also is sounded, and once, when the correspondence is enlivened with passages of colloquial Latin, after a holiday study of Ollendorff's hand-book, the ' Cub ' should have had no difficulty in interpreting the ' Dilectissime Scymne—Epistola tua me delectat, quod non jejuna est, sed notitiarum plena.'

Latin perhaps was not an alluring subject, but it must not be neglected, for ' whatever scientific line in life you enter upon, it comes into every examination. . . . If you go in for Botany, you *must* be able to write Latin easily and read it even when difficult, and only practice will enable you to do this.' Indeed he reminds the schoolboy that the minimum in any subject demanded by school hours is not of much worth when it comes to choosing a profession.

With school successes he is pleased ; with school disappointments he sympathises from his own old experience. For these his consolation is not perhaps novel, but is clear and sensible ; they teach patience, and should brace us up never to fall behind the place attained. At least there is the consciousness of doing one's best. ' You do not forget,' he asks once, ' the " talent de bien faire " ? ' He never pretended, as some pretend, that after-success has no relation to a school

career. Its foundation is laid in honest school work, though
success may be in some subject not taught in school.

A corollary to this is that his own teaching of his boys at
home was greatly in these very subjects omitted in school curricula, from Geography and Physiology onwards. In Botany
it had been chiefly by word of mouth and actual collection ;
as he remarks when regular botanical teaching is given the
boy at school :

> I have not bothered you yet with any introductory
> books on Botany, because I think you have plenty to do
> in the holidays in collecting and making up your Herbarium ;
> but without study you will make poor progress in the long
> run.

This ' study ' means the voluntary intending of the mind
and the physical powers with patience and hard work.

As to exercise and its relation to study, he praises swimming,
and considers cricket better than football, as being less violent
and unfitting for other work and mental exercise—not a
mere struggle.

Work and play bring up chance references to his own
early doings. Apropos of a good walk taken at the age of
14½ by Shoreham and Lancing, he writes :

> I hope you will inherit my powers of walking. When
> I was a youth of 20 I thought nothing of 30 miles and have
> done 60 in the day. My brother (who died in Jamaica)
> once did 80 !

Sending ' Lyra Heroica ' as a birthday present in 1891,
he suggests learning as much as possible by heart.

> I have often found it a great pleasure when alone with
> nothing to do, or when I cannot sleep at night, to recite
> to myself the poetry which I learnt when I was a boy.
> I dare say you do not know what sleepless nights are, but
> your turn will come !

And in 1895, when the boy was preparing for Little Go with
their old friend Mr. La Touche, a very uneven set of mathematical papers, where some of the work was good, some care-

less, was criticised as showing the need to go over and over
it all in his mind.

> I used to do this in bed. I liked it—it was a pleasure
> to me to go over a proposition in my mind. I spent hours
> of sleepless nights at sea going over some of the propositions
> and theorems of the 1st book. I was so fond of them—
> but then I always loved Mathematics—and what I did as
> a pleasure, you must do as a duty, if, as I fear, you have
> no love for the subject, and yet I was no Mathematical
> genius—quite the contrary. I do not think I have any
> more *aptitude* for the study than you seem to have—if so
> much.

To ' apply the mind ' was his one road to learning. It
can never be done too thoroughly, especially in taking up a
subject of which a little is known already. Thus he had
lightly taught his son a good deal of geology ; but is urgent
that at Cambridge

> you should take the Geology Lecture just as if you knew
> nothing about the subject, for after all what you did with
> me was superficial and you want the subject to be thoroughly
> impressed on your mind. It is a great mistake to begin
> thinking that ' you knew all this before.'

Application in acquiring knowledge, also, must be reinforced by practice in expressing it. As the Tripos approaches,
trial papers, especially in chemistry, must be worked through
under examination conditions.

> You must yourself have found out how different it is
> to know a thing, from being examined on it ; when alone
> you discover firstly how little you really know, and secondly
> how hard it was to put on paper what you did know—
> *hard practice* alone can overcome this.

The hardships of the Boer War, of which the young lieutenant in the Hampshires had ample share during the pursuit of
De Wet, suggest a comparison with the cruise in the Antarctic.

> Your food seems excellent, much better than I had
> at your age in the *Erebus*. We had no milk, bread, eggs,

jam, coffee or tea except what we bought for ourselves, and, except in harbour, only salt beef and pork, both as hard as wood, and salter than the sea—except twice a week a very little tinned meat. We had rum and Lime juice daily, and sometimes cranberries, made into a sort of doughy Swiss [?] tarts which we called cram-bellies. Yet we were always in perfect health.

The self-education of an officer should never be done. A reminder comes with the death of the gallant Lord Airlie.

His uncle, Lord Redesdale, says of him, 'Poor Airlie was one of the noblest of men, so chivalrous, brave and simple, a real Paladin. I loved him sincerely and we all are utterly miserable about it.'

I dare say that you have heard that Lord Airlie never shot or fished, though he had moors and rivers for his friends, and used to spend his leaves in reading and corresponding with a German officer friend on military problems. As you will soon find out, to be a good leader of men requires far more knowledge and thought than any education can give you. Manoeuvres are matters not only of skill and judgment, but of forethought, to be gained by reading and making plans of your own, and by getting others to set you military problems, and studying them together. Above all I would urge you to cultivate playing chess, and not to be the least discouraged by finding yourself perhaps an indifferent player—for it is the practice that chess gives that is so useful, and especially the habit of thinking what your adversary's plans are, as well as your own. This and Military History should employ your leisure hours, which too many spend in cards and billiards.

During active service he writes :

I wonder whether you keep a Journal—it would be wise to do so for your own improvement, and it would be so interesting for you in after life as recording your early military experiences.

At the end of the war, the young soldier, who desired anything but inactivity, obtained a transfer to the Indian

Army. The exchange was made easy by a word from Lord Roberts, who was a personal friend of Sir Joseph's. Soon he passed his Staff examinations and entered the Pioneers. Nothing could delight Hooker more than that his son should serve in the country with which he was himself so closely connected.

Your first object must be the interest of the Indian Government which you have undertaken to serve. Keep that always before you and you will never regret even a failure.

He threw himself, if it were possible, even more eagerly into his son's every interest. Hindustani must be learnt.

I well remember how helpless I found myself for the first year in India, and how little I understood of my surroundings, and especially of the Government of the country, and the position and duties of the political and civil officials, from Baboos up to Judges, into intercourse with whom I was thrown when travelling up the country, and down again, before going to Darjeeling.

But of course there was a great change since the day of John Company.

The Hookah was then quite in use at the dinner table— the meals were long and heavy, much wine was drunk, and people toasted one another in tumblers of beer! All this is, I hear, a thing of the past, but there is too much tippling of ' whisky and soda ' at all hours of the day.

But an abstract word of warning is not enough. Practical wisdom must suggest some feasible way to avoid being drawn into a habit which is doing great mischief. This comes in the experience of an old friend lately home from India.

He says that it requires great *firmness*, with a pleasant manner, to resist the invitations to join a friend in whisky and soda. He always did so, putting it off by having a pipe or cigar, or the soda without the whisky. Worst of all is the common habit of taking one before going to bed.

The enchanted land opened itself again to him through the eyes of his son. He has seen Parasnath.

> It was my first mountain climb in India, and I have still hanging in the hall the little marble head of the Jain idol which I found (with the Leper) in the ruined temple at the top.

He is yet more delighted that the lad has been to Darjiling and has found the ' Journal ' helpful ; and would have liked, had it been possible, to send him the pass for the railway— ' a little gold kukri that hangs to my watch chain—given by the constructor because of some information I gave him about 10 years ago.'

India is inexhaustible. ' It makes me giddy to think how much you have to learn of that wonderful country—its History, Peoples and productions, not to say also its languages.' To take an interest in all these things becomes a University man.

From old experience he can well understand the difficulty of grappling with Hindustani for entrance to the Staff and the ' appalling ' stiffness of the examinations, but heartens the young officer to grind away with his malodorous Moonshi by the reflection that unless they put stiff questions, the examiners have no means of picking out the men of highest ability. When one examination was passed after another, Hooker is delighted with his ' good head for lingos ' ; and bids him try for Persian next. To be a smart officer is not enough—' I do so want to see you taking the position of an exceptionally instructed officer.'

In this connexion he asks whether any of his brother officers are reading men.

> Is it true what all the people are now saying, that young officers never read and think it ' bad form ' to study, or talk on professional subjects ?

One of these brother officers is described as ' a quiet young man.'

> I hope [returns Hooker] that you make up to the ' quiet young man '—ten to one you will find something in him.

The months go by, but 'still I have 2000 questions to ask you,' one of which is:

How do your men compare with the Hampshires in conduct and bearing ?   It appeared to me that the soldiers of the Indian army were of a different type from ' Tommy Atkins,' more proud of their service, with more self-respect and better manners.

And later :

I shall be glad to hear how you get on with your men. Natives are just what Europeans make them.

Later again (1906) the point is put rather differently :

Native soldiers in India always look so dignified, as if they were models of self-respect, though individually they may be ' badmashes ' of a pronounced type.

Hooker's own philosophy of life comes out strongly in discussing the 'dulness' of a military routine career. His son was eager for greater activity ; meantime the alternative counsel was to make regimental life interesting.

This depends more on yourself than on your surroundings ; an active mind by seeking finds interests and occupations where a common mind sees nothing but boredom.   Please remember that your dullest duties can be made interesting to me.

And he believes that in most cases the dulnesses of regimental life are due to lack of intelligence or education in English boys and officers.

To this may be added another saying apropos of the need of frequent manœuvres.   True that generals thus learn their duties in war ; but much more

the best of all schooling for those beneath them, down to subalterns, is to witness their blunderings and mistakes on the field. . . . Wellington said he had any number of Generals who could get 10,000 men into Hyde Park, but not half a dozen who could get them out again without blundering and confusion! Whatever you do, do not encourage the feeling that nothing is to be learned from witnessing the evolutions of troops, however badly conducted. The wisest men reap their wisdom from their experience of the foolishness of others.

CHAPTER XLVI

No sooner was the Flora of British India finished than the Indian Government requested Hooker to undertake the completion of the Handbook to the Ceylon Flora, interrupted by the death of H. Trimen after the issue of three volumes. It would be, he thought, a relatively easy task coming immediately after the Indian Flora. It supplied occupation for nearly three years, Part IV appearing in 1898 and Part V in 1900.

From the long labour of the Flora of British India sprang two other works of substance. He was well aware of the imperfection of the material upon which he had worked. Some of the more difficult groups were still full of confusion, with genera and species not accurately compared and delimited by a careful monographer. One of the most perplexing was the group of the Balsams (*Impatiens*), and to this he devoted the chief part of his remaining years. Not only had new Balsams been discovered in the score of years that had passed since then, but he was not satisfied with the results he had obtained from the Herbarium specimens upon which he had worked for the Flora of British India. The species were very often only to be distinguished by very delicate differences in the shape of the flowers and the relations of their parts, which were too often masked or crushed in the dried specimens by the careless treatment of the native collectors. All his inexhaustible patience was needed in the tedious work of soaking the crumpled specimens from the paper to which they were heavily glued, and getting them into shape for drawing and examination

under the microscope. Even then they were not satisfactory.
But his friends in India came to the rescue, sending whole
herbarium sets from Madras, Calcutta, and Saharunpore.
They made drawings for him from the living plants, and Mr.
Duthie in particular made fresh collections for him in the
Western Himalayas, sending over new specimens most carefully
gathered and preserved by an admirable collector, Inayat
Khan, after whom Hooker gratefully named a new species.

He began re-examining the Balsams in 1898. By October
1899 he set steadily to work after the Ceylon Flora was off
his hands. In one form or another this continued to be his
principal occupation to the end of his life. In the main it
took shape in an Epitome of the British Indian species of
Impatiens, which occupied fifty-eight pages in the ' Botanical
Survey of India, 1904–6,' vol. iv. (1906).

Subsidiary to this was the identification and naming of
the specimens sent to him, and these were many, for gradually
all collections of Impatiens gravitated towards him, as the
one man who had anything like full knowledge of the 'Genus.
More than that, he published in 1904 a description of those
in the Wallichian collection at the Linnean Society, and in
1908 of those in the Herbarium of the Paris Museum.

He was then ninety-one, but continued to work in this rich
and inexhaustible field.   India completed, his botanical impetus
carried him on to the Balsams of all South-Eastern and Eastern
Asia. Thus in each of the three succeeding years he published
twenty-five drawings of curious Asiatic species in the Icones
Plantarum ; each year in the Kew Bulletin or the Botanical
Magazine he dealt with new species from Malaya or China or
the western Indian Peninsula, or reviewed the distribution
of the Order in particular localities such as the Philippines or
Chitral, while in the last year of his life also appeared his contribution of the *Balsamineae* to Lecomte's ' Flore générale
de l'Indo-Chine.'

These long and minute researches, however, did not use
up all his octogenarian energies. In the summer of 1901, being
now eighty-four, he was requested by the Indian Government
to draw up for the Imperial Gazetteer of India a succinct

survey of Indian Botany, a task which meant working anew over the exact distribution of the whole Flora and condensing the result into the compass of fifty-five pages. For this intricate piece of analysis and condensation he was at first allotted a bare three months, but exacted six ; when at last it was ready, he was told that publication must be held over : a serious thing, he complained, at his age. But happily it was not left for another hand to pass through the press. Though the Gazetteer was not published till 1907, a disheartening delay which made him regard his Essay 'with its corrigenda as quite behind time and hardly worth printing,' an advance issue of the ' Sketch of the Flora of British India' appeared in 1904, and he was able in the interval to work in various touches as they occurred to him.

More personal in the gratification it gave him was the record of his father's life and work which he published in the *Annals of Botany* in 1902, occupying more than 200 pages of the journal.

It was therefore on a very different scale from the sketch of Bentham's life-work in the *Annals of Botany* (December 1898, vol. xii. No. xlviii.), where he could not add much to what he contributed to *Nature* shortly after Bentham's death.

The memoir of his father was his principal occupation from March to September, though he had hoped to finish it for the April number. He finds it 'as laborious as it is a grateful work,' for biographical writing did not come easily to him. ' I wish I had your facility for writing biography ! ' he exclaims to Mrs. Lyell, mindful of her memoir of her father, Leonard Horner ; and here he was met by the further difficulty that his father had kept no diary or journal, and the threads of many matters had to be laboriously unravelled from his ' vast and voluminous correspondence at Kew, about 80 volumes,' some 27,000 letters.

To Inglis Palgrave

April 26, 1900.

I have just done glancing over my father's letters to Dawson Turner, 1805–1851. Except in as far as they throw

light on one's family affairs, they are very disappointing.
He seems to have confided everything to D. T. His affairs
and difficulties with publishers, booksellers, printers, European and colonists, are minutely detailed and advice asked,
but there is nothing in this that could interest the public.

His Scotch expeditions would have been interesting
had he described them or kept a diary, but you cannot even
trace his routes ; of one journey as far as the Hebrides,
before 1805, there is only an incidental allusion ! He must
have kept a journal of one of the long journeys, probably
with Borrer, for he consults D. T. about publishing it. Of
his several Continental journeys there are no routes or particulars, except in one letter to your mother, and no mention
even of the Botanists he saw, or Museums or Herbaria.

On his return from Iceland, several years before his
marriage, he was put by D. T. and Mr. Paget into the Halesworth Brewery to have the business under Mr. James Turner,
D. T.'s brother, who went stark mad.

My father seems to have spent most or all of his time
on his own botanical works (which cost him large sums, as
did purchase of books) and in trips to London. The Brewery
went down apace, and was sold at a great loss. Meanwhile
a large portion of his fortune, originally large (he sold the
land alone for £20,000) was put into the Spanish funds, and
realised in the long run from £200 in all.

The account of Mr. J. Turner's madness is gruesome
reading. My father was persuaded to take him to sea for
a month, with Mrs. T. and several gentlemen ; they hired
a pilot boat and the conduct of the lunatic was disgusting
to a degree. The weather being stormy they abandoned
the vessel and visited all the towns in Holland and some in
Belgium, dragging the wretched creature with them ! On
their return he was put into an asylum, which he never left I
believe.

Ruin was evidently staring my father in the face, when
he obtained the Glasgow Professorship ; for I should say
he had before his marriage been living on his capital. At
Glasgow he began to save, as much by his publications as
by the rapidly increasing classes, and improved Exam. fees
and Regius grants, but he was always seeking for a position
in London, and the letters are full to satiety of disappointed

struggles, hopes and castles in the air. Then came family sorrows, with the appointment to Kew.

Latterly his untiring efforts to get me on with the Admiralty, India Office, Treasury, Woods and Forests, and a host of Ministers, is quite bewildering. He thought far more of me than of himself.

These are mere hasty gleanings from 728 letters, which to read carefully would take a good many months. To typewrite them would cost a vast deal, and so many of the letters have no interest, or deal with better forgotten subjects, that they could not be intrusted out of the family. There is nothing before 1809 or after 1851.

The devotion to D. T. is quite touching throughout, and D. T. was very useful to him in many ways. There are large breaks in the correspondence and I suspect he gave many letters that he should have kept to D. T., especially from Ministers, about Kew and Glasgow College, and myself.

The letter to your mother is being copied, it could not be removed without breaking up the whole huge volume, the binding of which is like iron.

I shall look over my mother's letters to her father. Unfortunately she left orders that all my father's letters to her, indeed all her letters, were to be burnt.

Yours ever affectionately,
J. D. HOOKER.

To Mr. Duthie

The labour of picking the materials from his huge correspondence is very tedious. He was three years literally bombarding the Government Officials before he captured the Gardens—not till after the collections had been offered first to the Horticultural Society, then to the Regent's Park Gardens and been refused by both, upon which the Lord Steward ordered the Cape and New Holland collections to be destroyed ! and their houses to be occupied by vines. They had a very narrow escape.

To Inglis Palgrave
September 25, 1902.

The raking up of and co-ordinating his work has been a very heavy job. He tried to do far too much without

clerical aid, except of my mother. Thirty years of purely
editorial work without a Co-editor of any kind, was a hopeless task, and his ' Journals of Botany,' 31 vols., with useless
indexes, have been the despair of botanists. The amount
of curious, novel and instructive matter that they contain
is marvellous, and I have long appendices classifying their
contents as best I could.

These appendices were specially laborious, with their
chronological catalogues of all Sir William's writings great
and small ; the classification of the more important articles
and reviews in his various Journals of Botany, arranged under
the names of the countries involved, of the authors, of special
groups of plants or general works ; lists of economic plants
described or discussed ; of obituary notices, or articles physiological, morphological, and anatomical ; of Sir William's
principal botanical correspondents with the number of letters
from each.   But his affectionate admiration of his father was
built deep into the very foundations of his life, and it was
for him a labour of love as well as justice to raise a worthy
monument to his memory. To the last his pleasure was
always aroused by appreciation of his father's work ; witness
his letter to Sir Edward Fry in 1910 when he adopted the
plates drawn by Sir William for his own work on British Mosses.
(See p. 473 *seq.*)

This was not the only point where the cycle of time swung
round and brought up old interests of his again. The
Himalayas were to come before him again through Dr. Douglas
Freshfield's [1] mountain ascents and geological studies ;   Dr.
Arber's paper on Tasmanian fossil trees was to recall his own
first paper sent back from the *Erebus* ; the wonderful discoveries of the Antarctic voyage to be revived by the correspondence with Dr. Bruce and Captain Scott ; the Darwin
friendship to be lit up with a sunset glow by the publication
of the ' More Letters of Charles Darwin,' dedicated to himself,

[1] Dr. Douglas William Freshfield, D.C.L. (*b.* 1845), President of the Alpine
Club 1893–5, has written on his travels and ascents in the Caucasus, the Italian
Alps, and the Himalayas. Formerly Secretary of the Royal Geographical
Society, he has been President since 1914, and received the Gold Medal of the
Society in 1903, the year in which he published *Round Kanchenjanga.*

and the two celebrations, one the fiftieth anniversary of the
Darwin-Wallace communication to the Linnean Society ; the
other the centenary of Charles Darwin's birth.

Such were the main lines of his work.[1] Some details may
be filled in from the unceasing stream of his correspondence.
Thus even in November 1898 he would go three or four days
a week to Kew, being busy at the Botanical Magazine ; but
by way of avoiding the journey in the winter, he would bring
home the ' portable ' and less bulky Orders to dissect and
draw.

While at work on the completion of Trimen's Ceylon Flora,
he was consulted by Mr. Duthie as to the best method of drawing
up and printing local Floras.

A letter to Mr. Duthie of May 3, 1898, contains the first
reference to the work on the Indian Balsams. The Himalayan
Impatiens had ' worried ' him of late. As has been said, a
multitude of these had been discovered since the publication
of the Flora of British India, and some species had become
garden weeds in England. One of these (presumably *sulcata*)
in his own garden, was not even in Kew Herbarium. It differed
subtly from the type he had himself drawn in Sikkim. Specimens borrowed from Mr. Duthie's N.W. Himalayan collection
served to settle other points, but not this. As it came into
flower again in August, he resolved ' to figure it for the Bot.
Mag. whatever it be.' The usphot of long investigation, including the raising of young plants in Lord Redesdale's garden at
Batsford, was that *sulcata* as defined by Wallich was a collection
of extreme varieties of two other species. Such confusion
was an added difficulty in a genus already difficult by reason
of the extraordinary distortion of the parts of the flower.
Determination of a species was as difficult as analysing an
herbarium specimen. And most herbarium specimens were
unsatisfactory. The criticism which follows put Mr. Duthie on
his mettle and brought about a revolution in the mode of
collecting.

[1] Here it may be noted in passing that in 1901 Hooker devised a very
practical form of micrometer for use in botanical dissection, which was afterwards manufactured in Cambridge and then in Edinburgh under the name of
the Kew micrometer.

With regard to your herbarium specimens, the fire is
the best place for many of them, and so it is with all the
collections of the genus hitherto made ; and I would strongly
urge you to begin again, using a portfolio in collecting and
making notes on the spot, and above all getting capsules
and seeds carefully ticketed. If possible, drawings should
be made from fresh specimens.

Prain hopes to have this latter done for the Sikkim
species. He is sending me the whole Calcutta collection !
and I hope after the Ceylon Flora is finished to set to work
on revision of the Indian species of the genus.

Without better materials than exist in herbaria it will
be exceedingly difficult to unravel the Himalayan species.

By mid October, all the available Indian materials of
any importance had arrived ; and he exclaims to Mr. Duthie,
October 21, 1898 :

To tell you the truth, I quail before the task of tackling
them. . . . I think it is clear that my first job will be to sort
them geographically, as the only way of matching the hosts
of specimens which are either flowerless, or have the flowers
too damaged to be recognised. That done, I must get
them into natural groups according to Fl. Brit. India, and
so on.

It was a year, however, before he had finished off the
Ceylon Handbook, down to the indexes of the five volumes,
and next day was able to begin seriously on the Balsams.

Classification was not easy. ' I must confess,' he reports
to Mr. Duthie on October 18, ' that the outlook is far from reassuring and I quail before it.' After being for some weeks
' deep or rather shallow in *Impatiens* ' he found them more
difficult than ever. The capsule offered the best primary
division, but the want of fruiting specimens was the greatest
difficulty, while many points for study could only be followed
on living plants. Some indeed very uncharacteristically appeared not to burst elastically with resilient valves. But
seek as he might for a better co-ordination of the species
than that of the Flora of British India, he could, save in a few
particulars, see no material grounds for forming good groups,

and foresaw with regret that he must in the main stick to the Flora of British India for an intelligible arrangement, though the specimens and drawings that poured in meant ' much to correct and more to add to both species and descriptions ' in the Flora of British India now a quarter of a century old !

In the summer of 1900 he attempted to make use of the lip for classification—the saccate form in contradistinction to the funnel-shaped—at the same time making great use of the sepals and some use of the bracts. But this did not meet the worst difficulty. ' Without the wings I am all at sea, and the attempt to ascertain their forms is heart breaking.' To make out their characters was simply impossible except where each organ was dried separately, as had been done by Mr. Gamble for a good many species. Flowers preserved in alcohol broke at the least touch. The best help came from drawings. After eight months' steady work on the North-Western specimens, interrupted only by a bout of influenza and holidays to recruit, ' the result is akin to despair.' The re-examination of Wallich's Herbarium only proved

> that it was not safe to accept the distributed specimens as if they were the types. The specimens are in a frightful condition ; almost impossible to dissect with any confidence, and yet without dissection nothing can be done. The mixture of species is incredible.

However, Mr. Duthie lent valuable aid, especially in laying out the petals of new specimens and preserving them carefully in spirit, and before long the wings yielded some of their secrets to skilful method and patient handling, and by September 13, 1900, he had come to ' some sort of an ending ' of his long work on the W. Himalayan *Impatiens*. On that day he sent Mr. Duthie ' a crude sketch ' of the result, adding the types.

His practical difficulties are eloquently described in the following :

> Since writing last, I have made the distressing discovery that the wings afford most important characters, which it is impossible to ascertain on dried flowers, or even guess at

until you moisten and dissect; and I need not say that
this is a most laborious process and one destructive of
specimens, owing to the extreme delicacy of the tissues, the
shocking state of the specimens in so many cases ; owing too
to the grievous carelessness with which the specimens are
glued down, it often takes an hour to get out a flower from
under the leaves ! and two or even four to dissect it. . . .
Of course all has to be done under water on the stage of the
microscope.

I must now examine Brit. Mus. specimens, re-examine
Gamble's, and get specimens or notes on De Candolle's
species so as to identify them, for Wallich's individual
names apply to 3 or 4 plants—so awful is the mess in his
herb.

So, too, the Calcutta collection of Sikkim Balsams was

deficient in species, the specimens carelessly collected, very
badly dried, and literally ruined by the glue pot. . . .   The
fact is, that except myself, I doubt if any collector in Sikkim
laid in the specimens [i.e. into drying paper] with his own
hands, and no one has drawn a species except a few by Cathcart's artist, and more by myself.   [Of *Roylei*] though found
by me in abundance more than 50 years ago, there was not
a specimen in Herb. Calcutta !

However, he had been able to write a little earlier, ' Happily
my eyes are as good as ever, and my hand as steady ; patience
ought to be inexhaustible,' and he was not to be disappointed
in his expectations of the material that was collected under
Mr. Duthie's new instructions, though ' I should value even
more some observations, which only a botanist like yourself
could make, on the variation of species on the spot.'

When these arrived in November 1900, Hooker was
enthusiastic.

The specimens are splendid, quite enchanting, and the
floral detached organs all that could be desired.   Mr. Inayat
has indeed done well.

He set to work upon this ' inestimable ' material ' with
rare pleasure,' all day and every day except one a week when
he went to Kew for the Botanical Magazine.

I am mounting them on slips of white paper, after floating them in water, and taking out every fold or crumple. I add a sketch of each organ to the specimen in all cases except the manifest duplicates. I find the petals forming excellent characters, and many specimens have ripe fruits which I had not before seen.

Naturally it was good news that the same collector, Inayat Khan, was to be sent to Kashmir, while [Sir D.] Prain at Calcutta was sending another to Sikkim. The existing Kashmir material was in parts very incomplete ; Hooker therefore determined not to publish until the arrival of the new collections, which moved him to assure Mr. Duthie, ' It is indeed an immense service that you have done for me and I cannot thank you enough.'

As to the need of full monographs such as this to supplement the Flora of British India, he was very emphatic. Mr. Duthie's edition of Strachey's Kumaon Plants (1906), the result of a single season's collecting, showed eighty plants not in the Flora of British India. Apropos of the threatened suppression of the Saharunpur herbarium and botanist, which seemed specially serious since the question was mooted of giving the Forest Officers more botanical education, he writes (January 24, 1901) :

That N. W. India should be without the means of naming a plant by reference to a good Herbarium, would be a great blow to Indian botany. As to the Fl. Brit. Ind. providing for this, it is absurd. That work is a hurried sweeping up of nearly a century of undigested materials, and is in no sense a Flora like Bentham's Australian. It had to be carried out in a reasonable time, and except myself and Clarke none of my coadjutors was really well up in Indian botany, or authorities, or works, or climate, or geography. It is merely a crude guide to the extent and variety of the native vegetation of India. To have done *Impatiens* as it should have been, would have taken the time occupied by any one of the volumes.

And as he wrote later to Gamble (March 2, 1903) asking for the loan of a collection of Sikkim *Impatiens* :

Every collector seems to have got uniques, and the want of better specimens of many fairly well-known species is lamentable.

I am now writing up detailed descriptions of the E. Nepal and Sikkim species, and still find errors of identification in my last review of them ; and Fl. B. Ind. is past praying for in the matter of errors in detail, from bad identifications, bad specimens and bad examination. I feel as if I were now only beginning to realise the difficulty of my undertaking such a genus at all.

My great object is now to put the species into shape for the use of Botanists in India who would take up the genus ; for as to naming ordinary herbarium specimens by descriptions it is almost impossible ; for this good drawings are indispensable, and some species would take two or three plates to give an idea of its variations.

This pause in the work on the Balsams until the new collections arrived was busily filled up from the summer of 1901 by

an article on the Flora of India that I am preparing at the request of the Governor of India for the new edition of the Imperial Gazetteer. I am restricted to some 20 pages and am cogitating the dealing with the subject : as to the area of the Flora, composition, relation to climate of areas ; relation of whole to Floras of bounding countries, and a brief digest of the characters of each of the Nat. Ord. as regards the genera being temperate or tropical, European, Oriental, Malayan, S. Indian or Himalayan. Those are my ideas. (To Mr. Duthie, July 28, 1901.)

The idea of the sketch, in short, is that the botanical features delimiting areas are best expressed by the dominant Natural Orders.

This ' boiling down the Indian Flora '—' the work of a lifetime in 20 pages '—he found ' desperately hard work,' for he made out at least 16,000 known Phanerogams in India, of which about 4000 occurred in his favourite Sikkim, ' perhaps the richest flora in the world *for its area.*'

It was helpful to discuss the subject with [Sir D.] Prain, then in England on leave, and to read the proofs of his ' admirable introduction to the Bengal Flora.'

Much correspondence also passed between him and Gamble, especially in regard to the Peninsular and Forest botany.

All through November he describes himself as being ' still in the agonies of the little job for the Indian Gazetteer.' It was absolutely necessary to begin by tabulating from the Flora of British India, seven volumes of small print, as well as from later works, the species of all the provinces under which he purposed to discuss the Indian Flora, before he could get a clear comprehension of any, or compare one with another.

In one case at least, when consulting a recent list showing the distribution of the ten chief Natural Orders in India, he suspected it to be founded on his own Flora, and therefore could teach him nothing.

But nature resisted the concise and clear definitions demanded by such a survey. It was easy to recognise (October 25, 1901) the three subordinate geographic floras of the W. Peninsula, viz. Malabar, Coromandel and the central table-lands, ' India vera,' but ' I do not see how to draw their geographic limits with any approach to accuracy.'

Again he finds himself in danger of not observing due proportion in treating provinces of which he knows much or little. There is need for great reductions. Endless questions arise that must be smothered on birth. (January 8, 1902.)

So too the maps and forestry data with which Gamble supplied him were most valuable ; but again he found himself exceeding all bounds in introducing these data.

When one considers that the smallest weed like *Stylidium* has as much right to recognition in a phytogeographic point of view as a tall *Dipterocarp*, the question of balance of weeds against trees becomes delicate. (Jan. 17, 1902.)

Subsequent work in the same field he welcomed with but one regret.

I wish [he writes to Mr. W. A. Talbot [1] in 1906, when

---

[1] William Alexander Talbot, late India Forest Department, joined the service in 1870, became Assistant Conservator 1876, and Conservator 1901, retiring in 1906. Elected F.L.S. 1884. He published a *Systematic List of the Trees, Shrubs, and Woody Climbers of the Bombay Presidency*, 1894, second edition 1902 ; and the *Forest Flora of the Bombay Presidency and Sind*, 1909.

thanking him for his paper on the distribution of the Bombay and Sind Forest Flora] I had seen it before I wrote my sketch. . . . All is put so clearly with the evidence throughout of personal knowledge—whereas my sketch betrays the hand of a mere compiler.

Similarly he was greatly exercised over the proper use of such a term as Indo-Malayan as presenting a botanical type. Taking the whole Flora of tropical India as Indo-Malayan, with the Malayan element varying in extent in the different areas,

we know what Malayan types are—but in what shape do the Indian come in? To what type would you refer *Impatiens*? Really I suppose it is typically Indian, for comparatively few species are found in the Malayan area proper, and the Indian species are almost unexceptionally endemic. You could hardly regard *I. Noli-me-tangere* as a case of an Indian type in England without being laughed at; and yet we unhesitatingly regard *Stylidium uliginosum* as an Australian type in India. It is a case of ‘first catch your type.’ (To Gamble, January 17, 1902.)

To J. S. Gamble

March 2, 1902.

I am beginning to doubt if there is any Indian type Flora pure and simple at all! and that Indo is a geographical, not a botanical expression. Prain’s ‘ India vera ’ comes nearest to a real Indian Flora, but is that not really African? where not Malayan. Of the genera you mention as Indian: Tectona has its Philippine species; *Anogeissus* an African; *Chloroxylon* is monotypic; *Phœnix* abounds from the Canaries to the Cape; *Borassus* is all over tropical Africa; and *Impatiens* is now found to swarm in tropical Africa! We talk glibly of Indo-European, Indo-Chinese, Indo-Malayan, Indo-Arabian, and Indo-Oriental Floras, add Indo-African, and where does India proper come in? The bottom is knocked out of the vocabulary except as a geographical expression.

I think I shall have to formulate a note to this effect for the Gazetteer. Of course it is conceivable that India is the mother country of all the Indos, and I rather cotton

to the idea. ' India vera ' of Prain reappears in Burma
I believe.

On January 23, 1902, he tells Mr. Duthie :

I have finished the Gazetteer work, after a fashion that
I fear will satisfy nobody. The material I had would fill
a volume of 1000 pages—20 allowed me ; I have extended
it to 40, which will, I expect, be met by an order to cut
down.

As to publication, When ? is the question. They gave
me three months to finish it, I took six and now they tell
me that ' some considerable time may elapse before it is
put into print.' At my age this is serious. (To F. Darwin,
February 1, 1902.)

Impressed by the scientific breadth of the scheme, his
friends urged him to enlarge, not to condense his sketch.
He objected that an enlargement would not be suitable for the
Gazetteer, which was not intended for botanists, but for the
intelligent general reader. The elaborate work they desired
would have involved labour in collating the Indian genera and
species with the Chinese, Malayan, and African, and of working
out all the habitats of Mr. Duthie's Himalayan specimens in
the herbarium.

When you see the Gazetteer article, you will be amused
with yourselves for supposing that its author was capable
of writing such an account of the Indian Flora as you would
like to see.

Moreover, even if he were able, he did not think the time
had come for it. With so many regions from Nepal to Burma,
from Orissa to Bombay, still botanically unexplored, to add
to what we know of the known provinces would not throw
more light on the broad features of Indian phyto-geography.
On the other hand, a fuller geographical treatment of the
forests, he tells Mr. Gamble, would be most fruitful and useful—' but I am not up to it. Only a man like yourself could
grapple that.'

To J. S. Gamble

March 24, 1902.

What is now most wanted in the way of clearing the ground for a better dealing with the provinces of India, is a list of the genera, and another of the species common to all tropical and sub-tropical Indian Provinces, and to two or more of them, and so forth. This done we should have the characteristics of each province in high relief. As matters stand, it appears to me that the making my sketch fuller would consist mainly in piling on each province the names of species common to most or all of them.

I am glad that we agree as to the use of Indo.

Your speculations on India as the alma mater of the old world vegetation are very enticing. I have long dreamed over the condition of India before the elevation of the Himalaya, and always been brought up abruptly by the question, ' What was the nature of the vegetation of the area now occupied by the Himal., and which has been ousted by the elevation of the latter ? ' As also by the indisputable fact, that the Himalaya powerfully affects the climate of all India. The older you make the Deccan, the more dissimilar must its climate and vegetation be to those now existing. Elevations of sea bottom may help to the understanding of migration, but not the nature of by-gone vegetations.

You assume that the Himalaya was under water when the Deccan was still dry land, but the range must have taken thousands of years to have reached its present height, and if the European and Arctic Floras were developed originally on the range, they must have had ancestors from somewhere else.

Then again, what was the vegetation during the period of the gigantic Mammals now hoisted up 12–14,000 feet by the Himalaya ?

To the Same

April 21, 1902.

The Introductory Essay was all very well for half a century ago, when all was darkness as far as a knowledge of the Indian Flora *as a whole* was concerned, but I do not see that subsequent collections have furnished material for any very great advance on it. Had the Himalaya East of Sikkim, or Nepal or Burma, or the N. Hindustan been explored since that period ; or had the West Himalaya been analysed, or had any good local Flora been brought out in the interval, there would be more or less promiscuous feeding for digestion ; but except by the Forest Department, I fail to see any great light thrown on the Flora of all India, since the said Essay was published.

Were it otherwise, how can you think it possible that a man in his 86th year (next June), living 20 miles from Kew Library and Herbarium, could face the task ?

I agree with all you say of the ignorance, supineness, obstinacy, wrong-headedness, parsimony and indolence sciencewards of the political powers that be ; but I cannot think that in the case of India, the best use has been made of the money and opportunities that have been granted for botanical research, but that is a long story, better talked over than written about.

As to an alternative issue in some Indian scientific publication should it prove too long for the Gazetteer, he desired this to be one which would enable it to be circulated among Forest Officers ' who could indicate its errors and supply the hideous lacunæ it shows,' a procedure which produced but meagre results.   He writes on January 28, 1903, to Gamble :

I must be an ' Old Man of the Sea ' to you.   I have told King that my sending the article to the Calcutta Annals is out of the question.   He and Prain inordinately over-estimate it, never having seen it !   The Annals are rightly intended for plates.   The article would look ridiculous in Imp. Quarto, thick paper, broad margin and big type.   It should appear in most modest form (I had myself thought of the Records) for wide circulation in India, calling attention to the lamentable want of material for bettering such a lame goose.   If it has some good and new points of use and interest, that is the best that could be said for it.

In revising the Sketch during the autumn of 1903, he received much help from both King and Gamble in delimiting and individualising the sub-provinces into which he proposed to divide Burma.

To the latter he writes (November 11, 1903) :

> I am glad of your letter for I have wanted to let you know
> how grateful I am for what King tells me you have done
> towards correcting the sketch.   I have adopted all your
> suggestions, but King tells me that I had better go over some
> passages with you in respect to making them clearer to
> Forest Officers.   The oftener I re-read my Essay the less
> I like it, but I do not see what more I can do.   The fact is,
> that except at Calcutta the Botanists of India have been
> asleep since the days of Wight, Beddome, Law, Stocks,
> Dalzell and a few others.   The men of Bombay and Madras,
> ' Professors of Botany ' though they be, have done nothing
> at all.

As he wrote on receiving Mr. Bolus's ' Sketch of the Floral
Regions of S. Africa ' (October 15, 1905) :

> Your sketch is all as clear as a bell, and I cannot help
> comparing it ' longo post intervallo ' with my sketch of the
> Flora of British India, much to the disadvantage of the latter.
> My excuse is that we have no one area in India so well botani-
> cally explored as *any* of your Regions.   Vast areas indeed
> are actually unexplored, though everywhere accessible, and
> many of them peopled by Forest Officers, all of whom have
> had some botanical teaching.

> The Kashmir collections arrived before he was quite ready
> to turn to them.   He had to restrain his eagerness.   But he
> writes to Mr. Duthie, August 20, 1902 :

> At last I have got so far through my sketch of my father's
> life, that I dared to look at Inayat's Kashmir *Impatiens*.
> I make 8 species out of them, after a very careful analysis.
> The specimens are splendid, and the separated organs in-
> valuable. . . .   It is quite a pleasure to draw the sepals,
> standard, wings, lip, from the detached specimens.

By the middle of February 1903 he was sufficiently free
of other work to return to the Balsams.   ' I must now go to
Impatiens,' he tells Gamble (February 9, 1903), ' which really
terrifies me.   I cannot get good groups, and they keep me

awake.' If the first systematic difficulties had been overcome, he found that to complete his descriptions he had to analyse most of the species afresh, and even then the descriptions were ' vague and loose,' for ' every organ is variable except such as afford no character at all : it is like making species out of the waves of the sea. Still the species are stable enough, however loose on their pins ' ; and he plodded on, in his pet phrase ' groaning over *Impatiens* ' at the rate of about two species a day.

New specimens continually arrive, each with its own problem to raise or to solve. One beautifully preserved set from Gamble displays characters that had been destroyed in the more roughly handled herbarium specimens, and ' clear up a great long standing puzzle.' ' They settle *I. Gardneriana* definitely ! but what a queer beast it is ! the total dissimilarity of several states is very striking.' (November 25, 1903.)

Of one curious *Impatiens* indeed (*I. tingens*), he had written to Sir F. Darwin (June 11, 1903) that it ' differs so greatly from all its congeners that it may be called a case of evolution *per saltum.*'

Thanks to another set he makes out in the following June that the *I. scabrida* of the F.B.I. should be broken up into three species, among which he owed the determination of De Candolle's original *scabrida* to a photograph of the specimen in the De Candolle herbarium. ' I cannot tell you,' he exclaims, ' the worry these three have cost me ; due to the bad specimens, the confusion in Herb. Wallich, and my own *carelessness* or stupidity, or both.'

Long after, Gamble sent him for identification a Balsam which appeared to be *I. scabrida*. He placed it as one very like this, which had already figured under six names, adding (September 28, 1909) :

As to your finding that yours does not properly fit the description of either, that is the normal condition of every described Balsam. I am now revising my original clavis of the Chinese species with the help (hindrance I should say) of additional specimens and duplicated analyses, and the operation is most disheartening.

November 19, 1904:

I am having an awful time over the varieties of *I. Balsamina*, which are legion and strangely diverse—would you like me to collate yours with Kew Herbarium? . . . I have no faith in a single character of any one form; most of these being taken from single specimens. As, however, I indicate the locality of each form, future collectors may be led to investigate them in their homes.

I have written asking Talbot for the loan of his Balsams (of the Deccan), a suicidal proceeding, for if he has collected *I. Balsamina* with care, his specimens will be sure to upset my beloved varieties.

His prophecy as to the result of examining a whole new collection was duly fulfilled. After dissecting every one of the sixty and more species from the Madras Herbarium, he writes: ' In every case I have to add to or subtract from the previous description.'

Then arrived another little collection made by Inayat Khan in the Himalayas, including apparently three new species. Patient analysis showed these to be but one, which he proposed to name after Mr. Duthie himself; but with the usual perverseness of the genus, it did not fit into his sectional divisions.

And being tripped up by what seemed to be an entirely new species from the Peninsula, but which turned out to be a variety he had already described, he exclaims to Gamble: ' The fact is that Balsams are " deceitful above all plants and desperately wicked," and I am no match for them.'

I am quite in despair [he tells Mr. Duthie in January 1905] and should like to show you my attempts. The Eastern and Western Himal. species require a different treatment and the Malabar ones a totally different kind.

In short, ' the attempt to name Balsams by comparison of herbarium specimens is folly, except in a very few local cases.' (March 24, 1905.)

The Epitome of Indian Balsams, finished in 1905, was printed in the Records of the Botanical Survey of India in 1906. It was of only temporary finality. He continued 'constantly at work on the inexhaustible Balsams of India,' for the very thoroughness with which he had stimulated research into the genus everywhere, continually brought in more plants, and revealed the need of additions and corrections, especially in the W. Himalayan section. Species belonging to this, which were in no other herbarium, unexpectedly turned up in European collections, especially in the Vienna herbarium, and by 1909 he had detailed descriptions of some thirty British Indian species. To complete this section was 'almost an impossibility.' Kew did not possess half the species ; the district was botanically unexplored, because, excepting Inayat Khan, the collectors had been uninstructed men. Many species, also, had been founded on single specimens, perhaps found in one spot only by a single collector ; some in turn were finally reduced to varieties of the most diffused species, that 'terror' to botanists, which was named after Royle. Were he but younger, he declares at ninety-one, he would go and pick out species at Saharunpur.

Still, when he has to reject other workers' identifications among the Balsams, as in Mr. Duthie's paper in the Records on Chitral plants, he confesses :

I take to myself the blame, for you had nothing but the Fl. Brit. Ind. to refer to, and that is utterly unsatisfactory, full of imperfections and errors. In fact, it was not till after the publication of Vol. I. of that work that I essayed a critical study of the Indian species by moistening and analysing every specimen where there could be any doubt. The consequent labour has been trying, for within my experience no genus of Phanerogams approaches *Impatiens* in difficulty of analysis, description and classification of species. Except by geographical areas it is impossible to bring the species under control ; any attempt to bring all under one classification as in Fl. Brit. Ind. ends in chaos. (November 10, 1909.)

Having done so much since he was eighty, his one regret was that he could not complete everything down to the last detail. 'If I were 10 years younger,' he adds on the 29th, 'I would offer to re-name the whole Impatiens Herbarium of N. India.'

Moreover, the further he proceeded in this favourite field, whether surveying Indian botany as a whole or monographing one corner of it, the more he was disappointed by the slow progress that seemed to have been made since the work of the pioneer botanists and his own first effort to co-ordinate this. Brilliant exceptions there were, and Calcutta especially had upheld its old reputation,[1] but 1907 and 1908 find him lamenting that botany is almost dead in India. To the fellow labourers in his uphill task he deplored time after time the general lack of enterprise that left large areas of India still botanically unexplored, the futility of entrusting collection to the 'mere haymakers whom you magnify into botanical collectors,' the imperative need of taking a given area and working out the varieties and distribution of important species on the spot.

It is really too bad that the few Palms of the most populous and accessible parts of India should be botanically in confusion. One would suppose India to be an inaccessible country. (1901.)

The unexpected appearance of no less than five new Balsams in a supplementary collection from the north made in 1905 shows

how carelessly the country had been herborised, I cannot say *botanised*, for really the mere collecting without notes of any kind, even of colour of flower, is not botanising. (February 1906.)

To Captain Gage [2] he writes on March 8, 1909:

[1] 'Beccari's *Calami* will be a magnificent tribute to the energy of Indian botanists in having such diabolical plants collected in a condition for accurate description.

'The Annals of Calcutta Garden are magnificent; they are an imperishable record of the energy of King in starting the series.' (To Capt. Gage, April 17, 1909.)

[2] Andrew Thomas Gage (1871), M.A., M.B., B.Sc., F.L.S.; Major I.M.S.; Director of the Botanical Survey of India; Superintendent of the Royal Botanical Gardens, Calcutta, since 1906. Educated at the Old Grammar School and University of Aberdeen, he was Assistant Professor of Botany there 1894–6; entered the Indian Medical Service 1897, and was Curator of the Calcutta Herbarium 1898. He has published various papers on botanical subjects.

Except perhaps Sikkim, no part of the Himalaya has been systematically explored as far as Balsams are concerned, nor will they be till European eyes are employed in the Survey, as old Wight's and the Peninsular Missionaries' were nearly a century ago.

So, too, in an earlier letter to Captain Gage (July 12, 1906) :

I do hope that this season will get us some Balsams from unexplored territories, E. Burma especially. It is many years since any new country in India has been botanically opened up, and the contrast between India and China in this respect is deplorable, especially as I am sure that many of the new Chinese plants will be found in the E. Himal. While India is lagging behind in exploring even accessible regions by botanists, as the Katmandu Valley and Tenasserim, or the Shan States, splendid collections are being made in the Philippines—Australia is explored throughout and New Zealand and E. and W. tropical Africa and Rhodesia, and New Guinea beginning. India lags behind or if anything is done at all it is by ignorant natives, who do not give precise habitats, or dates, or even colour of flowers, still less such characters of growth as a botanist does.

Your little exploration of Nimbo in Burma is in one sense an exception, but was it possible to investigate a dry country vegetation in the dry season ?

Surely some effort should be made to obtain the means of redeeming the credit of India which is monstrous low at present. There is no more curious field of research in the world than the passage from the Burmese to the Malay Peninsula Flora, but it must be done by a botanist, not by ignorant natives.

Excuse my growl. I do love Indian Botany. I long to see another Griffith.

'I do love Indian Botany !' This is the keynote alike of his strictures and his corresponding delight in any achievement, such as that of Burkill, whose fruitful enterprise in 1907 dispelled the myth of the inaccessibility of Nepal. This it was that made him urge the co-ordination of effort, to organise the training of collectors, to obtain reports from the Forest Officers as to what botany had been done in their

respective districts, to stir the Government to interest itself
in the work of the botanists, scientific and economic, and the
means at their disposal.

Nor was he satisfied with the botanical publications of
India. Officially published works were indifferently distributed
to botanical workers ; and the writers themselves often did
not reach his own standard of fidelity in reference and thoroughness of re-examination, such as provoked his regretful criticism
to Gamble (June 13, 1904) :

I am writing to —— about that *Phoenix* which he boldly
refers to *P. robusta*, without having seen flower or fruits of
the species he has discovered, or a ghost even of *P. robusta* !
That is the way Botany is done in India.

And as for central organisation, he repeats to Gamble
(January 15, 1907) :

Botany seems to be dead in India ; some reform is needed.
There seems to be no organised scientific force which the
Govt. would respect and listen to, as the Home Govt.
does to the Royal Society—a position which the Bengal
Asiatic Society should take, and which would not allow of
the Lhassan Expedition being sent out without a Botanist,
Geologist, Zoologist and Agriculturist—without a remonstrance which would have reached England and engaged
the R.S. in its favour.

After the Indian Balsams, the African. These he worked
at in the same way from February to May 1905, while awaiting
some of his Epitome proofs from India, ' dissecting and drawing
the flowers of every species, but for which it would be impossible
to match future specimens.' Here were the same difficulties,
the same impossibility of good results from herbarium specimens
unless specially collected and preserved. ' Orchids are child's
play in comparison.'

It amuses me [he continues to Gamble on May 10, 1905] to
find that I only discovered all this at the end of my botanical
career ! The fact is that the genus will be a curse to systematists for many a long year.

Then in 1907–8 he dealt with the Chinese Balsams, from Kew, Calcutta and Berlin, from St. Petersburg, Christiania and Paris. They were overwhelming in number and variety. Of the 130 species of Kew and Paris combined, only thirty were common to both. From St. Petersburg came 350 sheets, without counting those of N. China. While he was correcting the proofs of his article in the Kew Bulletin on the Balsams of Indo-China and the Malay Peninsula, two new species came in from the Fribourg Herbarium, and as a crowning touch, he tells Mr. Duthie (January 28, 1909) :

> Only this morning I get a letter from the Paris Herb. telling me that they regret having overlooked some 40 sheets of Indo-Chinese specimens when sending me the lot I had described ! This is like a stroke of paralysis to a man approaching his 93rd year, but it is no use grumbling, my eyes are as good as ever, and my fingers as agile as ever, and I am indeed thankful.

In February 1909 he was ' still grovelling among Balsams,' finishing his monograph of the Indo-Chinese species for Lecomte's ' Flore Générale de l'Indo-Chine,' which took the longer to complete as individual comparison was necessary with two other collections. If the other specimens had been hard to handle, these had the sad pre-eminence of being by far the most troublesome he ever handled. By the end of this time his detailed descriptions amounted to nearly three hundred species.

It is a small but characteristic point that Hooker offered not only to defray the cost of translating his paper into French, but to provide special drawings up to £10 if desired by the editor of the Archives. Twelve of the new species had been figured in the Icones. Kew also offered to lend the drawings made from the Chinese specimens from the Paris collection to be copied or photographed, over seventy sheets. Hooker strongly recommended the acceptance of this offer,

> for species of Impatiens are inconceivably difficult of determination and identification from herbarium specimens ; and considering that the great majority of the Chinese species are described from single herbarium specimens, my determination of them must be often very faulty.

CHAPTER XLVII

INDIA, which opened out so wide a field of activity for Hooker's old age, brought up a few problems of economic botany during this period.

One was the question of tropical forests. The re-afforestation of parts of England was much discussed in 1909 ; in July he sends Gamble one of the contributions to the debate, remarking :

> It recalls a notion I have long held (always laughed at where expressed) that it is rather to tropical forestry than to temperate that we shall have to look for a check to the timber famine, and that this may seriously affect the hoped for profits from British forests.

The rapid growth of tropical timber, like teak, must tell in the long run, and scientifically arranged transport should make cheap use of the vast waterways of the Amazon basin. But an illuminating statement of the case in reply convinces him that this cannot come to pass for many a long year, when the over-population of the temperate zone will have left no space for forests.

Another was the revival of attempts to introduce fodder plants into the dry alkaline districts of the North-West.

To J. S. Gamble

January 14, 1903.

I am in correspondence with Sir W. Wedderburn on the subject of introducing drought-resisting plants on the Reh and Usar tracts. As Director of Kew I had seeds of various Cape and Australian plants supposed to be suitable sent to India for experimentation in the N.W., but my impression is that the results were disappointing. Of course I have no intention of taking up the subject, but in thinking over it I am tempted to ask whether there may not be other causes for failure than the Alkalis in the soil. On the one hand some of these Australians have taken at once to the Alkaline soils of California. On the other, Australia seems to have itself produced endemic drought-resisting plants by the score on its alkaline soils. Per contra, India neither accepts what Australia produces, and California greedily utilizes, nor does she induce any of her own alkali-loving *Chenopodiaceae* to spread over her alkaline tracts. Hence it appears to me to be probable that other, and possibly more potent conditions than the Alkalis may be found to obstruct the attempts to clothe the Usar with fodder plants, whether native or introduced.

I have put the question to Duthie, suggesting that the incidence of the rains may have something to do with it.

What was of closer interest, however, to the people at home, was the revival of the West Indies especially due to the wise application of botanical science. Hooker had seen the emancipation of the slaves, the rise and fall of the sugar industry, the growing poverty, discontent, and demoralisation. He was deeply conscious of English responsibility towards the Islands, lest the population in our hand should lapse into the condition of Hayti. He had long urged the introduction of other crops, tobacco and coffee, oranges as well as bananas, and though progress was often deplorably slow, he was delighted when far-sighted governors and merchants put the principles of botanical science into practice.

The wave of depression which passed over the Colony in 1897, just when he thought it was well on its feet, caused him the deepest concern.

To Rev. J. D. La Touche

November 14, 1897.

I am interested greatly in this W. Indian sugar question; it is the most serious look out by far that ever occurred in our

Colonial History. I shall be anxious to know how Chamberlain will deal with it. I had so much to do with the vegetable industries of the W. Indies when I was at Kew, that I cannot but feel deeply interested. It was very much due to Kew that Jamaica was rescued from bankruptcy, and has now a surplus revenue, and I see nothing for it but the establishment of small cheap Botanical Gardens, confined to economic plants, in the other colonies, and in the meantime we must make both grants and loans, or we shall have bankruptcy—and that means inability to return the Indian coolies, or even pay their work ; the throwing thousands of blacks out of employ and—civil war in some of the Islands ! I have no faith in Sir W. Norman's (the ablest of the Commissioners !) plan of granting bounties on our Colonial sugar, under the idea that we shall thus compel foreign Governments to reduce or abolish theirs ; granted that it did so— the measure would only be a temporary palliation. It is a curious fact, that never struck me before, that sugar is the easiest and cheapest to produce of all articles of diet (except we include salt as such) and this by the beet, not the cane. Curiously enough this state of things in the W. Indies is *directly* brought about by our two great boasts, the abolition of slavery, and free trade, principally the first. Had we been wise in our method of liberation, and not at the same time hustled the white planter out of the Islands with the mark of Cain on him, the Islands would not have been ruined off hand, and free trade, when it came, would not have hurt them. As it is we must now pay for our two luxuries, and there will be a howl in the Commons.

To W. E. Darwin

March 22, 1898.

Politics are in a muddle ; the W. Indies interest me most. I can see my way a little there, but not elsewhere. We, by a most iniquitous system of slave catching dumped down a population of Blacks in our W. Indian Islands. After netting several millions by the use of them in manufacturing and growing coffee, sugar, &c., we suddenly give them their liberty, paying their *owners*, and them nothing ! Well, the natural consequence is, that the planters bring their ill-gotten gains to England, thus robbing the Islands of both capital and labour wages. We send out a lot of Governors, not one in ten of whom knows anything of tropical products or can promote old or new industries.

We have incurred a gigantic obligation in having to rule, educate, police, and provide sanitary and medical aid for the poor of a huge population for which there is no labour provided!

The fact is, if we do not expend largely and soon, some of the islands will lapse into the state of Hayti and St. Domingo.

To Rev. J. D. La Touche
March 2, 1898.

England has brought this about, and England must pay to set the matter right—not by bounties or duties, but by advancing money, some to clear off large standing loans, some for promoting new industries—the latter chiefly in the way of establishing Gardens for new profitable vegetable products —and some loans to encourage capital.

Jamaica has prospered ever since, under the stimulus of good Governors, she encouraged new industries. Thirty years ago, sugar formed something like ¾ of her wealth, it is now I believe nearer ¼, and instead of a chronic debt she has an annual surplus, and is making roads, railroads, &c., &c., &c.

Sir J. P. Grant and Sir A. Musgrave were the chief instruments, and the abuse they got from the sugar planters was scathing. That this danger of a relapse of some of the Islands into black barbarism is a reality cannot be disputed, it is a terror to the whites of some of them.

Yet great progress had already been made in organising West Indian industries.

To Dr. Fawcett [1]
April 1, 1897.

It astonishes me to read of the extension of your useful work in agriculture, botany and horticulture, since my old

---

[1] William Fawcett (1851), B.Sc. Lond., F.L.S. to 1916. He was assistant in the Botanical Department at the Natural History Museum 1880–86, when he became Director of the Public Gardens and Plantations in Jamaica. He held various positions in the Jamaica Institute, 1887–1907, and retired in 1908. He has contributed several papers to the *Bulletin* of the Botanical Department, Jamaica, and in addition to a Guide to the Gardens, has published on the Flora, Woods and Forests, and Economic Plants of Jamaica.

friend Sir J. P. Grant started the culture of other plants besides the coffee in the Island. He came and spent a couple of days at my house at Kew before he went out, with the view of getting all the information he could that might tend to increase the value of the vegetable resources of the Island.

Except the United States, there is no country in the world where the teaching (practical) of tropical produce is so efficiently taught as in Jamaica, certainly no other British Colony.

He was constantly on the alert to seize some point where better organisation would mean better progress. After reading [Sir D.] Morris's [1] account of West Indian matters in 1902, he notes (July 6):

Under Agricultural Institutions, p. 78, I find no fewer than five Agric. and Hortic. institutions in Jamaica alone, besides Fawcett's Garden Bulletin, and the West Indian Bulletin. Could not these, or some of them at any rate, be combined? As it is, they must entail a lot of waste of time, material and expense of housing and administrative work. The burden of botanical, horticultural and agricultural literature is becoming insupportable—the costs of binding and space occupied in shelves are hideous, for all that one gets 99/100 of the pages are never referred to.

The history of tobacco in the Island is told in a letter to Dr. Fawcett of October 15, 1905.

Thank you much for your letter of the 13th ult. and the box of excellent cigars, which Mr. Arthur Farquharson has been so good as to send me. Please thank him cordially from me, and tell him that I believe I was the 'Deus ex machina' through whom the manufacturing of tobacco into good cigars was introduced into Jamaica.

[1] Sir Daniel Morris, K.C.M.G. (b. 1844). He was educated at Cheltenham, the Royal College of Science, South Kensington, and Trinity College, Dublin. First-class honours Natural Science, Gold Medallist; Assistant Director of Royal Gardens, Ceylon, 1877; Director of the Botanical Department, Jamaica, 1879; Assistant Director of Kew Gardens, 1886-98. He went on special missions to the West Indies, Bahamas, &c., and was instrumental in furthering trade and agriculture in the West Indies, and published many papers on the subject. Adviser in Tropical Agriculture to the Colonial Office, 1908-13. Resides at Boscombe, and is President of the Bournemouth Horticultural and Natural Science Societies.

It was in this wise : when Sir J. P. Grant was appointed Governor, he, being an Indian friend of mine, asked me what he could do for the introduction into the Island of useful vegetable products, and he came and stayed with me for a couple of nights to talk it over. I reminded him that it was a scandal to our Government that with the East and West Indies in our possession we had not a good cigar from either ; that in India (excellent cigars are now made in India) nothing but Manilla cheroots were smoked by Europeans, and in England Havannas. I suggested my getting seeds of various good kinds from Cuba, Manilla, &c. through our Consuls, and histories of their manufactures ; and that he could get some enlightened Jamaica proprietors to commence the cultivation. This we carried out, and in the third year Sir J. P. Grant [1] sent me a sample of tobacco grown in Jamaica from their seeds, which I sent to an expert in London who pronounced it as most promising. I also suggested to Sir John offering prizes for the best cigars. His successor Grey, also an old Indian, would not trouble himself to encourage the experiment, but my old friend Musgrave [2] did, and sent me several boxes of excellent cigars. Afterwards the quality of the article went down, some boxes I ordered arrived full of weevils and were carelessly packed, and I gave up smoking the Jamaicas. Quite lately I have had good accounts, and the box you send bears testimony to them.

It was not enough, as he insisted to Dr. Fawcett in 1897, to be content with raising smokable tobacco and then taking no more trouble. The best flavours must be obtained, by scientific experiments on the various kinds raised in different soils, and the selection of the best seedlings. Here the difficulty lay in the time spent in reaching results after

[1] Sir John Peter Grant, K.C.B. (1807–93), who bore the same name as his father, Chief Justice of Calcutta, was Secretary to the Government of Bengal at the time of Hooker's visit. He played a distinguished part in Indian administration both before and during the Mutiny, and became Lieutenant-Governor of Bengal, leaving India in 1862. In 1866 he succeeded Eyre as Governor of Jamaica, and in the seven years of his administration effected a total reform, both legislative and economic.

[2] Sir Anthony Musgrave (1828–88) was a successful Colonial Administrator, whose career lay in North America, South Africa, the West Indies, and South Australia before he became Governor of Jamaica in January 1877, afterwards proceeding to Queensland. He published *Studies in Political Economy*, 1875.

innumerable failures. And later he asks whether experiments
have been tried in the way of making Navy twist in the form
Jack Tar likes, or óf introducing cheap cigars into regimental
canteens.

In 1897 he asks Dr. Fawcett, 'Has Jamaica as yet sent
oranges to England ? Surely it is time that an orange trade
with this country should be established.' Indeed, he finds
it most difficult to get an eatable orange in any shape. Jaffas
and Floridas are disappointing ; there was no sign of a delicious
orange, as big as two fists, which he had met with at the Cape
Verds in 1839. 'In my young days it was very different
when the St. Michael's came in shiploads.' By 1900 : 'At
last we are getting really good oranges (Californian ?) at the
Stores at a moderate price. The W. Indies must look to
their laurels if they do not soon begin to supply our markets.'
By Christmas 1901 he is glad to see that the imports of bananas
have doubled and oranges have largely increased, though none
come up to the Californians. Somehow also the Jamaica
bananas are not favourites ; and his own experience pronounces
them inferior to those from the Canaries. It is the same story
in 1902 ; the oranges are too full of pips and core ; they are
poor in appearance and not fit to be seen on the table. Some
scheme is needed for grading them according to quality. But
even then the packing sometimes was not careful. Later in
the year, after trying a most unsatisfactory chestful, Hooker,
who left no stone unturned to improve matters, wrote to one
of the chief importing firms ; their reply was that although
they had already written repeatedly, it had been to no effect—
they are afraid the West Indians never seem to learn
wisdom.

Hooker sent this answer with a sharp comment to [Sir D.]
Morris, whose wife at the time was in England, and had just
visited The Camp.

Really and truly the Jamaicans do not deserve the sacrifice England is making in respect of its fruit trade. They
want a rousing up—a good stinging *series* of articles in a
good Jamaica Newspaper or Journal—they have had too
much pity. A smart scold and a showing up of their idle-

ness and folly might induce other Islands to do better. On the table to-day, there were even worse specimens than what Mrs. Morris saw, trash in fact, as the best have 28 seeds! I wish I could continue to patronise Jamaica oranges, but it is impossible.

But he could not throw up a cause he had championed so long, and when staying at Bexhill in the following January, he ' kept his eye on the fruiterers *in re* oranges,' and learnt that these now arrived in good condition, wrapped in papers, but not graded. The grading was done by the fruiterers, who declared the best grade to be usually excellent save for their colour. A fresh supply coming in the day before he left, Hooker sampled a dozen of the best, and ' at the risk of being thought a bore,' reported them as larger and better coloured than before, seedless and thin-skinned, but disappointing in flavour and substance.

The pity is, that in buying a Jamaica orange, you do not know what you will get—and the Jamaica growers should, I think, have a governing syndicate of experts to direct their efforts.

He paid no less attention to detail with regard to other W. Indian products, all of which he faithfully tried in his own household. Thus taught by experience he notes that the public should be warned that sweet potatoes, unlike the common potato, will not keep long.

His triumph was great, however, when success followed.

To Sir D. Morris

July 19, 1904.

It is indeed good of you to send me those splendid Barbados Bananas, which are quite the finest I ever ate in England. I am delighted to hear from my daughter in London that W. Indians in good condition are now hawked about at $\frac{1}{2}d.$ each! They are thus becoming what I have long hoped to see, a food for poor people. For this they are mainly indebted to you and you may well be proud of it.

To Sir D. Morris

The Beacon, Sidmouth : January 15, 1911.

I have occupied this forenoon in reading your admirable paper,[1] word for word, with Lady Hooker participating and pronouncing it to be very able.

Before commencing I was impressed with the necessity for such a review of the labours of your Department, but failed to grasp the vast extent and multiplicity of the aims to be sought and the amount of almost superhuman knowledge, experience, energy, tact and endeavour required to cope with the situation.

Please accept my most hearty congratulations and thanks for your all too appreciative mention of me. How I wish that my father could have been present. As founder of Kew in an economic sense, he was the great originator, and you are the most brilliant of his successors in the tropical field.

Perhaps to you your greatest reward is the confidence of the Colonial Office.

Most sincerely yours,

Jos. D. Hooker.

[1] On ' The Imperial Department of Agriculture in the West Indies,' read before the Royal Colonial Institute, January 10, 1911, and issued in March as No. 75 of ' Colonial Reports—Miscellaneous ' (Cd. 5515). This Department was established by Mr. Chamberlain and placed under Sir D. Morris. Nearly twenty-eight years before (June 13, 1883) he had read a similar paper before the same body : ' Planting Industries in the West Indies.'

CHAPTER XLVIII

HOOKER'S POSITION AS BOTANIST

By Prof. F. O. Bower

For several decades before his death Sir Joseph Hooker occupied a position unique among living botanists. A glance at the list of distinctions awarded to him, as set out in the official list of the Royal Society, will show the catholicity of his appreciation in countries other than his own. Within the British Empire the leading position had long been his without question. Thus contemporary science gave its verdict in no uncertain way. But the opinion of a period is not necessarily the opinion of posterity. There are, however, solid reasons in the present case for believing that the two will not diverge in any marked degree. In attempting to analyse and appreciate those qualities which gave Sir Joseph Hooker his assured position among his contemporaries, it may be possible at the same time to recognise the permanent features in his work. For it is these which will secure for him a prominent place in the History of the Science, as it may be reviewed from some vantage point in the remote future.

What first strikes the observer is the mere superficial fact of an unusually long life, zealously used. In the year 1837, while still a student, he described three new species of mosses. In 1911 he established several new species of the genus Impatiens. Thus his published record covers a period of three-quarters of a century. Doubtless this was a factor, but only a minor one. What is more important is that to the very end he never grew really old. He never outlived his freshness of interest in a new discovery, whether his own or that of his younger contemporaries. Doubtless the extraordinary length of his productive period made the great volume of his work possible. But it is not upon the mere quantity of the output that his title to fame is to be based. It is the critical quality, the originality, and the diversity of the work that are its outstanding features. Throughout it all runs the golden thread of acute observation. He knew his plants personally. As a boy he absorbed specific knowledge almost unconsciously in his father's house in Glasgow, and in the Botanic Garden there, which, as a source of novelties, was at that time without its equal in this country or probably in any other. As a young man he travelled the world over, to see plants in their native surroundings. As a veteran he lived among them in the great Garden at Kew. Few, if indeed any, have ever known plants as he did. Such knowledge comes only from growing up with them from earliest childhood.

But he was not only a botanist. His interest extended into kindred spheres. He shared with Darwin that wider outlook upon the field of Science that gave a special value to the writings of both. The best sample of his work as a geographer is embodied in his 'Himalayan Journals,' a book which ranks with Darwin's 'Voyage of the *Beagle*,' and Wallace's 'Malay Archipelago.' These form a veritable trilogy of the Golden Age of travel in pursuit of Science. The data collected on his journey in Sikkim and Nepal formed the basis of a map published by the Indian Topographical Survey. By its aid the operations of various campaigns and political missions have since been carried to a successful issue. If Hooker were not known as a botanist he would still have an assured place as a geographer.

Similarly in the Science of Geology he made solid contributions to knowledge. He was early in the field in the microscopic examination of plant-tissues preserved in coalballs. These were studied by sections, a method then newly introduced by Witham, and since greatly developed in this country. He may be said to have himself originated another line of study, since largely pursued by geologists. For he

examined samples of Diatomaceous Ooze from the ocean-floor
of the Antarctic, and so initiated the systematic treatment
of the organic deposits of the deep sea. These, together
with his observations on glaciers and on sub-aerial denudation,
were all carried out in his earlier years of travel. The quality
and the rapidity of the work showed his mastery in a science
not specially his own ; while the problems which he handled
were all nascent at the time when he worked upon them.

But though such excursions into the sphere of the kindred
sciences illustrate Hooker's natural power and the breadth
of the basis of observation upon which he worked, his fame
rests upon his purely botanical writings. The most important
of them fall into three groups, though these naturally over-
lap : viz. the works of *Systematic*, of *Morphological*, and of
Philosophical character. His greatest Systematic Works were
the 'Antarctic Flora,' the 'Flora of British India,' the 'Genera
Plantarum,' and the 'Kew Index.' As all of these four differ
in scope and character, each demands separate notice and
analysis.

On the Antarctic Voyage Hooker had the opportunity
of collecting on all the great circum-polar areas of the Southern
Hemisphere. His 'Antarctic Flora' was based on the collections
and observations then made, supplemented by those of other
travellers. It was published in six large quarto volumes.
They describe about three thousand species, of which over
one thousand are depicted, usually with detailed analytical
drawings. But there is more in them than reports of explorations or descriptions of new species. All the known
facts that could be gathered were incorporated, so that they
became systematically elaborated and complete Floras of
the several countries. Moreover, in the last of them, the
'Flora Tasmaniae,' there is an Introductory Essay, which in
itself would have made Hooker famous, for it contains a discussion of the permanence of species, to which we shall return
later. It contained also his first enunciation of broad theory
of Geographical Distribution of Plants. While it was still
in preparation Darwin wrote to him in terms of prophetic
enthusiasm : ' I know I shall live to see you the first authority

in Europe on that grand subject, that almost keystone of the laws of Creation, Geographical Distribution.' Never was a forecast more fully justified. Hooker the traveller had prepared the way for Hooker the philosopher.

What he did for the Antarctic in his youth he continued in mature life for British India. While the publication of the 'Antarctic Flora' was still in progress, he made his Indian journeys. The vast collections amassed by himself and Dr. Thomson were consigned by agreement with Government to Kew. Thither had also been brought the herbaria of Falconer and Griffith. Such materials, with other large additions made from time to time, formed the foundation upon which Sir Joseph Hooker was to base his 'Magnum Opus,' the 'Flora of British India.' Though conceived, he says with regret, upon a restricted plan, it ran to seven volumes, relating to 16,000 species. It is, he says in the Preface, a pioneer work, and necessarily incomplete. But he hopes it may 'help the phytographer to discuss problems of distribution of plants from the point of view of what is perhaps the richest, and is certainly the most varied botanical area on the surface of the globe.' This great floristic work was fitly rounded off by his completion of the 'Ceylon Flora,' left unfinished on the death of Dr. Trimen. His last contribution to the Flora of the Indian Peninsula was in the form of a Sketch of the Vegetation of the Indian Empire, including Ceylon, Burma, and the Malay Peninsula. It was written for the Imperial Gazetteer, at the request of the Government of India. No one could have been so well qualified for this as the veteran who had spent more than half a century in preparation for it. It was published in 1904, and forms the natural close to the most remarkable study of a vast and varied Flora that has ever been carried through by one ruling mind.

The third of Hooker's great Systematic Works is the 'Genera Plantarum,' produced in collaboration with Mr. Bentham. Its three massive volumes contain a codification of the Latin diagnoses of all the genera of Flowering Plants known at the date of publication. It is essentially a work for the technical botanist, but for him it is indispensable. The only real predecessors of this monumental work were the 'Genera Plantarum' of Linnæus (1737–1764), and that of Jussieu (1789), to which may be added that of Endlicher (1836–1840). Both Bentham and Hooker had felt the inconvenience of the want of a Genera Plantarum founded on actual observation, to replace the already antiquated ones of Endlicher and Meissner, both of which, especially the latter, had been in great measure mere compilations. In view of the gigantic nature of the task they joined forces. But the authors specially wished that the whole should be considered as the joint production of them both. The characters embodied in the diagnoses were drawn from the actual examination of specimens. Such data could only be derived from a reliable and rich herbarium such as Kew had then become. Thus the book is not in any sense a compilation from the work of earlier writers, but it contains a redrafting of the diagnoses on the basis of personal observation. Probably into no work on Botany is there condensed so wide a field of personally recorded fact, expressed in such precise terms. The authors were both mature observers. But while Hooker was at home in the forest and the jungle, Bentham was rather a denizen of the herbarium. His education as a conveyancing barrister gave point to his naturally acute mind in the exact wording of diagnoses. The difficulty of the task of Bentham and Hooker was greater than that of their predecessors by reason of their wider knowledge and the great increase in the number of recognised genera, consequent upon the activity of collectors the world over. But their 'Genera Plantarum' was on that account a nearer approach to finality. Hitherto its supremacy has not been challenged. On the other hand it has formed the source from which diagnoses have been liberally borrowed.

In the arrangement of the contents the 'Genera Plantarum' followed the prevalent custom of the time. This may puzzle generations that come after. For they may say it is true that Hooker took the first step towards a phyletic classification, by adopting the view of mutability of species. He was the first Systematic Botanist who did this. They may ask ' Why, after making this important advance on the older methods, did he

maintain so nearly the grouping of orders and genera prevalent
before ?   Should it not have been a logical necessity to attempt
some grouping more nearly in accordance with the probable lines
of evolution than that retained in the " Genera Plantarum " ? '
Those who would urge these grounds for criticism long after
the event of publication should remember two essential facts.
The first is that the work was, as its name conveys, a work on
genera, not on the grouping of genera. Its value does not lie
in the order of the arrangement of the diagnoses, but in the
strictness of their definition. It deals with the cutting of the
gems, as apart from the plan of their setting. Gems these
diagnoses certainly are, and it is probable that they will not
be improved in the cutting for long enough to come. The
second essential point to be remembered by critics is the immensity of the task of arriving at any phyletic grouping of
Angiosperms, and the uncertainty of the methods to be used.
Moreover the time was not ripe. For this work was planned in
1858, and the first part was published in 1862, within three
years of the production of the ' Origin of Species.' Even if the
authors had attempted a grouping according to some theory
of descent, they would have courted disaster. They knew
as well as any men of their time the complexity of the inter-
relations of Seed-Bearing Plants : the nicety of the distinctions, and the vastness of the number of closely related forms.
To those who appreciate this, the wisdom of retaining the old
groupings is manifest. A quarter of a century later an attempt
was made by Engler and Prantl to attain a more satisfactory
arrangement (Die Natürlichen Pflanzenfamilien. II. Embryo-
phyta Siphonogama, 1889). They altered the sequence of orders
and genera, with results which are no doubt beneficial in the
main, though certainly not final. But the relatively brief
diagnoses there given are in no sense a substitute for those of
the ' Genera Plantarum,' which remains, and will probably long
remain, the ultimate court of appeal.

The ' Kew Index ' was produced under the personal supervision
of Sir Joseph Hooker. The expense of it was borne by Charles
Darwin, and by his family after his death. The scheme originated in the difficulty Darwin had found in the accurate naming

of plants. For 'synonyms' have frequently been given by different writers to the same species, and this had led to endless confusion. The object of the Index was to provide an authoritative list of all the names that have been used, with reference to the author of each, and to its place of publication. The correct name in use according to certain well-known rules of nomenclature was to be indicated by type different from that of the synonyms superseded by it. The only predecessor of such an Index was Steudel's ' Nomenclator Botanicus,' a book greatly prized by Darwin, though long out of date. Hooker was asked by Darwin to take into consideration the extent and scope of the proposed work, and to suggest the best means to have it executed. He undertook the task, and it was carried out by Dr. Daydon Jackson and a staff of clerks. The plan of Dr. Jackson was that it should be based on the ' Genera Plantarum ': that it should be carried out at Kew : and that reference to the source of origin should be given for each species. The work was started in 1882, and took almost ten years. It extended to four large quarto volumes, with 2500 pages, bearing about 375,000 specific names. Hooker read and narrowly criticised the proofs, supplying himself the statements on geographical distribution. Surely no greater technical benefit was ever conferred upon future generations of botanists by a veteran of the science than this Index. It smooths the way for every systematist who comes after by sweeping aside the superfluous weight of effete names, and guiding those who consult it directly to the proper designation of the species referred to. The Index stands as a monument to an intimate friendship. It bears witness to the munificence of Darwin, and the ungrudging personal care of Hooker.

While such purely scientific activities as these of Sir Joseph naturally claim attention first, his effective administration should not be lost sight of. Its most tangible result is the great botanical establishment at Kew. Three generations of the Hooker dynasty—Sir William, Sir Joseph, and his son-in-law Sir William Thiselton-Dyer—made Kew what it is. In 1840, when taken over from the Crown by the Department of Woods and Forests, the Garden at Kew left much to be desired. It was small in extent, and without adequate library, museum, or herbarium. During the rule of the Hookers it became a great scientific establishment. The living collections, which necessarily fluctuate in quality with the skill of the gardening staff, attained the highest degree of success. But the more permanent parts of the establishment, the herbarium, library, and museums, form the basis upon which finally the systematic study of plants must be pursued. Their framework consisted of the Hookerian collections themselves : first those of Sir William, acquired by the State after his death ; and afterwards those gathered by Sir Joseph in the Antarctic, and in India. These were largely added to by gift, by purchase, and by exchange, so that for the botany of the world, and for that of the British Empire in particular, Kew became the centre for reference and study. It grew into a great co-ordinating machine for systematic comparison. It was the source from which a series of Floras of the British Dominions and Colonies has been officially issued, many of them planned by Sir Joseph himself.

While this is what Kew means to the Systematic Botanist, it is to the general public a place of the purest delight. The living collections, and especially the Arboretum, on the perfecting of which both of the Hookers spent their best efforts, give pleasure and instruction to the serious student and the artist, as well as to the masses. This the public owes in great measure to the administrative capacities of the first directors.

There is only one other family record in European Botany which can compare with this of the Hookers at Kew. It is that of the De Candolles at Geneva. For three generations they also were in the forefront of Systematic Botany. The greatest of them was A. P. De Candolle. He was a most versatile writer on physiology, and on geographical distribution. But his greatest work was the 'Prodromus Systematis Naturalis,' in which all known plants were to be arranged according to his natural system, and described at length. He initiated this stupendous work, but did not live to complete it. It was based chiefly upon his own collections, still preserved in the family house in the Place St. Pierre, at Geneva. We visit it with interest and pious respect. But it is evident that the active science of the present day has drifted elsewhere. The dynasty of De Candolle, brilliant and effective as it was, has left behind no co-ordinating machine like that of the establishment at Kew.

The year 1841 was notable in the History of Botany. It witnessed the death of A. P. De Candolle, and the move of Sir William Hooker to Kew. It may be held as the year of birth of the new establishment there. We may then pause and consider the position of Botanical Science in Europe at this date. The glamour of the Linnæan period had faded, and the Natural System of Classification of Plants initiated by De Jussieu had fully established its position, taking its most elaborate form in the 'Prodromus' of A. P. De Candolle, the continuation of the unfinished work being left in the hands of his son Alphonse. In England, Robert Brown was in the full plenitude of his powers, and, in possession of the Banksian Herbarium, was evolving out of its rich materials new principles of classification, and fresh morphological comparisons. In fact Morphology was at this time being differentiated from mere Systematic as a separate discipline. Nothing contributed more effectively to this than the publication of 'Die Botanik als inductive Wissenschaft,' by Schleiden, the first edition of which appeared in 1842; for in it development and embryology were indicated as the foundation of all insight into Morphology. But notwithstanding the great advances of this period in tracing natural affinities, and in the pursuit of morphological comparison, branches which would seem to provide the true basis for some theory of descent, the Dogma of Constancy of Species still reigned. It was to continue yet for twenty years to dominate botanical thought.

Meanwhile great advances had been made also in the knowledge of the mature framework of cell-membranes in plants. Anatomy, initiated in Great Britain by Hooke, Grew, and Malpighi, had developed in the hands of many 'phytotomists,' the series culminating in the work of Von Mohl. But it was chiefly the mere skeleton which was the subject of their interest. Eight years previously, it is true (in 1833), Robert Brown had described and figured the nucleus of the cell, and had approached even the focal point of its interest, viz. its relation to reproduction. But the demonstration of the cytoplasm in which it is embedded was yet to come. In fact, the knowledge of structure omitted as yet any details of that body which we now hold to be the ' physical basis of life.'

The period immediately succeeding 1841 was, however, a time pregnant with new developments. The study of protoplasm soon engaged the attention of Von Mohl. Apical growth was investigated by Naegeli and Leitgeb. The discovery of the sexuality of ferns, and the completion of their life-story by Bischoff, Naegeli, and Suminski, led up to the great morphological generalisation of Hofmeister. Thus Morphology in its modern development was initiated. On the other hand, Lyell's ' Principles of Geology ' had appeared, and obtained wide acceptance. Darwin himself was freshly back from the voyage of the *Beagle*, while Sir Joseph Hooker was at that very time away with Ross on his Antarctic Voyage, and shortly afterwards started on his Journey to the Himalaya. These three great figures—the forerunner of Evolution, the author of the ' Origin of Species,' and Darwin's first adherent among biologists—were thus in their various ways working towards that generalisation which was so soon to revolutionise the science of which Kew was becoming the official centre. Well may we then regard this date as a nodal point in the History of Botany not only in this country, but also in the world at large.

It was into such an atmosphere of development and change that Sir Joseph Hooker entered on his return from the Himalayan Journey in 1851. His first care was to work out his results systematically. Two volumes of the ' Antarctic Flora ' were already out, and the ' Flora of British India ' soon took form. These works show how fully he was imbued with the old systematic methods : how he advanced, improved, and extended them, and was in his time their chief exponent. Not only did he add greatly to the genera and species recorded, but he co-ordinated previous results and defined the limits of distribution of species : thus giving more coherent treatment to the vegetation of vast areas of the earth's surface. It is interesting to compare his systematic method with that of his father. The elder Hooker, true to his generation, treated his species as fixed and immutable. He did not readily generalise from them. His end was attained by their accurate recognition, delineation, description, and classification. His attitude towards microscopic detail is noteworthy. He remarks in his ' Genera Filicum ' that Presl ' has laid too much stress on the number and other circumstances connected with the bundles of vessels in the stipes which in the herbarium are difficult of investigation.' Occasionally he gave his reasons for this opinion, as in a notable passage in his ' Species Filicum ' (vol. iii. p. 3), where he explains that a grouping based on the microscopic details of the annulus in Ferns ' would be inconvenient to retain in a work whose main object is to assist the tyro in the verification of genera and species : and natural habit is often a safer guide than minute microscopic characters.' Thus we see that for the elder Hooker convenience of diagnosis was more important than details of structural similarity.

But the younger Hooker, while he was not a whit behind the best of his predecessors in the recording and tabulation of detail, saw farther than they. He was not satisfied with the mere record of species as they are. He sought to penetrate the mystery of the origin of species. To the elder Hooker species were units. The younger contemplated the summing of those units into progressions, which would thus in a sense make visible the changes in descent. To the elder Hooker the study of plants was static. In the hands of the younger it became dynamic. Development and microscopic detail, used according to the methods of Schleiden and Hofmeister, became then of the first importance. Such enquiry we see illustrated in those of Sir Joseph's writings which may be styled *Morphological*. The great outburst of systematic work in Britain in the middle of the nineteenth century had had a deleterious effect on those of lesser breadth of view than he.

Anatomy and Physiology were in danger of being atrophied
in the very land of their birth. Hooker himself formed a link
between the herbarium and the laboratory. In his own work
he held the balance between the two by a series of Memoirs
which were 'morphological' in the modern sense. Already he
had pursued the microscopic study of the fossils *Lepidoden-
dron* and *Pachytheca* in a fashion in advance of his time; later
he made similar investigations on living plants. Examples
of such work are found in his Memoir on the Balanophoraceae,
and in his study of the development and structure of the
pitchers of *Nepenthes*. The physiological significance of these
and other organs of carnivorous plants formed the subject of his Address before the British Association in 1874.
It was in 1863 that the great Monograph appeared on that
most remarkable of all Gymnospermic plants, *Welwitschia*.
This may be held as the best example of his morphological
work, and compares favourably with any similar Monographs
of the period. The material came from a very limited area
of dry country inland from Walfisch Bay, on the South-West
Coast of Africa. It was supplied by Dr. Welwitsch and
others. The plant differs from any other known type, but
after a full examination of the structure of its vegetative and
reproductive organs, it was referred by Sir Joseph to the
Gnetaceae. The analyses of the propagative organs were
carried out by him with minute care. The whole plant is of
so unusual a character that it was a real triumph to trace
the comparisons leading to the systematic position which he
assigned. Much modern work, by the aid of refined methods
of fixation and the use of the microtome, has only served to
confirm his classification of one of the most bizarre plants in
the Vegetable Kingdom.

Such works bore the character of a time later than when
they were produced. They tided over the period when in
Britain investigation in the laboratory by means of the microscopic analysis of tissues was almost throttled by the overwhelming success of systematic and descriptive work. The
revival dated from about 1875. But we see in Hooker one of
the few who, prior to that date, pursued microscopic enquiry

side by side with systematic and floristic work. It is another indication of the breadth of his scientific outlook. This revealed itself later in the lively sympathy which he showed in the anatomical and physiological enquiries of younger men.

But above and beyond all this systematic and morphological work lies Hooker's pursuit of *Biological Philosophy.* In his hands the former necessarily led onwards to the latter. It is indeed his use of facts rather than the acquisition of them that constitutes his highest title to rank among botanists. He fully grasped that ' the battle of the evidences will have to be fought out on the field of Physical Science, and not on that of the Metaphysical.' This was the difficult lesson of the period when Evolution was born, and Hooker learned it early. He cleared his mental outlook from all preconceptions, and worked down to the bed-rock of objective fact. Thus he was able to use his vast and detailed knowledge in advancing along the lines of induction alone towards sound generalisations. These had their very close relation to questions of the mutability of species. That subject was approached by him through the study of Geographical Distribution, in which as we have seen he had at an early age become a leading authority.

The fame of Sir Joseph Hooker as a Philosophical Biologist rests upon a masterly series of Essays and Addresses. The chief of these were the Introductory Essay to the ' Flora Tasmaniae,' dealing with the Antarctic Flora as a whole ; the Essay on the Distribution of Arctic Plants, published in 1862 ; the Discourse on Insular Floras in 1866 ; the Presidential Address to the British Association at Norwich in 1868 ; his Address at York, in 1881, on Geographical Distribution ; and finally, the Essay on the Vegetation of India, published in 1904. None of these were mere inspirations of the moment. They were the outcome of arduous journeys to observe and to collect, and subsequently of careful analysis of the specimens and of the facts. The dates of publication bear this out. The Essay on the Antarctic Flora appeared about twenty years after the completion of the voyage. The Essay on the Vegetation of India was not published till more than half a century after Hooker first set foot in India. It is upon such foundations that Hooker's reputation as a great constructive thinker is securely based.

The first named of these Essays will probably be estimated as the most notable of them all in the History of Science. It was completed in November 1859, barely a year after the joint communications of Darwin and Wallace to the Linnean Society, and before the ' Origin of Species ' had appeared. It was to this Essay that Darwin referred when he wrote that ' Hooker has come round, and will publish his belief soon.' But this publication of his belief was not merely an echo of assent to Darwin's own opinions. It was a reasoned statement advanced upon the basis of his ' own self-thought,' and his own wide systematic and geographical experience. From these sources he drew for himself support for the ' hypothesis that species are derivative, and mutable.' He points out how the natural history of Australia seemed specially suited to test such a theory, on account of the comparative uniformity of the physical features being accompanied by a great variety in its Flora, and the peculiarity of both its Fauna and Flora as compared with other countries. After the test had been made, on the basis of study of some 8000 species, their characters, their spread, and their relations to those of other lands, he concludes decisively in favour of mutability and a doctrine of progression.

How highly this Essay was esteemed by his contemporaries is shown by the expressions of Lyell and of Darwin. The former writes :

I have just finished the reading of your splendid Essay on the Origin of Species, as illustrated by your wide botanical experience, and think it goes far to raise the variety-making hypothesis to the rank of a theory, as accounting for the manner in which new species enter the world.

Darwin wrote :

I have finished your Essay. To my judgment it is by far the grandest and most interesting essay on subjects of the nature discussed I have ever read.

But besides its historical interest in relation to the Species Question, the Essay contained what was up to its time the most scientific treatment of a large area from the point of view of the Plant-Geographer. He found that the Antarctic, like the Arctic Flora, is very uniform round the globe. The same species in many cases occur on every island, though thousands of miles of ocean may intervene. Many of these species reappear on the mountains of Southern Chili, Australia, Tasmania, and New Zealand. The Southern Temperate Floras, on the other hand, of South America, South Africa, Australia, and New Zealand differ more among themselves than do the Floras of Europe, Northern Asia, and North America. To explain these facts he suggested the probable former existence, during a warmer period than the present, of a centre of creation of new species in the Southern Ocean, in the form of either a continent or an archipelago, from which the Antarctic Flora radiated. This hypothesis has since been held open to doubt. But the fact that it was suggested shows the broad view which he was prepared to take of the problem before him. His method was essentially that which is now styled ' Ecological.' Many hold this to be a new phase of botanical enquiry, introduced by Professor Warming in 1895. No one will deny the value of the increased precision which he then brought into such studies. But in point of fact it was Ecology on the grand scale that Sir Joseph Hooker practised in the Antarctic in 1840. Moreover it was pursued, not in regions of old civilisation, but in lands where Nature held her sway untouched by the hand of man.

This Essay on the Flora of the Antarctic was the prototype of the great series. Sir Joseph examined the Arctic Flora from similar points of view. He explained the circumpolar uniformity which it shows, and the prevalence of Scandinavian types, together with the peculiarly limited nature of the Flora of the southward peninsula of Greenland. He extended his enquiries to oceanic islands. He pointed out that the conditions which dictated circumpolar distribution are absent from them ; but that other conditions exist in them which account for the strange features which their vegetation shows. He

extended the application of such methods to the Himalaya
and to Central Asia. He joined with Asa Gray in like enquiries
in North America. The latter had already given a scientific
explanation of the surprising fact that the plants of the Eastern
States resemble more nearly those of China than do those of the
Pacific Slope. In resolving these and other problems it was
not only the vegetation itself that was studied. The changes
of climate in geological time, and of the earth's crust as demonstrated by geologists, formed part of the basis on which he
worked. For it is facts such as these which have determined
the migration of Floras. And migration, as well as mutability
of species, entered into most of his speculations. The Essays
of this magnificent series are like pictures painted with a full
brush. The boldness and mastery which they show sprang
from long discipline and wide experience.

Finally, the chief results of the Phyto-Geographical work
of himself and of others were summed up in the great Address
at York. As President of the Geographical Section he chose
as his subject ' The Geographical Distribution of Organic
Beings.' To him it illustrated ' the interdependence of those
Sciences which the Geographer should study.' It is not enough
merely to observe the topography of organisms, but their
hypsometrical distribution must also be noted. Further, the
changes of area and of altitude in exposed land-surfaces of
which geology gives evidence, are essential features in the
problem, together with the changes of climate, such as have
determined the advance and retrocession of glacial conditions.
Having noted these factors, he continued thus :

> With the establishment of the doctrine of orderly evolu-
> tion of species under known laws I close this list of those
> recognised principles of the science of geographical distribu-
> tion, which must guide all who enter upon its pursuit. As
> Humboldt was its founder, and Forbes its reformer, so we
> must regard Darwin as its latest and greatest law-giver.

Now, after thirty years, may we not add to these words of his,
that Hooker was himself its greatest exponent ?

But before all else it was the part which Hooker played in

the drama of the birth of Evolutionary Theory which gives him a permanent place in the History of Human Thought. He was an almost life-long friend of Charles Darwin. He was the first confidant to whom the Species Theory was entrusted. Excepting Wallace, he was its first whole-hearted adherent. He was also Darwin's constant and welcome adviser and critic, drawing upon his unrivalled knowledge of botanical detail as evidence for, or check upon, the advancing theoretical position. The published letters of Darwin reveal in a way that none of the completed works of Darwin or of Hooker could have done, the steps in the growth of the great generalisation, and the part in it which Hooker himself took. We read of the doubt of one or the other : the gradual accumulation of material facts : the criticisms and amendments in face of new evidence : and the slow progress from tentative hypothesis to assured belief. We ourselves have grown up since the clash of opinion for and against the mutability of species died down. It is hard for us to understand the strength of the feelings aroused : the bitterness of the attack by the opponents of the theory, and the fortitude demanded from its adherents. It is best to obtain evidence on such matters at first hand ; and this is what is supplied by the correspondence between Darwin and Hooker. From the letters it is clear that his friendship, advice, and alliance were of incalculable benefit to Darwin himself, who summed this up in the words : ' You have represented for many years the whole great public to me.' But while this in itself gives Hooker his natural place in history, it must never be forgotten that he himself upheld in the 'Flora Tasmaniae' the mutability of species, and based his opinion, as Darwin stated, on 'his own self-thought.' Among botanists Hooker was in fact the Protagonist of Evolution. His influence during that stirring period, though quiet, was far-reaching and deep. His work was both critical and constructive. His wide knowledge, his keen insight, his fearless judgment were invaluable in advancing that intellectual revolution which found its pivot in the mutability of species. The share he took in it was second only to that of his life-long friend, Charles Darwin.

When we review these varied activities, extending throughout the long life of Sir Joseph Hooker, it is not difficult to
account for the pre-eminent position which he held among
his contemporaries. This estimate will be an enduring one.
For the quality and extent of the systematic work is such
that its effect must be felt wherever Flowering Plants are
defined and classified. On the other hand, the originality
of the generalisations on Geographical Distribution, and on
the Species Question, has lifted current opinion into new
channels, and so altered it that his place in the History of
Human Thought is for ever assured.[1]

[1] Foreign opinion has been aptly called a court of contemporary posterity.
Recalling this, it is interesting to record that soon after Hooker's death he was
selected by the Japanese as ' one of the twenty-nine Heroes of the World that
Modern Time has produced.'—L. H.

[*From a Photograph by W. End, Sunningdale.*

Sir J. D. Hooker in his Study at the Camp.

CHAPTER XLIX

In 1898 the completion of ' a monumental work in botany, the " Flora of British India," ' was chosen by the Linnean Society as a fit occasion for commemorating Hooker's eminent services to biological science. A gold medal was specially struck, and presented to him on May 24 at the Anniversary Meeting of the Society, with which he and his father and his grandfather before him had been so closely connected.

In his reply, Hooker recalled the fact that the Linnean was the first scientific society in which he was enrolled, fifty-six years before. It was perhaps due to his family record that he was elected as the youngest Fellow on the list with no more solid scientific claims than that he was serving as naturalist in the Antarctic under Captain Ross, who was himself a Fellow, and had a copy of the *Transactions* in his cabin, which proved a godsend to the young naturalist. The ships were at the Falkland Islands when the election took place, and nearly a year and a half elapsed before Captain and Naturalist knew that they were fellow-Linneans. Now he was the only Fellow who personally knew four of the 169 naturalists who, 110 years before, formed the nucleus of the Society.[1] He concluded with these words :

[1] 'Of these four, I knew two in my later teens ; they were the Rev. W. Kirby, the author, with Spence, of the immortal *Introduction to Entomology* ; and Dr. Heysham, of Carlisle, an excellent entomologist and ornithologist. The others were Aylmer Bourke Lambert, a former President, and the last, as I have been informed, who wore in the chair the presidential three-cornered hat ; and Archibald Menzies, who as naturalist accompanied Vancouver in his voyage in the Pacific, and who introduced the *Araucaria imbricata* into England. These all died very near the year of my election.'

It remains, sir, to thank you cordially for coupling my father's name with my own in this award, but for which, indeed, I could not have accepted it without a protest. I inherited from him my love of knowledge for its own sake, but this would have availed me little were it not for the guiding hand of one who had himself attained scientific eminence ; who by example, precept, and encouragement kept me to the paths which I should follow, launched me in the fields of exploration and research, liberally aided me during his lifetime, and paved for me the way to the position he so long held at Kew with so great credit to himself, and benefit especially to our Indian and Colonial possessions.

At home the summer brought Hooker its share of trouble. His son William had just pulled through a serious illness, and he was looking forward to spending a happy week at Batsford, when he was called to the death-bed of his sister, Mrs. Lombe,[1] who had long been an invalid. The tie of affection between them was very close, and maintained by regular correspondence. 'We had been fast friends for well-nigh 80 years,' he exclaims to Lord Redesdale, and now that the last of his own generation was gone, he felt deeply the loss of a lifelong love and friendship.

Other memories of the past, however, were kept warmly alive. [Sir] Francis Darwin, with the collaboration of Professor A. C. Seward, was preparing to bring out a collection of ' More Letters of Charles Darwin.' Hooker responded with delight.

To F. Darwin

February 1, 1899.

MY DEAR FRANK,—I will gladly help you all I can ; so have no scruples. By all means send me any of my letters you think I can throw light upon.

You are right to make the book uncompromisingly scientific. It will be greatly valued. I am getting so old and oblivious that I fear I may not be of much use.

Ever affectionately yours,

JOS. D. HOOKER.

The exchange of unpublished letters brought some surprises.

[1] Her husband, Dr. Evans, had taken the name of Lombe.

To the Same

February 24, 1899.

I had no idea that your father had kept my letters. Your account of 742 pp. of them is a revelation. I do enjoy re-reading your father's; as to my own, I regard it as a punishment for my various sins of blindness, perversity, and inattention to his thousand and one facts and hints that I did not profit by as I should have, all as revealed by my letters. I do not think I gave my mind as I ought to have —but I had always my head and hands full of all sorts of duties, and my correspondence with your father was the sweet, amongst many bitters.

Yes, I will gladly go down at some future time and confab with you.

To the Same

March 21, 1899.

I enclose copies of your father's letters to mine. The first refers to his testimonial towards my candidature for the Botany Chair of Edinburgh University. If you care for a copy of this I will send it, though it savours of vanity to offer it.

Ever affectionately yours,

Jos. D. HOOKER.

P.S.—You are most welcome to the originals of my letters to your father. If I had them I should be tempted to burn them!

For he was, as ever, very critical of his bygone letters, as he dipped again and again into the four red portfolios of them now at his elbow: 'From what I read of them, I thought they were very poor stuff' (February 1, 1901). He preferred his present rôle of throwing light where it was needed on Darwin's current interests, and again insisted, 'Do not hesitate to ask me for any information I can give you.' Going over the slip proofs in May 1902 was no burden, but a pleasure: 'To me the letters are most refreshing—they bring *all* Down home to me.'

The crowning pleasure came as the book neared completion, and the authors proposed to dedicate it to Darwin's closest friend.

To F. Darwin

July 18, 1902.

I can imagine nothing that would greet my declining years with anything approaching the pleasure of having the letters dedicated to me, and I do heartily thank you and Mr. Seward for thinking of me. I do feel as if it would add years to my life.

The first page of the book bears these words :

DEDICATED, WITH AFFECTION AND RESPECT, TO

SIR JOSEPH HOOKER

IN REMEMBRANCE OF HIS LIFELONG FRIENDSHIP WITH
CHARLES DARWIN

' You will never know how much I owe to you for your constant kindness and encouragement.'—*Charles Darwin to Sir Joseph Hooker, Sept.* 14, 1862.

The revival of the Darwin interest was intensified by the inauguration of a Darwin statue in the Oxford Museum on June 14, 1899. This, the work of Mr. Hope Pinker, was the gift of Professor Poulton to the University. Hooker attended the ceremony, and spoke, being asked

to give some little account of that long and intimate friendship with which he affectionately honoured me. Of course I can do little more than repeat what I said at Shrewsbury, except you can give me a hint as to any other topic. (To F. Darwin, June 7, 1899.)

This speech (a report of which appears in *The Times* of the following day) he prefaced with an apology for possible distortions of memory, for ' Narrators of an advanced age are proverbially oblivious and too often victims of self-deception in respect of what they think they remember.' Beginning with the parallelism of their early careers and their common friendship with Lyell, he told in much fulness the history of the origin and growth of their friendship, especially in the ' inaccessible house ' at Down ; his first sight of the sketch of Darwin's theory ; and his retort to the friends of a later day who asked why he had not shaped all his own researches upon the lines of that illuminating sketch : It was confidential.

Of his character and peculiar power of work he repeated the impressions given in preceding letters, but added that when Darwin claimed for himself only a fair share of 'invention,' he meant the quality that Hooker would define as originality, the exercise of imagination in critical experiments. And referring to the reception of Darwin's Primula paper at the Linnean, he told the story of how an ardent supporter of Darwin's compared previous students of the flower to Peter Bell with his view of ' A primrose by the river's brim.' On being told of this, Darwin exclaimed : ' I would rather have been the man who thought of that on the spur of the moment than have written the paper that suggested it.'

A sketch of his reading in these days shows among other things the unending interest in Indian religions.

To Lord Redesdale

January 25, 1899.

I am glad that you have taken up Buddhism, a favourite subject with Huxley and myself. I have a few good books on the subject ; shall I send you a list of them ? You can then have what you please of them. I regard the Essenes as a branch of Buddhists, tinctured some with Greek, others with Jewish ideas (Philosophy so called), and that Christ's teaching was one outcome of the movement. I shall be glad to know if you come across in your reading any rational explanation of the identity in ritual ceremonies, offices, vestments, &c., &c., &c., of the Buddhist and Roman Churches. I have proposed this question to many a learned churchman on one hand, and Buddhist scholar on the other, without obtaining the smallest satisfaction. That it was all accidental is the answer I generally get, at which I scoff. I have my own ideas on the subject, but do not suppose they would be accepted without more evidence than I can offer. My friend, Brian Hodgson, was an arch Buddhist scholar, and we spent many a long evening in the Himalaya over Buddhism ; but his knowledge was too profound to be communicated intelligently to a novice. I have his works. I fancy he did more by the collection of materials than by his dissertations, to advance the study.

My reading of late has been all but demoralising, for its

variety and, to a great extent, *vacuity*. Novels of sorts, intersected between fits of Spencer's last ponderous volume, wherein the old matter interests me more than the new. Travels I devour and only partially digest. Metaphysics I cannot abide. I was disappointed with Tennyson's Life, made up of snippets in too great proportion. I have read Prescott's Cortes, Pizarro and Philip II. with renewed pleasure. Also Motley's Ferdinand and Isabella, all stale viands, but the two former still appetizing.

The Illustrated Edition of Green's History is just come. I ordered it for Dick, with whom I am reading Huxley's Physiography and Pope's Odyssey.

It is high time I ended this fatuous gossip.

On April 16 he sends his friend a batch of his own books on Buddhism, adding with perhaps unnecessary emphasis :

My memory is now so bad that the whole subject is a blur in my brain—a confusion of Thibetan, Japanese, Singhalese and Burmese developments of the creed.

Follows a reminiscence of Dartmouth, where he had just spent a fortnight :

We went over the *Britannia*, very interesting as you know. I was astounded at the multiplicity and variety of subjects crammed into the 15 months' course. It is a grand education. I was amazed at the size of the lads' sea-chests, quite thrice the size allowed in my time ! Dartmouth Harbour is charming, but the town beastly, swarming with dirty children and an undersized population of loutish men and distressingly plain women. The predominance of dirty little lolly-pop shops is the feature of the place.

On May 7, enclosing a page from a book circular with two Buddhist works which Lord Redesdale might care to get secondhand, he relates his own fondness for such advertisements.

I get book catalogues almost every day and run my eye through every one, not with the idea of purchasing, but because it keeps up my memory of my father's and grandfather's fine libraries.

The love of pictures, also, was common ground between him and Lord Redesdale.

The Camp, Sunningdale : March 28, 1900.

MY DEAR OLD FRIEND,—Our last letters crossed. I was delighted to have news from yourself and especially to know that you had congenial work with the Wallace trust. The collection must be a glorious one. I saw a portion of the pictures at Bethnal Green, when exhibited there many years ago, and believe that I recognised a few pictures there from my Grandfather's collection, which was sold during the Crimean War. If I mistake not, one was a small Titian, Europa and the Bull, and there were one or two old Cromes, small pictures. I have a privately printed volume of outline lithographs of my Grandfather's small collection, with full accounts of each picture. Of these I have two, a *very* slight Crucifixion by Van Dyck, of no value ; and a magnificent enamel on copper, by Bone, of L. da Vinci's Christ blessing the world, 12 by 9 in., taken from the (then) Leigh Court collection.

Also I have a very interesting picture by Beechey, which passes as a Rubens, the history of which is that my Grandfather accompanied Beechey to see the Rubenses in Whitehall, and the latter, on returning home, painted the subject of one of the panels from memory, which so pleased my G.F. that he purchased it from him on the spot.

My only other painting of any value is a small Vincent, whose works are very rare ; but for knowing its history it might be a Stark, Nasmyth or Stannard.

I am very busy trying to get my huge heterogeneous correspondence into some order. I have nearly completed the Benthamian, which is extraordinarily rich. B. was in the full swing of Society in France as a young man ; and his diaries are full of interesting matter, from 1810 onwards. I let the Brit. Mus. have (some years ago) all his uncle Jeremy's MSS., an enormous bulk, that will I fancy never be consulted. Maunde Thompson has had them all arranged and catalogued.

It is time I put my house in order, and so good night.

Ever, dear Mitford, Affectionately yours,
JOS. D. HOOKER,

Among his other interests, that in Wedgwood ware continued undiminished in his later years. It was the interest of the connoisseur rather than the collector pure and simple. As he tells W. E. Darwin (July 6, 1900) :

> We now make all our marriage presents in Wedgwood plaques, chiefly ' the Hours ' or ' the Muses '—framed and glazed, and you would hardly believe how much they are prized, and how distinguished they look amongst the fish-slices, paper-knives, salt-cellars and egg-spoons of the bridal gifts.

More than this, the beautiful plaques included many portraits of great men of the past. These cameos, with their historic significance, their memorial to genius as well as their artistic perfection, appealed to him beyond all. He would record the discovery of any which he had not seen before, and if given a photograph of the rarity, offer a copy to W. E. Darwin or Lord Redesdale, his fellow enthusiasts, or send his duplicates. The absence of such portraits he found a blemish in an otherwise magnificent show of Wedgwood ware in 1905.

> It was a show of Jasper Ware and copyists' skill in reproducing and adapting classical figures, &c., but a score or two of Wedgwood's common cups and saucers, teapots, and such articles would have better shown the genius of the man in adapting these to their uses and as being faultless in modelling, ornamentation, and all the best attributes of manufacture and material. So would a collection of medallions and busts have shown his appreciation of learning and genius and great services rendered to the country. (To W. E. Darwin, August 24, 1905.)

A private collection offered for sale in 1907 which ' swarms with cameos and portraits I never saw before ' fills him with proportionate enthusiasm and regrets that he must not commit the extravagance of buying it.

When these memorials had slipped out of memory, his rare knowledge found happy use in reviving them. Thus in 1900 he corresponded with Etruria about the Herschel cameo. Having found by chance that neither Miss Herschel nor her brothers had ever heard of it and were all most anxious to obtain it, search was made for the mould, and a rubbing sent ' of an old gentleman as like Herschel as me.' The mould was identified finally from Hooker's own medallion, which had been made for the 1851 Exhibition, and turned out to be a fine piece of Flaxman's work.

Similarly he suggested that the Wedgwoods should supply the Linnæan Jubilee at Upsala in 1907 with the Linnæus medallion, with the result that ' the firm joyfully respond, and will also send Capt. Cook, Banks, Solander, Bergman, Queen Christina, Charles XII, and Gustavus III.' To complete the matter, he wrote to his correspondent, the Professor of Botany at Stockholm (Professor Wittrock),

> asking him if he could introduce at the Jubilee the subject of the Linnæus Medallion portrait being the work of the famous Swedish sculptor Inlander ; and that Dr. Solander, a pupil of Linnæus (afterwards Banks's Librarian), declared it by far the best likeness of his old master. Also if he could recommend for the Etruria Firm a good agent for the disposal of the medallions, the firm having no correspondent in Sweden. (To W. E. D., January 1, 1907.)

The memories of old times, often curiously re-echoed in the present, are often warmly renewed in the letters to his remaining contemporaries, Mrs. Lyell, whom he had early known as Katherine Horner, and Mrs. Paisley, who, as Sabina Smith of Jordan Hill, had been his playmate in childhood.

To Mrs. Paisley

February 4, 1899.

My dear Sabina,—Your kind letter of the 15th gave me very great pleasure. You are now the oldest of all my friends ! the only one antedating 1830, so that when my mind wanders back and back, ever so far, your name comes as the first and last in the long list of old companions, and always with unclouded associations.

Do you remember our ' black-bide ' [i.e. blackberry] hunts in the hills above Helensburgh, our games in the conservatory at the Baths where Bell's steam-engine lay ? the Amethyst ? the dogs, Copper and Combie ? and the wonderful apparatus for kindling a match by a stream of gas upon platinum (I think), which your mother used to show us ? That reminds me that none of the great scientific discoveries of the century have been more utilised than the progressive ones from the tinder-box flint and steel of our earliest days, to the ' strike a lights ' of the present.

The season here has been quite exceptional—as every season is according to my experience in every part of the world that I have visited ; in this year doubly exceptional, in not being for the worse ! We have had frost at last for two days, but it is passing over and threatens to snow.

I should indeed like to visit you at Helensburgh. The last time I was there was on a visit to Mr. Buchanan at the Baths, some 30 years ago ! The time before that at Ardincaple, when your mother was still alive, and Archie and I paddled about in his skin canoe.

Mrs. Paisley had an hereditary interest in Polar exploration. Her second name was in honour of Douglas Clavering, who commanded the *Griper* which took Sabine to Greenland and Spitzbergen on magnetic work in 1823. Surveying an unexplored part of the coast, he bestowed many Scottish names on his discoveries. One of these was the familiar Jordan Hill !

To the Same

December 12, 1899.

You will be interested to hear that the measures for another Antarctic Expedition [1] are progressing favourably. It will not be on the scale of the last, not being undertaken by Government, which however grants some £45,000 towards it. The contract for building the ship is all but signed, and it will absorb the Government Grant. I am on two Committees concerning it, the general and biological, so I shall end my active life as I began it, in the interest of Antarctic discovery ! Mr. Rücker,[2] one of the Secretaries of the

---

[1] Under Captain Scott, in the *Discovery*.

[2] Sir Arthur William Rücker, M.A., LL.D., D.Sc., F.R.S. (1848-1915); Fellow of Brasenose Coll., Oxford, of London University ; Prof. of Physics, Yorkshire Coll., Leeds, 1874-85 ; R. Coll. of Science, London, 1886-1901 ; Royal Medal, 1891 ; Secretary to the Royal Society, 1896-1901 ; Principal of London University, 1901-8 ; knighted 1902. Sir Joseph's son Reginald married the only daughter of Sir A. W. Rücker, in 1911.

Royal Society, a very able mathematician, is taking the part that Archie did in devising the arrangements for magnetic work, which, as in the former voyage, is one of the chief objects of the expedition. What with steam and a better sailing ship, the coming Expedition ought to do far more work than did the *Erebus* and *Terror*.

Do you remember my father and me breakfasting at Jordan Hill, when your father kindly invited us that I might be presented to Captain Ross, as an applicant for a berth with him ? I well remember that Ross took his place in a separate table with you and your sisters and amused you all, and I longed to be there too ! The expedition is not to sail till 1901, so I cannot expect to see it return and perhaps not even see it sail !

Answering further questions in 1910, he tells Mrs. Paisley how little of a ship's doctor he was.

To the Same
September 13, 1910.

The *Erebus* was my ship when I met Ross at Jordan Hill in 1838, and he promised me (or my father) the appointment of naturalist to his expedition. I had no idea of going as a medical man, but Ross would not take me in any other official capacity, and I had to gallop through a medical degree at the last hour : happily for the crew we had no sickness and hardly an accident to either ship throughout the voyage and we had three other Medical Officers, hence my time was devoted throughout to my natural history studies, in some of which Ross took a keen interest.

To the Same
March 29, 1901.

Yes ! this Antarctic Expedition occupies much of my time and mind. As I am (for now a good many years past) the only surviving officer of Ross's Expedition, I am consulted a good deal, and with the Hydrographer and Sir A. Geikie,[1] had the final revision of the orders to the Captain and the head of the Scientific Staff. I am looking forward with the greatest interest to see the ship when in the Thames.

---

[1] The geologist, President of the Royal Society.

The Captain and head of the Scientific Staff [Dr. E. A. Wilson] both came here and looked over my Antarctic sketches. I liked much what I saw of both.

In a discussion at the Royal Society on an Antarctic Expedition (February 24, 1898), speaking of the unknown origin of the Great Barrier, where no landing seemed possible on its precipitous ice cliffs, he said :

It probably abuts upon land, possibly upon an Antarctic Continent ; but to prove this was impossible on the occasion of Ross's visit, for the height of the crow's nest above the surface of the sea was not sufficient to enable him to over-look the upper surface of the ice, nor do I see any other way of settling this important point except by the use of a captive balloon—an implement with which I hope any future expedi-tion to the Antarctic regions will be supplied.

Add to this its possible use in recovering a lost party, and finding open water. There were several occasions when Ross could have used it when coasting along the Barrier, and more when it would have helped navigation in the Pack. Hence in sending a subscription for the purpose to Captain Scott, Hooker put it neatly :

May 19, 1901.

DEAR CAPTAIN SCOTT,—As I was the first to suggest the use of a captive Balloon in Antarctic discovery, so I ought to be one of the first to respond to your appeal, which will, I do hope, prove successful.

Very sincerely yours,

J. D. HOOKER.

Enclosed cheque £10 10s.

The fact that the German Expedition under Dr. Drygalski in the *Gauss* had at once taken up the idea no doubt aided its adoption here ; but when he had finally seen the ' cumbrous gasometer ' on the *Discovery*, he was fain to confess that if he had known the space the apparatus would occupy on board he might not have been so insistent. For after his first visit to look at the *Discovery* in July 1901, he strongly urged the utility of a balloon upon Sir C. Markham, and advised him to

appeal to the public—using Hooker's name if need be in stating
that without this instrument the Expedition might lose half
its means of accomplishing its end.

With the fund thus raised, two small captive balloons and
their equipment were provided, which were duly used on the
Barrier. (See the 'Voyage of the *Discovery*,' i. 197 *seq*.)
Thanks to the sympathy of the War Office, two officers and
three men of the Expedition had been trained for the work in
advance.

The other point on which he specially dwelt in his remarks
at the Royal Society was that the Antarctic offered endless
investigations to the naturalist, for the South Polar Ocean
swarms with animal and vegetable life. The large collections
made under Ross, i.e. chiefly by Hooker himself, had never
been examined, except the Diatoms.

> A better fate, I trust, awaits the treasures that the hoped-
> for expedition will bring back, for so prolific is the ocean
> that the naturalist need never be idle, no, not even for
> one of the 24 hours of daylight during a whole Antarctic
> summer, and I look to the results of a comparison of the
> oceanic life of the Arctic and Antarctic regions as the herald-
> ing of an epoch in the history of biology.

His regrets over this stifling of scientific results were most
strongly expressed in a letter of January 10, 1901, to Dr. Bruce,
of the *Scotia* expedition, already quoted (see i. 56).

Captain Scott set sail on the last day of July 1901. Sir
Joseph, accompanied by Lady Hooker and his youngest son
and their friend Dr. Smallpiece, had paid a farewell visit to
the *Discovery* on the previous day. When Scott returned
three years later, no one gave him warmer welcome than the
veteran explorer, to whom was brought a renewal and enlargement of the vision of the South which till but three years
before no living eye but his had seen. The photographs, so
much more adequate than the drawings he himself had brought
back, stirred his memories ; across the gap of sixty years he
recognised and named every point in the scenes shown to him,
and pronounced the most interesting fact for science to be the

retrocession of the Barrier, in some places as much as twenty
or thirty miles since Ross's visit. He remembered the ice
reaching the slopes of Mt. Terror, where now stood bare dark
cliffs, while the remains of Barrier ice on the shores of the
continent go to show that in a recent geological epoch it must
have covered the whole of the Ross Sea. He found in Scott's
book ' an indescribable charm '; ' his observations on the
great ice sheet are pregnant with new and sound views.'

To an appreciative letter raising these and other points of
critical detail, Captain Scott replied (November 5, 1905) :

> 56 Oakley Street, Chelsea Embankment :
> November 5, 1905.

MY DEAR SIR JOSEPH HOOKER,—No criticism of my book,
public or private, has pleased me so much as your letter.
My reviewers have been kind and in some cases discriminating, but nothing they have said can reward my literary
labours so fully as the thought that I have really brought
vividly before you those scenes of ice and snow which you
once knew so well.

I can see how carefully you have read, and that you
should have done so with appreciation more than repays me
for the difficulties and trouble of writing.

It is very interesting for me to have a written confirmation of the verbal account you gave me of the condition of
C. Crozier at your visit. I have thought a good deal on this
matter, and cannot bring myself to believe that any great
thickness of ice can have disappeared in so comparatively
short a time. It is possible that during your visit a heavy
summer snowfall may have temporarily covered the bare
land shown on the photograph, page 164, and thus the appearance of a complete ice-cap may have been given. You will
see from the photograph what a large tract of uncovered
land there is at present.

As to the retrocession of the Barrier, he wrote :

It is ridiculous of course to suppose that Ross's latitudes
can have been in fault. One of the most satisfactory points
in connection with the proof of the retrocession of the
Barrier edge is that the evidence rests on Sights for latitude.
Had there been a question of longitude one might reasonably

doubt. I suppose it is rather too much to expect the ordinary reviewer to understand what a lot of difference this makes.

For a sailor it is easy to understand what Ross's position was when sailing along the Barrier. One knows well how careful he was obliged to be and one never ceases to wonder that he accomplished so much in his unhandy ships.

Two other points may be quoted :

The whole question of scurvy is bewildering—the history of the disease seems full of contradictions. The account you give of your provisions is extremely interesting and shows that there were certainly no very elaborate precautions taken in your case. It would almost seem well that your supplies did not undergo the test of a polar winter.

I remember that it was quite news to me to hear from you that Ross was coldly received on his return. At first it seems inexplicable when one considers how highly his work is now appreciated. From the point of view of the general public however I have always thought that Ross was neglected, and as you once said he is very far from doing himself justice in his book. I did not know that Barrow was the *bête noire* who did so much to discount Ross's results. It is an interesting side light on such a venture.

I find however that Ross did put Barrow's name on his Chart. You will find C. Barrow on the Chart in vol. i. of my work ; it is just North of Cape Adare.

Other appreciations of the work of the Expedition will appear later.

To return to the sequence of events, a wave of influenza devastated England in the opening months of 1900.

To Lord Redesdale
January 3, 1900.

The column of deaths in the *Times* is appalling, day after day ! Not a few of my old friends appear in it, chief amongst them Paget, which depresses me much. We had botanised together in our teens. I shall go to the Abbey to-morrow.

Paget and myself were two out of the first Board (of 4) that was appointed by Government to examine on the competitive system. It was for the E.I.C. Army and Navy Medical Department. I took all comers in Science for 12 years—during the first few not 50 per cent. knew the freezing and boiling points of water! This was nearly half a century ago!

All at The Camp fell victims to the epidemic. Hooker in particular was badly hit, and with Lady Hooker went to recruit at Bournemouth. ' I have neither taste nor smell,' he wrote, ' per contra my hearing is improved.' But other sequelæ followed, and at midsummer they took the waters at Harrogate.

My Rhododendrons [he tells Lord Redesdale on June 14] have just burst into full bloom, and I was meditating the proposal that you should come and see them when a bad attack of eczema determined me to lose no time in starting for this place. I shall thus lose the sight of my place in its glory : but on the other hand be spared the horrid sight of seeing the Rhododendrons go into widows' weeds, or rather commence wearing the green willow for eleven consecutive months on end.

Meantime [he tells Mrs. Lyell, on April 22] Kew still claims about one day of the week, devoted to the Botanical Magazine, and I occupy my days here chiefly in dissecting plants for the good of Kew Herbarium, and drawing the analyses on the sheets for the use of those coming after me. This work, dissecting flowers, fruits and seeds, has been a lifelong passion with me. I often think of my dear father working on his Ferns with unabated energy up to the very week of his death.

Then in October and November they were at Weston. He had had bronchitis, his youngest son whooping cough. Having thrown off the effects of the influenza, he was able to winter at home, escaping with no more than a touch of bronchitis in February, which temporarily kept him from going to London to attend the Antarctic Committee. Till summer returned, he had to be cautious about visiting far afield, but in June attended the Jubilee of Glasgow University, and renewed his old memories of the Clyde and Edinburgh, now so different.

June 29, 1901.

My DEAR MRS. LYELL,—It is indeed kind of you to
remember my birthday, and to send me the beautiful slippers,
worked by your own hand too. I do not deserve them—
having let your natal day pass over in ignorance of its
date.

The Jubilee of Glasgow University was well carried out,
and I enjoyed it very much, though I could not undergo
all the festivities. Of the city itself I have no great loving
memory. My happy days in Scotland were spent in the
Highlands, and especially at Helensburgh on the Clyde ;
and these were delightfully recalled by a visit of a sweeter
character to Mrs. Paisley, *née* Sabina Smith, one of the 7
daughters of Mr. S. of Jordan Hill ; and the only remaining
one. We were playmates as children in Helensburgh, where
Mr. Smith kept a yacht, and the revisiting the scenes of my
youth there was most pleasurable. The quondam village had
grown into a town, but the neighbourhood is little changed,
and is as beautiful as before, though the Firth of Clyde is
rendered hideous by the black smoke of steamers, of which
there are myriads, plying in every direction, and all vomiting
clouds that literally stretch right across the Firth from shore
to shore !

Edinburgh [1] is as attractive as ever, though enormously
extended on every side. What struck me as even more
remarkable than the dilatation of the city is the number
of magnificent buildings springing up everywhere in the very
heart of the old town. The Botanical Gardens are now in
the centre of a magnificent Arboretum commanding beautiful
views of the city, and adjoining an equally beautiful public
park. The collection of plants in the Gardens is enormously
increased and is kept in perfect order—all are well and legibly
named. The walks in the Arboretum are most skilfully
laid out, and beautifully kept, and the number of rare and
attractive herbarium plants in the Students' department
is really astonishing. There is a good Herbarium, Library
and Museum for instructional purposes and a class of 300 to
400 pupils annually ! who work in a Laboratory, supplied
with microscopes, and all that is needful for research in

[1] ' We had four very pleasant days in Edinburgh with Professor and Mrs.
Balfour, at the lovely Botanical Gardens.' (To Mrs. Paisley, July 11.)

Botany, besides attending the lectures of the Professor—a son of the late Professor Balfour.

To Mrs. Paisley

July 11, 1901.

I cannot express to you the pleasure which my visit to you gave me, chastened though it was by memories—not regrets. Then, too, the many familiar scenes of Helensburgh and the Gareloch were more welcome to me than I could have believed possible. The fact is that, beyond my own family, your family and Helensburgh are the dearest of my memories of Scotland, kept up as they were at Kew by my intimacy with Archie, in his home, his office, and at our fortnightly meetings at the Philosophical Club of the Royal Society.

I should indeed like to have seen the fleet of yachts. I once saw them assembled at Rothesay, and they reminded me of a flight of white butterflies in a lake in a tropical forest, dancing and dipping on the surface of the water.

It was on this occasion that he was taken round the Glasgow Botanic Gardens by the curator, Mr. Christopher Sheney, who, writing in 1912, thus describes the visit :

I need scarcely say that he took a remarkably keen interest in the various collections of flowering plants in the greenhouses. It was, however, on his reaching the Moss House that he expressed his keenest delight, as he evidently never before saw such a large group, nearly four hundred living species of mosses, together, and he was anxious to know what induced me to cultivate them.

I explained that being successful in securing the prize offered by Professor Bayley Balfour for a collection of British Musci and Hepaticae, I thought of trying the experiment of growing them, in which I was more or less successful. I was previously aware of Sir Joseph's vast knowledge of flowering plants of all kinds, but was scarcely prepared to find that his knowledge of these comparatively insignificant members of the vegetable kingdom was, if possible, more vast. Nothing came amiss to him. The water moss, *Fontinalis antipyretica*, *Hookeria lucens* (named in honour of his father), the various species of the apple moss (*Bartramia*), *Splachnum sphaericum*, and that other Alpine species

Andraea ? Alpina), *Lygodon Mongeoltii* and *Leptobryum pyriforme*, seemed to be all quite familiar to him, and he recognised them at once. He seemed particularly anxious to know what species or genus adapted themselves permanently to confinement. The genus *Fissidens*, all species ; *Bartramia*, all species ; *Minum*, many species ; *Dicranum*, many species ; and *Hypnum*, many species.

His wonderful knowledge of these plants seemed nothing short of amazing, and came to me as a great surprise.

Looking back upon the benefits he had received from the great scientific societies during his long career, Hooker was anxious to make some substantial recognition of this. Thus, having compounded for his subscription to the Linnean Society in 1842, he had had all the privileges of membership and had received the volumes of Transactions for nearly sixty years in return for what had turned out to be in comparison a ridiculously small sum. Then the Society had given him its Gold Medal ; it had also struck in his honour a special gold medal bearing his likeness, the work of Mr. Frank Bowcher ; had had his portrait painted by Herkomer, and published papers by him at considerable cost for expensive illustrations. Sentiment added an hereditary tie with the Linnean : both his father and grandfather had been Fellows.

Furthermore he wished to leave a sum to the Royal Society fund for the relief and support of distinguished scientific men and their families, for which, when President, he had officially received large sums. Accordingly in 1901, by a codicil in his will, he bequeathed £100 each to the Royal and Linnean, free of legacy duty, declaring to his cousin, Sir Inglis Palgrave,[1] the constant friend and business adviser, with whom he kept up an unbroken correspondence mainly on family and business matters, that neither he nor his wife would feel comfortable for the rest of their days if he did anything less.

[1] Sir Robert Harry Inglis Palgrave, F.R.S., third son of the late Sir Francis Palgrave, K.H., Deputy Keeper of the Rolls, and of Elizabeth, second daughter of Dawson Turner, and sister of Maria, Lady Hooker, was born 1827, and educated at Charterhouse ; is a J.P. for Suffolk, a Director of the banking firm of Barclay & Co., a Knight of the Order of Wasa of Sweden, a Freeman of Yarmouth, and Lord of the manor and patron of the living of Henstead ; author of several works relating to banking ; edited the *Dictionary of Political Economy*.

In 1902 his eighty-fifth birthday was celebrated in diverse ways. By a happy coincidence, as has been recorded, it was closely followed by the dedication to him of the ' More Letters of Charles Darwin.' The German Emperor, a little before-hand, sent him the highest Prussian decoration, the order ' Pour le Mérite,' to the huge excitement of the German governess at The Camp, who sentimentally kissed the ribbon of the order.

' It makes me a Ritter,' he tells his son, ' and if ever I go again to Germany, the soldiers will present arms to me ! '

Of this he writes to [Sir] F. Darwin, June 15, 1902 :

Thanks for your congratulations. I well remember the pleasure which the recognition gave to your father. I have refused all foreign orders, and only accepted this on the assurance that the King permitted its being given and worn —being a Civil Servant I am bound by rules of ' Orders in Council.'

You will think me a sad growler when I tell you that I have two faults to find with the thing—the French title—and that the badge is a reminder of a school medal with ' Virtue's Reward,' or ' For Good Conduct ' on it. This between ourselves.

' The badge,' he tells Lord Redesdale, ' is rather insignifi-cant, but the collar ribbon is that of the Black Eagle ! '

On the day itself, June 30, arrived a sundial for the garden, presented by a number of his friends. Of this he writes to Mrs. Lyell, July 2, 1902.

I do indeed deeply appreciate your affectionate regards so long granted me, and now so touchingly expressed. I do indeed thank you heartily. The Sundial was a great surprise and no tribute ever paid me has given me such pleasure, and your name at the top of the list of the 42 ! I did not want a reminder of you here, for I never pass the *Linnaea* without thinking of you.

The year 1902 saw the Coronation of King Edward VII. As a G.C.S.I. and distinguished Civil Servant Hooker found it necessary to break in upon his usual quietude and attend this and other functions. Thus on July 1 there was the Astronomer Royal's garden party. ' Having, as P.R.S., been chairman of the Board of Visitors for 5 years, I felt bound to go, and met only two persons known to me ! '

Then he continues to Mrs. Lyell :

On Thursday (the 7th) we go to the grand Indian affair. I shall think of you and wish you could renew your sight of the grand Indian Chiefs. As I was at the Waterloo station yesterday, 4 Indian regiments filed past me—they sent the blood tingling to my finger tips, such grand fellows, and such gentlemen, such proud yet pleasant faces, such an air of dignity and self-respect.

On the 27th, for the Coronation was still a fortnight ahead, he tells Mr. Gamble :

What with the ' Nature Study ' exhibition and the ' Chelsea Garden Jubilee ' and the dinners given to the ' Most Meritorious ' [i.e. in honour of the members of the newly created English Order of Merit] I have been in a whirl last week, and greatly obstructed in dragging the lengthening chain of my father's life and works.

His solace lay in being transported to his beloved India as he read the proof sheets of Gamble's Malayan Botany, while

economising the time spent in toasting my toes, which even in July I cannot keep warm without a fire. . . . Your sheets have been Godsends, for the moment after I get them I fling myself into my easy-chair and thoroughly enjoy the memories they stir up of collecting, preserving and working up such a lot of old friends in the shape of specimens, and localities of India, and above all old friends of botanists, there and at Kew.

The Coronation took place on August 9.

I have to wear a voluminous blue silk mantle with a huge gold star worked on it, and shall feel the ' lean and slippered pantaloon ' that I am, in doll's clothes. (To Mrs. Paisley, July 20.)

His impressions of the ceremony, taken from letters to

Mrs. Lyell (August 12) and Mrs. Paisley (August 20), are worth recording.

You may like to hear our experiences of the Coronation, which we managed with much less time and trouble than I had anticipated. We had to sleep in town, of course, and be up by 5 A.M. to get the dressing and breakfast over by 7¾ when we started, and after many stoppages reached the Abbey at 9.30 to a minute. The crowds of orderly people in the streets were really imposing, and the troops of soldiers of all countries and of sailors, drawn up along the whole length of Whitehall, with dashing officers of all ranks capering about, were a wonderful spectacle for colour. On being set down we were at once conducted to seats in the nave, almost within hand-shake of the procession. The Ceremony itself we could not see, as the central area of the Abbey was crammed with officials and Peers and Peeresses, &c. The Procession was most imposing, stately and dignified—everything in perfect order. (L.) We were deeply impressed with the solemnity and stateliness of the whole proceedings, broken only once, and then by the volley of cheers for the Crowned King, the effect of which in the Abbey was a spasm of wonder, love and awe. The multifarious, many-coloured garments of the officials were striking, but their gorgeousness almost suffocating, and the width of gold lace tawdry— especially on the clericals! (P.) Bishops, Deans, Canons, who, as it appeared to me, could hardly stagger under their resplendent mantles ; had they been all in pure white robes the effect would have been far more to my liking and more effective too. (L.) The darkness of the Nave was a great drawback—the sky was all but black, and the windows were blocked by the tiers of woodwork for the accommodation of those invited. It was difficult to recognise the most familiar faces. (P.) In the theatre, as the central area is irreverently called, electric light was turned on. As it was, in the nave the jewels did not sparkle, not even in the King's crown. The Archbishop and Dean were both evidently very frail, the latter literally tottering along, and the D. of Cambridge was rather dragged than walking. The King and Queen bore themselves with quiet dignity. (L.) I had no idea that the coronets of peers were so hideous when on their heads (as the procession left the Abbey), and the ' Cap of Maintenance ' was little better. The crowns of King and Queen were particularly elegant, and fitted well, but from want of light their ' unspeakable ' gems did not sparkle ; which was a disappointment to me, who love jewels (but not for myself) and fancy I am a connoisseur, mineralogically at least. (P.) The music was of the best, and most admirably selected (L.) but did not gratify me though for ' time-keeping ' it was marvellous. The voices drowned the organ, and to my ears were harsh—but the Abbey is notoriously a bad building for music. (P.) We could not hear the words of the ceremony, but could time them with our books, so that nothing was really lost. (L.) The voices of the Heralds' silver trumpets were lovely, at least I thought so. Curiously enough I had from childhood wished to hear them, no doubt from some dim recollection of the Coronation of William IV—so I was prepared to greet them and be gratified. (P.) Lastly as to finding your carriage, every one of them was numbered and the drivers came up one after another in rotation calling out each his number ; if you were not ready he passed on and came round again in his turn to pick you up.

The solemnity of the whole ceremony was most impressive, and I am glad I went, though I was bothered by my gorgeous, voluminous sky-blue satin mantle of a G.C.S.I. with a gold star on it as big as a soup plate, and a heavy gold collar on my shoulders.

Part of December, and nearly all January 1903, were spent at Bexhill. In the spring he was crippled by a return of his old enemy eczema, and in the middle of May went again to Harrogate. By the end of June, ' though the demon was not yet exorcised completely,' he was able to walk about, and spent his birthday in taking his youngest son and a grand-daughter from Cirencester over York Minster, where ' a most civil official showed us many things not usually seen,' from the Archbishop's crozier, a magnificent affair 6 feet high, to a sketch of poor Martin, who *conflagrated* the Minster in 1829.

By the autumn some trouble was still to be felt in ankles and instep, which hindered his walking or standing about over his botanical specimens.

Two books read this autumn carried him back, one to the progress of Victorian science, the other to his own explorations round Kinchinjunga.

To Mrs. Paisley

October 7, 1903.

I have just finished reading Sidney Lee's ' Life of Queen Victoria.' It is most interesting, but depressing. She was indeed a good woman though with many imperfections. From a political point of view it is very difficult to judge her on Sidney Lee's showing, one sways backwards and forwards in estimation or the contrary. Her indifference to all the great discoveries in Science during her reign, and especially the Medical and Surgical, strikes me as abnormal. This is not pointed out, and must go with her neglect of Ireland, as being under my view the great drawbacks to a warm appreciation of her reign.

To D. Freshfield

December 3, 1903.

You have indeed sent me a crowning present, of your really monumental work [' Round Kangchenjanga '] with its dedication, which I regard as the greatest honour by far that my Himalayan Journals have received.

I have arrived at a time of life when my contributions to Sikkim Geography might well have been forgotten, and to find them fresh in the memory of those most capable of appreciating them is a greater satisfaction by far than I can express in words. I shall read with keen interest your admirably got up work. The reproductions of the photographs are perfect, which contrasts to my impotent attempt to represent similar objects. The two Lepchas opposite page 36 almost upset me when I remember how kindly helpful the poor fellows were to me.

December 16, 1903.

MY DEAR FRESHFIELD,—I have just concluded my reading ' Round Kangchen ' with absorption, with pleasure that I cannot express in words. Never since reading, as a boy, Franklin and Richardson's journey to the Polar Sea, have I been so fascinated. You have brought to me visions of my happiest early days that I never hoped to see : for your descriptions are as happy as they are truthful ; so much so that they have set me dreaming by night of the Teesta, Zemu, Jongri, and above all Jannu. In your mention of my work you have gone far beyond justice, and I thank you heartily for this.

The sum of work done and light thrown on the structure of Sikkim by your journey is indeed great, and remembering the terrible snowfall is indeed surprising. I was not prepared for your being able to hug the great massif so closely, i.e. at such heights, nor for so complete a chart of the origin and course of the glaciers. This is really a fine piece of work. The photographs giving the sculpturing and structure of both snow and rocks are of the greatest beauty, and remind me of every detail impressed on my mind when studying the realities. Appendix A. has almost upset me. I had no idea that my geological work had any value ; no one hitherto had paid any attention to it, and I had myself forgotten it— I may say *utterly*. Mr. Garwood's resurrection of it, and his most liberal appreciation of it, is I need not say an extraordinary gratification. He has turned it to great purpose in his original views of the origin and building up and sculpturing of Sikkim, and his speculations are of very great interest and promise for the future.

As regards my blundering between the Zemu and Thlonok I make no doubt but that I was intentionally misinformed by the Raja's people, who leading me to believe that the Zemu led into Tibet hoped that its jungles and snows would sicken me.

As to the spectacular effects of Jannu versus Mont Cervin, you are right. Though I have seen the latter often since, it has not the hold on my memory and imagination that Jannu keeps and which you have greatly emphasized I am glad to say. The view of Mont Cervin I alluded to was taken from a shoulder of Monte Rosa on my way to the *old* Weiss-thor pass of grim memory by which I descended to Macugnaga just 50 years ago.

Again thanking you heartily for your book and for my rejuvenescence, believe me,

Sincerely yours,

Jos. D. Hooker.

To Inglis Palgrave

December 23, 1903.

[In reply to some remarks on Herbert Spencer's works.]

I am not surprised at what you say regarding Spencer and his work. I attribute much of your dislike to the effect of his style and diction, which Huxley and I often discussed and regretted ; but more I think may possibly be put down to the stern face set against scientific thought, method and teaching in the educational system of your early years. You somehow acquired an appreciation of scientific methods by the light of nature, and showed it in those of your early writings, which in the opinion of your more scientifically minded friends, induced them to urge your claims *on these very grounds*, for election into the Royal Society, but this appreciation went no further than your own professional work and habits of thought, as far as you were concerned. Then too, may there not be a little of the *odium theologicum* in your dislike of Spencer's system of philosophy ?

As it is, I do not think that any one, except a deeply read man, can appreciate the immensity of Spencer's converse with all that man has done in the spread of knowledge, and of its influence in the development of every phase of his advancement from the savage to the highest civilisation.

I am wholly unable to draw the line between Bacon and Spencer ; I feel that I do not know enough of the work of either, though I have everything that Spencer wrote, up to his last volume, all gifts from himself.

My mother read his little work on education, and was much taken with it, though thinking it was too highly pitched for practical purposes. She told me it was the best book ever written for *bachelor's children*.

Did you happen to read Rücker's address to the ' Modern Languages Association ' in to-day's *Times*, p. 5 ? It is very good ; but there is one matter affecting early education that I have never seen discussed, it is the adverse effect of the modern boy being in point of self education so enormously in advance of what boys were 50 years ago. By self education I mean all that he gets by contact with his surroundings, social, political, commercial, and everything else, especially penny papers. It was comparatively easy for an empty-headed boy 100 years ago to get up his classics and mathematics. Now a boy's head at 12 and 14 is already loaded with knowledge of sorts, that had for his grandfather's boyhood no existence.

But I am maudling, so no more from

Yours ever affectionately,
Jos. D. Hooker.

In March 1903 another revival of early interests was heralded by a note to Professor D. H. Scott : ' I was much amused the other day on finding my infant attempt upon a fossil plant christened in the Geological Journal as a new (genus ?) of plants.'

The sequel appears from the following note by Professor E. A. Newell Arber :

In January 1904 I published in the Geological Magazine (decade v. vol. i., p. 7) a short description of a fossil tree trunk from Tasmania, which had been described by Sir Joseph in 1842, in what I believe to be his first scientific paper. The tree was brought to England for the Great Exhibition of 1851, and was afterwards presented to the British Museum. It remained in the cellars of that Institution until the removal of the Natural History Museum to South Kensington and eventually, in the early 90's I think, was mounted and exhibited in the fossil plant gallery of the Geological Department, where it remains to this day. On the publication of my paper (a copy of which is enclosed) I naturally sent a copy to Sir Joseph, and his reply is appended.

To Prof. E. A. Newell Arber

Bath : January 30, 1904.

My dear Sir,—I am really very much obliged to you for the copy of your paper on the Tasmanian Fossil Tree. I had seen it in the Geolog. Magazine, which Mr. Winwood here kindly sent me, and it came to me like ' Bread cast upon the waters—found after many years ! ' I am indeed gratified by your generous treatment of my virgin attempt at fossil botany. My paper has a history, it having been read in Lady Franklin's drawing room after dinner, quite privately in 1840, the occasion being the embryo meeting of her endeavour to found a scientific society in Tasmania, which subsequently blossomed into the Tasmanian Journal of Natural Science, now I think the Royal Society of Tasmania. Sir John Franklin was Governor of Tasmania at the time, and my only audience was the Governor and his Lady, the private Secretary, Captain Ross, and the Surgeon of the *Erebus.*

It is good news that you will undertake the study and arrangement of the fossil woods in the B.M. When next I can get to London I shall visit the Geological Gallery and hope to make your acquaintance.

Believe me, Yours faithfully,

Jos. D. Hooker.

A much appreciated gift was one from Charles Eliot Norton, President of Harvard, in the shape of a little book he had written on ' The Poet Gray as a Naturalist, with Selections from his Notes on the Systema Naturæ of Linnæus and facsimiles of some of his drawings.' This was brought over by W. E. Darwin on his return from a visit to the States.

To W. E. Darwin.

December 3, 1904.

Many thanks for bringing me C. Norton's present; I am gratified exceedingly by his recollection of me. His Memoir of Gray is charming. How beautifully he writes, and how accurately he sets forth the poet's power as a Naturalist. I wish that your father could have seen the little book. How like and unlike Gray was to the Selborne Naturalist.

To Mrs. Lyell he adds :

It is a revelation to me. Gray seems to have devoted his life to Natural History, all for himself, for he had not even a correspondent ! Shall I send it to you ?

This year, 1904, there was an accession to the household of the young life in which Hooker so much delighted. Of the sons of his second marriage, one was in the Army, had served in the Boer War, and was now in an Indian regiment ; the other had started his schooldays in 1899, and was to enter

Emmanuel College, Cambridge, in the autumn of 1905, to study medicine,[1] a choice on which Sir Joseph remarks :

> I cannot understand it.   I never cared for it, and took it up solely in the view of travelling.   My brother Willy was passionately devoted to it.   (To Mrs. Paisley, August 8, 1905.)

The newcomer was a grandchild from Australia, his son Brian's daughter. ' She is, as you suppose,' he tells Mrs. Paisley, ' a great interest to me.   I give her half an hour of Geography every morning before breakfast, and find her a very apt pupil ; she is further of a happy temperament.'   And when at Christmas in this year he also had with him three of his sons and his unmarried daughter, he called himself ' especially joyful.'

The end of the year brought him fresh echoes of his earliest travels.   Captain Scott returned in the *Discovery*, and the letter of December 3 to W. E. Darwin, already quoted, continues :

> I have been but once in London since you left for America —to see the sketches of the Antarctic Expedition, exhibited in Bruton Street.   The Doctor of the Expedition (Wilson) is a first-rate water-colourist, and his drawings, of the birds especially, are I think unrivalled.   His landscapes, sea-scapes and ice-scapes are most interesting, including extra-ordinary sunsets.

To Mrs. Lyell, on the same day, he is even more emphatic : Above all his drawings of birds are superb : all alive oh ! '

Dr. and Mrs. Wilson were soon among his guests.   Of others he writes to Mrs. Paisley, December 29, 1904 :

> Two days ago I had a call from Col. Younghusband of the late Tibetan Expedition.   He was staying for 2 days at Ascot and most kindly, knowing my interest in Tibet, came over to see me.   He was much amused at seeing, framed and hung up, a telegram which he and his Expedition sent to me on its first arrival in Tibet ; [2] it was prompted by the fact that they had followed my footsteps of 1849, since which

[1] He afterwards abandoned Medicine for the Law.
[2] See i. 275.

year no other Englishman had crossed the frontier ! Nor
in fact had any Englishman for 50 years preceding my
doing so.

To-day I am expecting Capt. Scott of the Antarctic
Expedition, with Admiral Wharton, the late Hydrographer
of the Admiralty.

To F. Darwin

January 4, 1905.

[Answering an enquiry as to what plants are represented
on the Darwin Medal.]

MY DEAR FRANK,—I have botanised over the reverse of
the medal and make out :
1. At the bottom *Dionaea,* followed on each side by
2. Primula,
3. A confused group of leaves and flowers of some tropical
Orchid—I cannot remember its name, though I recognise
the flower. It is not figured in your Father's two works,
i.e. Forms of Flowers and Orchids. I will run it down.
Neither *Nepenthes* nor *Drosera* are thus.
4. *Ampelopsis.*

January 6, 1905.

The Orchid on the medal is *Phalaenopsis Schilleriana.*

To W. E. Darwin

February 19, 1905.

[The ' Letters of Emma Darwin,' edited by her daughter,
which have since been published, were privately printed in
1905.]

I have read every word of Henrietta's interesting volume
with great pleasure ; and with emotion in respect of what
relates to your parents. I often recall with deep feeling your
Mother's winning reception of me on my first visit to Down in
1843. It was followed by your father, who was earnest in
acquiring botanical information, inviting me to come and
stay for a week at a stretch, bringing my own work ; his
reason being that he could not (owing to his head symptoms)
discuss scientific matters for more than one half hour a day ;
and that my shorter stays would involve endless corres-

pondence. On these visits your Mother did everything to make me feel at home. Often I worked in the dining room, (latterly in the billiard room) through which your mother often passed on her way to the store closet in the end, when she would take a pear, or some good thing, and lay it by my side with a charming smile as she passed out. Then in the evening she always played to me, and sometimes asked me to whistle to her accompaniment of some simple air! Those were happy days to me. Your father and I never discussed scientific questions except for the half hour after breakfast and even that always fatigued him. At other times we had long chats by which I profited enormously, especially during the forenoon and afternoon *sand walks*,[1] for which he invariably summoned me.

I cannot express the pleasure that your sister's work has given me.

To Mrs. Paisley

August 8, 1905.

I have kept very well indeed throughout the spring and summer—always at home—I have only occasional attacks of my trouble and these are always bearable. I read a great deal, especially the Lives of eminent men. I have just finished the Letters of Sir Walter Scott to Mrs. Matheson[2] [?] with deep interest. They would, if anything could, raise my admiration for Scott. I wonder whether your father knew him. I remember his son when with his regiment in Glasgow. I cannot comprehend the positive distaste that the present generation of young folk show for the Waverley Novels, and stranger still for the Minstrelsy, which latter especially, as having been in a measure learnt by heart, are as fresh and charming to me as they were when in my youth.

To A. R. Wallace

The Camp, Sunningdale : November 12, 1905.

My dear Wallace,—My return from a short holiday at Sidmouth last Thursday was greeted by your kind and

---

[1] A dry sand-walk had been made round a certain coppice in the grounds at Down, and on this Darwin used to take his appointed measure of daily exercise—so many times round.

[2] Query :—Mrs. Hughes' *Letters and Recollections of Sir Walter Scott*, edited by Horace Hutchinson, 1904.

welcome letter and copy of your ' Life.' The latter was, I assure you, never expected, knowing as I do the demand for free copies that such a work inflicts on the author. In fact I had put it down as one of the annual Xmas gifts of books that I receive from my own family. Coming as it thus did quite unexpectedly, it is doubly welcome and I do heartily thank you for this proof of your greatly valued warm friendship. It will prove to be one of four works of greater interest to me than any published since Darwin's ' Origin ' ; the others being Waddell's ' Lhassa,' Scott's ' Antarctic Voyage,' and Mill's ' Siege of the South Pole.'

I have not seen Clodd's Edition of Bates's ' Amazon,' which I have put down as to be got, and I had no idea that I should have appeared in it.[1] Your citations of my letters and their contents are like dreams to me ; for to tell the truth, I am getting dull of memory as well as of hearing, and what is worse, in reading, what goes in at one eye goes out at the other. So I am getting to realise Darwin's consolation of Old Age, that it absolves me from being expected to know, remember, or reason upon new facts and discoveries. And this must apply to your query as to any one having as yet answered De Vries. I cannot remember having seen any answer, only criticisms of a discontinuous sort. I cannot for a moment entertain the idea that Darwin ever assented to the proposition that new species have always been produced from mutation and never through normal variability. Possibly there is some quibble as to the definition of mutation or of variation. The Americans are prone to believe any new things, witness their swallowing the thornless Cactus produced by that man in California, I forget his name (Harland ?)[2] which Kew exposed by asking for specimens to exhibit in the Cactus House.

I have been for years working at the Indian species of *Impatiens*, the distribution of which is unparalleled amongst Indian phanerogams. One species alone, the indigenous Garden Balsam, is found in most parts of India. Of the rest, some 200 species, most by far are strictly limited to geogra-

[1] It had escaped his memory that he had furnished Mr. Clodd with this material.

[2] *New Creations in Plant Life : an Authoritative Account of the Life and Work of Luther Burbank*, by W. S. Harwood, 1905. Among these ' new Creations ' was a thornless *Opuntia*.

phical areas. The species are wonderfully constant though insect fertilised and are most demonstrably (to appearances) dependent for their creation, variation, &c. on Insect action. I am sending you the published part of an epitome of the species, the object of which is to draw collectors' attention in India to the necessity of observing as well as collecting.

To W. E. Darwin

January 7, 1906.

I knew Lady Dorothy Nevill very well, and had many invitations to her hospitable house. Her narrative does not do justice to herself. She was not the frivolous character she paints. She was thoroughly interested in the rare plants of her noble garden. Her exertions in the hopeless endeavour to establish a silk culture in England were earnest and long continued—and her efforts to improve Donkey breeding and other industries of a like nature were as intelligent as useful. I ought to go and see her, as she made me welcome in London too, but have not for years.

To Mrs. Paisley

July 20, 1906.

[After his eighty-ninth birthday.]

My dear Sabina,—I cry shame on myself for so long delaying to acknowledge and send grateful thanks for the welcome congratulations of one who is by many years my earliest friend. Of the pile of tokens of affection that lies on my table, yours is surely the only one that carries me back to childhood's years and to memories that have cheered many an hour of sad and serious as well as happy thoughts. The ' Baths ' and ' Jordan Hill '—your father and mother and all your sisters and above all Archie and yourself are no mere dreams to me, but vivid realities. Louisa comes next after you. Of the few companions I had in Glasgow all have gone years ago, and I really think that you are by good 20 years or more the earliest living friend, as you are the most prized, on many accounts.

It is good news that you are so well and above all free from pain. No doubt much vigour cannot be hoped for at our ages, and I do get a little stiff ; but except for flying fits of eczema which render walking troublesome, I have nothing but ' hardness of hearing ' to complain of.

My boys are all *in statu quo*, which a witty aunt of mine used to say should be translated ' always worse and worse.' Happily this does not apply to my children. Willy, the eldest, is a merchant, deeply engaged in East African affairs. He made me a birthday present of a most beautiful walking stick of one piece of ivory nearly a yard long, with a heavy gold handle. I am at my wits' end to know what to do with it. I cannot buy a safe for it as I had to do for my diamond mounted star and collar of the G.C.S.I. Such honours are real burdens.

I have not yet read the Duke of Argyll's ' Life,' having quite an incubus of books to get through before taking it up. I am glad to see it is so highly spoken of. I well remember the Duke's sending to my father the cone of a Scotch pine which he saw dropped from the mouth of a corbie or raven, and which curiously enough was infected by a fungus never previously found in Britain ! It is still I suppose exhibited in the Kew Museum of Economic Botany.

I am ashamed to ask you to accept my tardy congratulations on *our* birthday.[1] I can truly say that I have never ceased to love the memory of you, and can still feel your little arms round my neck as I carried you ' pick-a-back ' up the deep road at Helensburgh, and I rejoice in the memory. Hyacinth who shares my joys sends best love.

[1] Mrs. Paisley's birthday was really on the same day, June 30 (though not the same year; she was the younger).

CHAPTER L

THE LAST YEARS

IN 1907 Hooker completed his ninetieth year. The even tenor of his way was only chequered by minor ill health ; his unabated work was little interrupted from outside.

> The fact is [he tells Mrs. Paisley, January 2, 1907] that I live almost a hermit's life, occupied with books and plants as of yore. Kind friends come to see me, but I rarely leave the house. I read a great deal, but the number of books worth reading is now so great that I cannot keep pace with the authors. India always occupies much of my thoughts, and the Durbar about to be held for the meeting of the Governor General and the Ameer of Afghanistan especially interests me.

It interested him none the less because his son Joseph with his Indian Sappers was engaged in preparing the ground at Agra.

In the spring of this year, as in 1906, he went to Bath, where the change of air and scene gave the refreshment desired, without either baths or waters, for the 'former experience of the efficacy of these vaunted cures was not encouraging.'

The ninetieth birthday itself was marked by several honours, described in the following letters.

To Lord Redesdale

(who had sent his congratulations a little prematurely).

The Camp, Sunningdale : June 30, 1907.

MY DEAR OLD FRIEND,—You are the first whom I thank for your welcome and affectionate congratulations, written

when approaching my 10th decade ; as to which I assure
you that their foreshadowing the event added to the great
pleasure they gave ; *bis dat qui cito*, I inconsequently exclaimed. They left me an interval in which to enjoy them,
undisturbed by the advent of the floods that have arrived
within the last two days, leaving me not an interval to dis-
criminate amongst them.

It was good of you to recall the close intimacy of our lives
and work at Kew. Your appointment to the Board of Works
heralded the termination of a long period of official indifference to the real objects of the establishment over which you
watched and in which you played so great a part with zeal
and success. But for you I should never have had an
assistant-director.

Now I have your kind postscript written from Nuneham,
where I was for a good many years an annual guest.

This month has brought me a cupful of honour. The
unique broad gold medal struck by the Swedish Academy in
commemoration of the bicentenary of Linnæus, has been
awarded to me as first of living Botanists, with the acclama-
tion of the host of scientific men there assembled ; and
presented by the Crown Prince to our Ambassador.

To-day I have been waited on by Col. Douglas Dawson
bringing the insignia of the Order of Merit, with a letter from
Lord Knollys informing me that he is commanded by His
Majesty to tell me that it is conferred in recognition of my
eminent services to science, adding the King's hopes that,
notwithstanding my advanced age, I *will* live long to enjoy
the honour.[1]

Excuse this volcano of vanity and believe me ever, dear
sympathetic Redesdale,

Your truly affectionate,
Jos. D. Hooker.

The cumbrous addresses I have received in Latin and
German are terrors to translate and stupefaction to answer.

To celebrate Hooker's ninetieth birthday and sixtieth year
of Fellowship, an address was presented to him by a deputation
from the Royal Society.

[1] 'Is it not curious that Lord Kelvin and I, who sat in the same class in
Glasgow College as boys, should both be recipients of this rare honour ?' (To
Mrs. Paisley, July 16, 1907.)

At [Sir] F. Darwin's request he promptly put together his remembrances of what he said to the deputation, though he would have liked to have a week to shape it. As he sent the draft on July 3, he added :

To F. Darwin

It was indeed a gleeful thing to me having your father's son as a member of the deputation.

I am overwhelmed with addresses, British, French, Swedish, Norsk, German, Dutch, Italian, Finnish ! Austrian, and Russian, all too elaborate to answer cursorily.

What really is wanted are portable phonographs that we could send by post, charged with our worded answers to such communications.

I hope I am right in saying I was the oldest living Fellow R.S.

The chief event of 1908 was the jubilee of the communication of the Darwin-Wallace paper to the Linnean Society in 1858. Hooker was the sole survivor of those immediately concerned, and though now ninety-one, accepted the Society's invitation to speak on the subject. He it was to whom Darwin, then in great distress over the illness and death of one of his children, had first confided Wallace's unexpected letter ; he had first suggested the joint publication and the obtaining of Lyell's judgment, and had offered to write to Wallace explaining matters. But one or two of the letters that then passed were missing ; and he wrote anxiously to [Sir] F. Darwin, who had gone over all the existing material in the 'More Letters,' for documentary confirmation of his recollections. But even when satisfied that his memory had not deceived him, his hatred of *réclame* raised other doubts, and he begged not only [Sir] F. Darwin, but his brothers William and George, to say if they 'entertain the *smallest* doubt of the expediency or propriety of telling the public of the part I took.'

The address was delivered at the afternoon meeting of the Linnean on July 1. Hooker had intended not to go to the evening meeting, ' remembering that the soirées of the

Royal Society were followed by an attack of ' phalangitis,' but
on the day felt strong enough to venture it.

To Mrs. Paisley
July 12, 1908.

For this occasion Lady Hooker took me up to London
(where I had not been a day for many months) on the previous
day, in great dread of my knocking up !   I was overwhelmed
with visitors and letters of congratulation on my *age* no less
than upon my apparently robust health, which continue to
pour in.   All this seems to be very vainglorious, but respect
and consideration for old age is very pleasing—whether
shown by the very few old friends that are spared to me or the
comparative strangers who compete with their kind feeling.

As for me, I am in rude health as far as appearance goes ;
can take little walks, read small print and use the microscope
as well as ever I did, but there is a skeleton in the closet ;
I am troubled with eczema, which, with the stiffening of age,
obliges me to have a nurse always hard by.   Of course I
am taken far more care of than I am worth—am not allowed
to go out if there is a drop of rain, or too much wind, &c.,
&c.   This damps all hope of ever getting to Scotland again,
dearly as I should like to see you and the Clyde again.   It
takes long to think out the fact that we shall not in human
probability meet again on this earth, and can only look
hopefully to a future existence.

I fear you will have trouble in your kind anxiety to read
this scrawl ; the first two pages were written by lamplight.
I have tried to be larger handed in the following.   All I can
say is what the porters at Broomielaw used to say as you
staggered up the gangway from the steamers : ' Tak your
time.'

I must exhaust my vanity.   I have just received a photograph of a drawing of my head made by command of the
King by the Countess Feodora Gleichen for his collection of
portraits of members of the Order of Merit at Windsor.   It
is a charming drawing but ' reproduction is prohibited *by
command*,' so I cannot have it repeated and send you a copy.

The great event of 1909 was the centenary of Darwin's
birth.   Of all the galaxy of notable men who saw the light

in the *annus mirabilis* 1809, Darwin, least in the public eye, came to have the profoundest influence in the world, transcending, beyond all others, the limits of his own country and his own lifetime. It was fitting that this honour should be paid to his memory and his enduring inspiration by Cambridge, his old University, where, if Darwin himself had profited little save by Henslow's direction of his bent towards science, science had since sprung up lustily under the Darwinian impulse, and a strong personal link with his name was kept up by the active work in the University of his distinguished sons.

The proceedings extended over three days, the 22nd, 23rd, and 24th of June; 1500 invitations were sent out. The first evening there was a reception by the Chancellor, Lord Rayleigh, in the Fitzwilliam Museum. Next morning, a presentation of addresses by delegates of Universities, Colleges, Academies, and Learned Societies, in the Senate House; in the afternoon, a garden party at Christ's College; in the evening, a banquet in the New Examination Hall, followed by a reception at Pembroke. On the Thursday, honorary degrees were conferred in the Senate House; the Rede Lecture delivered by Sir Archibald Geikie, P.R.S., and in the afternoon a garden party given by the members of the Darwin family in Trinity College. There was an exhibition also of portraits, books, and other objects of interest in connexion with Darwin, in the Old Library of Christ's, his own College.

It was a brilliant function, resplendent with the bright and many coloured academic robes of various distinctions from a hundred seats of learning in every quarter of the civilised world. Of the guests who represented science at large or some personal link with the Darwin tradition, over five hundred sat down to the great banquet, a polyglot assembly keyed to the highest appreciation, where the admirable interest of Mr. Balfour's historic speech was only eclipsed by the sense of personal charm in Mr. W. E. Darwin's reminiscences of his father. Simple, direct, instinct with the same rich, unassuming humanity that they affectionately depicted, his words seemed to reveal from a still living source the very qualities of his father. ' Now,' one who had met Darwin whispered to his neighbour, 'those who never saw him will be able to understand why Darwin was so much beloved by his friends.'

Writing to Mrs. Paisley on August 11, Hooker describes his share in the celebrations.

At Cambridge we stayed with one of the Darwin family, Horace, the youngest of Mr. Darwin's sons, a scientific instrument maker in Cambridge and F.R.S. (as are two other of Mr. Darwin's sons, George, Prof. of Astronomy, Frank of Botany). The celebration was most successful, and nothing could exceed the delight of the Delegate foreigners, some of whom were invited to bring their wives and daughters. The number of lady guests was remarkable, and added brilliance to all the functions, besides amazing the foreigners, who are not accustomed to see ladies at their Jubilees. The hospitality was boundless, and what struck me most was Mr. Balfour's address at the Banquet (at which I was not present) ; he grasped every salient point in Darwin's character, works, and their results on the progress of science and civilisation in a truly magic manner.

Of course H. took care that I took only corners and snatches of the intellectual food that was spread over every day and part of every night ; and living as I was in the heart of the Darwin family as a brother, I did indeed feel grateful and happy with what I had.

He tells also of their meeting with the famous Dr. Metchnikoff of the Pasteur Institute,[1] whose wonderful sour milk cure Lady Hooker had been trying, and of his amusement when Hooker introduced her as a patient who had benefited by his nostrum.

Of the public functions, he attended the presentation of addresses by the delegates, where the German orator, not yet by Imperial decree cursing where he had blessed, was among the most brilliant of the speakers ; he attended the garden parties and even the late reception at the Fitzwilliam, where the inward eye can still see him, robed in his LL.D gown, as he rested in a sheltered alcove, receiving the affectionate

---

[1] Dr. Elias Metchnikoff (d. 1916), F.L.S. 1880, was elected a Foreign Member of the Royal Society, 1905, and awarded the Copley Medal in 1906.

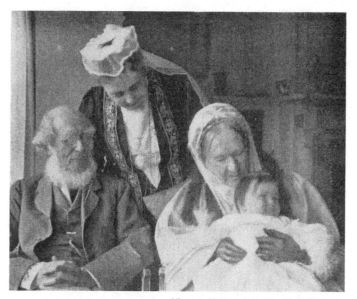

AT THE DARWIN CENTENARY.

Sir Joseph and Lady Hooker. Mrs. T. H. Huxley and Ursula Darwin.

ii. 468]

homage of his friends and admirers. The marks of a hale, serene, and dignified old age were upon him in the softly heightened colour of his face, encircled by a complete halo of silver hair and fringing beard ; in the enhanced prominence and luminous quality of his eyes, which shone very blue from under the veritable penthouse of his eyebrows. As he sat there, still firm and upright, it was hard to believe that he was ninety-two years old. Indeed the two figures which most strongly caught the general imagination as living links with all that those days commemorated, members of Darwin's generation and his close friends in the great days of the past, were such as might move men to love and admire the best gifts of old age. One was Hooker, the other Mrs. T. H. Huxley, then eighty-four. She also was staying in one of the Darwin households, and an historic memento of the reunion of the three families is the photograph here reproduced of the youngest and the oldest representatives of the living tradition : Sir Joseph and Lady Hooker, Mrs. Huxley, and, in her arms, Darwin's great-grandchild, Ursula Darwin.

The flood of congratulations which poured in upon him a few days later on his birthday prompts the reflection :

It is a curious episode in old age when a man gets letters of congratulation from all but strangers—the tribute being not to the individual but to the age he has attained ! Such old age. (To Dr. Bruce, July 13, 1909.)

During July he paid three other visits before settling down again at The Camp : to Cirencester for the marriage of his son Charles's eldest daughter ; to his daughter, Lady Thiselton-Dyer, ' in her pretty house and garden on the Cotswold Hills ' near Witcombe ; and thence for a few days to Pendock, a Worcestershire village where Lady Hooker owned ' a very out of the way property.'

To Mrs. Paisley

August 11, 1909.

My late father-in-law was Rector of Pendock, and the charm of our visit was the delight of the old peasants who had known my wife as a girl. We inhabited there a cottage, charmingly furnished, and rented by my son, the Captain in India, who is devoted to the place and dreams of retiring there 20 years hence! I tell him that he reminds me of the many Indian friends I knew, who dreamt of retiring to their old homes in the Highlands and Lowlands, and whom I found spending their last years in Bayswater and S. Kensington!

I rejoice that you can feel free from any chronic pain. I hope you may yet walk a little with a stick. It may amuse you to hear that my cousin, Mr. Inglis Palgrave, who was knighted the other day, wrote me previously in dismay, saying that if he had to kneel to receive the accolade he could never get up again! I told him to take a walking-stick, and lent him a nice ebony one that he used, and the good-natured King seeing his difficulty had him helped by some of the attendants. He is over 80.

Old interests were again revived by a letter from Mr. T. D. La Touche, son of his old friend, with descriptions of Sikkim and recent changes in the country.

To T. D. La Touche

July 8, 1909.

MY DEAR MR. LA TOUCHE,—I thank you very much for your long and interesting letter of the 4th inst. from Jongri. The contents have intensely interested me, recalling so many scenes once familiar to me. The Oscillations of the Glaciers must be very difficult to determine, for in most cases they debouche in narrow valleys, not as in Switzerland in open meadows or flats. I think that the Lachen and Lachung Glaciers would serve your purpose better than the Western ones.

What you tell me of the destruction of forests, the spread of cultivation, export of maize, the dying out of the Lepcha and his replacement by the Nepalese, and the rarity of Murwa beer, are all shocks to me.

The improvement of the roads alone gratifies me, and I could certainly put up with the bridges, and the diminution of the leech attacks. I hope too that the ticks which I most especially abominated, are less accessible to the traveller.

I enjoy your account of the Rhododendrons and fancy I
can smell the bruised leaf of the little *R. anthopogon,* which
you allude to, and which is the only species of the Himalayan
that stretches away north into the Altai.

I am sorry that my likeness is no longer in the Changa-
chelling portrait gallery.[1]

Your investigation of the little Sikkim lakes will be
very interesting. They are entirely different from those of
Eastern Nepal which I visited. That of Catsuperri is especially anomalous. I shall never forget the weirdness of itself
and its surroundings. You are I expect right in attributing
its existence to silt and landslips. It would be worth while
to have it surveyed and the depth of the soil all round
ascertained, as well as that of the water at various points.
I quite forget at what distance behind it the land rises.
I think it is too thickly clothed with forest to the water's
edge to see what is behind it.

Have you met the Rajah yet? I had him here for a
day [when Kumar], and was charmed with his appearance,
manners and conversation. An excellent photograph of
him, which he gave me, hangs in Lady Hooker's boudoir.

Do you know Mr. Charles E. Simmonds? a gentleman who
called on me last year with magnificent specimens of copper
ore and plumbago from spots which I indicated in my Journal,
and where he has opened mines, under a concession from the
Rajah, as he now writes to me.[2]

P.S. Should you be in Sikkim in the seeding time of the
Rhododendrons and could send me seeds of any, I should
be much obliged. They should be shaken out of the pods,
enclosed in paper capsules (I enclose a sample)—half full
is enough—and despatched *without delay* by post to my
address.

In the earlier part of 1910 Hooker was at Sidmouth escaping the cold winds of spring, a place whose only drawback in
Hooker's eyes was the absence of ships, their course up and
down Channel being far out of sight. In his unfailing birthday
letter to Mrs. Paisley, he tells her how from here one day some

[1] See i. 280.
[2] Besides this rediscovery, Mr. Simmonds found in Sikkim a living memory
of Hooker's visit fifty-nine years before. (See the illustration, i. 272.)

friends took him by motor the sixteen miles into Exeter to
see the statue of his 'Uncle by ancestry,' Richard Hooker,
presented to the city by a very distant relative,[1] and set up
on the grass of the cathedral enclosure, which struck him as
' really a very fine thing.'

Fortified by the good effects of Sidmouth, Hooker was
able to continue working at the Balsams, though he did not
feel equal to the more intense fatigue of journeying to London
to join in the ' send off ' to Scott's last expedition. Instead
Scott paid a farewell visit to The Camp. Moreover, in September he repeated the round of family visits to Cirencester,
for his son's silver wedding, to Lady Thiselton-Dyer, and to
Pendock. The following December he renewed his successful
experiment, and wintered at Sidmouth, his last absence from
home.

One of the greatest pleasures of this stay was seeing his
friend Colonel Cunningham,[2] who with his brother paid them
a visit of which he gives some impressions in the following
letter :

Col. Cunningham to Lady Hooker

Tor Mount, Torquay : January 26, 1911.

I must write a line to say how greatly we enjoyed our
visit of yesterday, and how grateful we feel to you and Sir
Joseph for having allowed us to make it ! My brother, in a
note which I had from him to-day, says ' Seeing Sir Joseph
made me feel quite youthful again ' ; and, though I don't
know that it produced exactly that psychic effect on me, I
came away from Sidmouth with very much the feeling that
I suppose many of my Indian friends experienced when they
returned from a successful ' Tirath ' or pilgrimage to a shrine
containing one of their special objects of adoration !

The following to Professor Oliver, who had barely escaped
from an accident at a railway station, may be quoted as reveal-
ing his warm affection for his old friend and fellow worker.

---

[1] Robert H. Hooker, of Amalfi, Weston-super-Mare.
[2] Colonel David Douglas Cunningham, C.I.E., F.R.S., I.M.S. ; sometime
Professor of Physiology in Calcutta Medical College, and Hon. Surgeon to
the Viceroy of India and Hon. Physician to the King ; retired 1898.

August 10, 1910.

My dear Oliver,—I have just been informed by a note from Mr. Gepp of the horrible danger you have encountered and the narrowness of your escape. It is enough to make an old friend's blood freeze. I and Lady Hooker offer to Mrs. Oliver and your family our heartfelt thanks for your providential preservation.

Ever affectionately yours,
Jos. D. Hooker.

Two letters to Sir Edward Fry, and two to Professor Judd,[1] display his earliest loves and interests maintained to the last. Sir Edward Fry, a botanist in his leisure moments, was writing on the Liverworts, and wished to illustrate his book with plates drawn by Sir William Hooker.

To Sir E. Fry

July 29, 1910.

My dear Sir Edward,—Nothing would have more pleased my father, or does me, than your intention of utilizing the plates of the ' British Jungermanniae.'

The work was a labour of love to its author, and I am very proud of him, and so with the Musci Exotici.

The plates of both works are etchings, nothing can exceed their truth and beauty.

Most truly yours,
Jos. D. Hooker.

The Beacon, Sidmouth : January 30, 1911.

My dear Sir Edward,—I cannot express to you the profit and pleasure that the perusal of your kind gift of ' The Liverworts ' has given to me. Absorbed as I have been for so many years in the study of Phanerogams, I have really lost all count of my ignorance of the higher Cryptogams and am

[1] John Wesley Judd (1840–1916), C.B. 1895; LL.D., F.R.S., F.G.S. He was educated at Camberwell and the Royal School of Mines. He was on the Geological Survey of England and Wales 1867–70 ; Inspector of Schools 1871 ; President of the Geological Society 1887–8 ; Professor of Geology 1876–1905 ; Dean of Royal College of Science 1895–1905, and Emeritus Professor of Geology in the Imperial College of Science and Technology 1913. He contributed many scientific memoirs to the *Transactions* of the Royal and other societies, and published various scientific books, on geology, &c., including *Volcanoes*, 1878 ; *The Student's Lyell*, 1896 and 1911 ; *The Coming of Evolution*, 1910.

consequently ' brought up all standing ' by the revelations of your admirable digest of the structure, classification and relationships of a group of plants known to me hitherto only as generic names, or little more.

The many allusions you make to my father's British Jungermanniae touch me deeply. What has always struck me as most remarkable in all his drawings of Mosses and Liverworts is, that they look alive. He had the gift of seeing nature as she is, and transferring her to paper.

I hope that Miss Fry will let me include her name with yours in my appreciation of the excellence of her analyses and her representation of these.

Ever, dear Sir Edward, sincerely yours,

Jos. D. HOOKER.

Of the two books by Professor Judd mentioned below, ' The Coming of Evolution ' is a succinct account of the lines of thought that expanded through Lyell and Darwin into a well-founded theory. It was suggested by the Darwin centenary.

' The Student's Lyell ' is a handbook to Geology adapted from Lyell's great work.

October 29, 1910.

MY DEAR JUDD,—I have twice read ' The Coming of Evolution,' and twice rejoiced that its authorship devolved on you.

I am impressed with the remarkable fulness and completeness of the narrative, and even more by the sense of proportion that is displayed in the treatment of its main features, their inter-dependence and strictly logical sequences. I like also the swing of your style ; you trip only in saying so much of me.

Lady Hooker shares my view, and intends making Xmas gifts of copies to her friends.

I am very glad to see prominence given to Scrope's labours and early views. I travelled over the scenes of his labour with Huxley and a copy of his book,[1] when I discovered the remains of ancient glaciers in Central France (see *Nature*, xiii. 1876, p. 31), the source of which he angrily

[1] See p. 185 *seq.*

disputed. These remains had, I believe, been actually recognised by a young French Geologist, only a few days before
my visit. Having no good map, I gave a wrong name to
the valley in which the remains occurred.

With Lady Hooker's kind regards,

Ever sincerely yours,

Jos. D. Hooker.

July 9, 1911.

My dear Judd,—The Student's Lyell arrived as a most
welcome congratulatory birthday gift, for which I cordially
thank you, together with congratulations on the new edition.
I have only just commenced reading it, my time having been
fully occupied with the Life of Sir Joseph Banks, which
preceded it on date of arrival. The historical introduction
is of course not new to me ; but at my age, memory plays
sad pranks, and I have re-read it with all the interest and
pleasure of novelty. It is a rare tribute to the memory of a
man, the scientific importance of whose labour cannot be
exaggerated.

I well remember first seeing him, when as a boy I was
staying at Kinnordy [probably in 1836], and looking out of
the window saw him wheeling a barrow of marl up to the
house from the pit [to search through for shells].

My earliest knowledge of The Principles was of the fifth
(1837) edition, in two volumes, which I took to sea with me,
and still have, and of which there is a copy in the Kew
Herbarium.

I have heard my mother talk of his visit to my grandfather, Mr. Dawson Turner, who published a memoir of
Dr. Arnold [1] in 1819.

Very sincerely yours,

Jos. D. Hooker.

Many years before, when Governor of the Cape, Sir Henry
Barkly had described in a letter to Hooker a giant *Mesembryanthemum* discovered in Namaqualand, the specimens of
which unfortunately rotted away during the return journey.
Time after time since 1904 Hooker had inquired of Professor

[1] Joseph Arnold, M.D., of Beccles (1782–1818), Surgeon R.N. ; traveller
and botanist, a friend of Dawson Turner. He died in the E. Indies when serving
as Naturalist under Sir Stamford Raffles.

Pearson, now at the Cape, whether he had come across this lost plant, and his final wish as Pearson started for a most interesting bit of botanical exploration in 1910, was that he might at last find General Barkly's huge *Mesembryanthemum*, and plenty of plants throwing light on the present or past distribution of the Flora. The result was of the happiest.

To Professor Pearson

Sidmouth : March 11, 1911.

Your most interesting letter of February 12 reached me here two days ago, and gave me a shock of pleasure. Your Namaqua third journey has been indeed a success, and I heartily congratulate you, as I do myself for having lived to read of the rediscovery of the giant *Mesembryanthemum*. So important an event cannot be hidden from scientific purview, and I felt compelled to communicate it to Colonel Prain for the Kew Bulletin, hoping that you will approve of my action.

Constant to the last in his claims for the scientific extension of practical botany, he enclosed a formal message of support to Professor Pearson's efforts in this direction.

Sidmouth : March 12, 1911.

MY DEAR PROFESSOR PEARSON,—I have read with deep interest your excellent pamphlet advocating the establishment of a Botanical Garden at Cape Town.

My long official connection with the Royal Gardens, Kew, the originator of so many Colonial Gardens and the active correspondent of so many more, leads me to hope that my voice may be heard in support of your appeal.

That our Colonies both temperate and tropical have profited exceedingly by their Botanic Gardens in economic and æsthetic points of view needs no demonstration by me, and there is not one of them known to me that its Government or people would dream of abandoning.

The South-Western African Flora is the richest and perhaps the most beautiful of any temperate one in the world, and must contain a great number of plants of a great economic value that can only be tested under continual cultivation : that none of these should be rubber yielding is inconceivable ! To test these a small laboratory should be attached to the Gardens, where duly qualified amateurs might work, as at Kew, Ceylon, Java, &c., &c.

Australia possesses five liberally supported Botanical Gardens—all I think with laboratories and libraries attached.

The Cape is now rivalling Australia in glorious fruit Gardens, and it is earnestly to be hoped that it will not remain long without a Botanical Garden.

Sincerely yours,

Jos. D. Hooker.

As the Antarctic had been the first great interest of his life, so, after the lapse of seventy-two years, it occupied his last correspondence. This was with his ' brother Antarctic,' Dr. Bruce, with whom he had formed a warm personal as well as scientific friendship, after his return from the South ; backing his first application, in 1909, for a Treasury grant towards the working out of his valuable scientific results, and when he set forth on a new expedition to Spitsbergen, speeding him with wishes for all success and safe return.

Then, in 1911, Dr. Bruce wrote his ' Polar Exploration,' and ultimately dedicated the book to Hooker, for the latter, having consented to look over the account of Ross's voyage, not only suggested various points from his unique knowledge of the circumstances, but offered to do the same for the rest of the proofs. The official account of the voyage hardly made clear, for example, that Ross and Hooker were the only collectors of marine invertebrate organisms throughout the Expedition.

Hooker also arranged to send a number of illustrations and mementoes of the Ross Voyage to a Polar exhibition which Dr. Bruce was getting up in Edinburgh, including a plaster medallion of Ross, ' an excellent likeness by a young artist, brother of one of the officers (Smith) of the *Erebus*, who died young in Australia, I think ' ; a medallion of Sir John Richardson, and portraits of Ross and Franklin, of Davis, the second master, and Lyall, the assistant surgeon, of the *Terror*, besides pictures of the perilous collision of the two ships, and scenes in the Ross Sea and off the Barrier, some of these framed in wood from the smashed rudder of the *Erebus.*

The following, taken from this correspondence, show his vigour still undimmed.

The Beacon, Sidmouth : February 20, 1911.

MY DEAR DR. BRUCE,—I return herewith the proof sheets which I have perused with extraordinary interest and an amount of instruction and information that I never expected to receive at my age.

The extent and amplitude of your personal experience amazes me, as does the use you make of it in clear exposition of the phenomena of Polar conditions, physical and biological.

I return also the 64 pages set up, and Mr. Perris' letter to you of 17th February.

As to the introductory note by me he wants by the *beginning of this week,* I cannot supply it. With me composition is a very protracted affair, I rewrite over and over again. Mr. Perris does not know, and you, I think, forget, that I am in my 94th year, far advanced, and that writing this note is no slight labour—nor is any such appeal to the public really wanted. ' Let every herring hang by its own head ' must be a familiar proverb to you. On the other hand, I should be proud of having your work dedicated to me. To tell you the truth, I habitually distrust such introductory notes by other than the author, they are really publishers' toutings.

My position under Ross was exceptional, my father's friendship with Franklin, Parry, Richardson, Irvine and others, had to be considered.

It does not, I think, appear in the narrative of the Voyage that I was the sole worker of the tow net, bringing the captures daily to Ross and helping him with their preservation, as well as drawing a great number of them for him.

Except some drying paper for plants I had not a single instrument or book supplied to me as a Naturalist, all were given to me by my father. I had, however, the use of Ross's library, and you may hardly credit it, but it is a fact, that not a single glass bottle was supplied for collecting purposes, empty pickle bottles were all we had, and rum, as a preservative, from the ship's stores.

Throughout the voyage I took hygrometer observations

twice or thrice daily, by Daniel's hygrometers, till these were all broken, then by wet and dry bulb.

The Camp, Sunningdale : March 19, 1911.

MY DEAR DR. BRUCE,—Referring to the publication of the results of your two perilous Antarctic Explorations and of the unpublished material, I regard them as of the highest scientific interest and importance in respect of Meteorology, Magnetism, Geography, Hydrography, Geology, Zoology and Botany.

I cannot therefore but expect a favourable answer from the Prime Minister to the application to the Treasury grant for £6,800 to enable you to complete the publication of the Scientific Results, and to repay the sums advanced by your friends who so liberally came forward in your aid.

It is now going on for 70 years since Sir Robert Peel, then Prime Minister, procured me a grant of £1,000 towards the publication of the ' Botanical Results ' of the first Antarctic Expedition (1839–43) in which I had the honour of serving.

Very sincerely yours,
JOS. D. HOOKER.

The Camp, Sunningdale : May 6, 1911.

MY DEAR DR. BRUCE,—' Polar Exploration ' has reached me and I have read it through with great interest and pleasure, greatly heightened by its kindly and flattering dedication to myself, for which I cordially thank you. It is an excellent digest to our knowledge of the Polar region, and was much wanted. As the precursor to your forthcoming ' History of Polar Exploration,' it will be widely welcomed. I have noted a few misprints of which you may be glad of knowing in the event of a new edition. . . .

The freedom from scurvy of the *Erebus* and *Terror* deserved mention. One case alone occurred in the *Terror*, who had it before embarking.

The only serious omission that I notice (if I have not carelessly overlooked it) is that of the marvellous retrocession of the Barrier since Ross mapped it. To me this appears the most momentous change known to be brought about in the Antarctic in little more than half a century. I have seen doubts thrown upon Ross's demarcation of the sea front of the Barrier—but that is ridiculous, he was a first-rate naval surveyor, he visited it in two successive years, and there is the *Terror's* log to confirm it and my sketches showing the East base of Mt. Terror at the junction with the Barrier clothed with white, where now black cliffs appear.

I hope that you can report a good attendance at the Exhibition.

Though he kept at work till but a little before the end, his physical strength began to fail in August. Yet his mental powers remained clear and strong ; till the last he was keenly interested in current topics and the latest contributions to natural science. On December 10 he passed away in his sleep, peacefully and without pain. The last honour of burial in the Abbey was offered by the Dean of Westminster, where his ashes would lie beside those of Lyell and Darwin, in death not divided from the beloved friend and inspirer, whom in turn he had strengthened by his affection and his knowledge. But Hooker's own wish had been to be laid to rest in the family grave at Kew, beside his father and close to the life-work of them both, which will ever be linked with their names. Here, in the churchyard at Kew Green, he was buried on December 17. In the church is a tablet to his memory, bearing his portrait medallion from a model by Mr. Frank Bowcher, reproduced in Wedgwood ware, and the inscription :

1817–1911

JOSEPH DALTON HOOKER,

O.M., G.C.S.I., C.B., M.D., D.C.L., LL.D.

ASSOCIÉ ÉTRANGER OF THE INSTITUTE OF FRANCE, KNIGHT OF THE PRUSSIAN ORDER 'POUR LE MÉRITE,' SOMETIME PRESIDENT OF THE ROYAL SOCIETY.

FOR XX YEARS DIRECTOR OF THE ROYAL BOTANIC GARDENS, KEW.

BORN AT HALESWORTH, 30TH JUNE, 1817.
DIED AT WINDLESHAM, 10TH DEC., 1911.

'THE WORKS OF THE LORD ARE GREAT, SOUGHT OUT OF ALL THEM THAT HAVE PLEASURE THEREIN.'

On the tablet also five plants are portrayed in Wedgwood, representative of some of Hooker's chief interests : *Aristolochia Mannii* (Africa), *Nepenthes albomarginata* (Malay Peninsula), *Cinchona Calisaya* (America), *Rhododendron Thomsoni* (Asia), *Celmisia vernicosa* (New Zealand).[1]

Kew has his personal memory, but Westminster Abbey enshrines another memorial for the nation. This also is from the hand of Mr. Bowcher. It is of marble, a profile medallion in high relief, slightly over life-size, set within an oblong frame—a presentment of him in old age, at once strongly conceived and delicately executed ; in form and expression admirably lifelike, save in the small point that the exigencies of sculpture demand a greater fulness of beard than he habitually wore.

It is placed in the north aisle of the nave, where the Abbey honours modern science. Here is the Darwin memorial, erected some thirty years before ; then a group of men famous beyond their own generation ; last the memorial of Hooker himself. But though this group includes other contemporaries and friends of his, the understanding eye overleaps them, and sees closest in commemoration, as closest in affection, those lifelong fellow-workers.

[1] These were designed by Miss Matilda Smith, already mentioned as successor to Walter Fitch, the Kew draughtsman.

APPENDICES

APPENDIX A

In Van Diemen's Land Jorgensen is seen; is sinking very low, for he is constantly drunk. He died in 1844 in the Hobart hospital, a sordid and unpicturesque ending for a wildly picturesque character—a modern Benvenuto Cellini in his mingling of genius, high spirits, and madly irresponsible audacity. Like Benvenuto Cellini he left an Autobiography, but on a smaller scale.

The son of a mathematical instrument and watch maker, he was born at Copenhagen in 1770. Love of adventure took him to sea first as apprentice on an English collier, then to the Cape, where he entered the Naval service, and as midshipman under the famous Captain Flinders, shared in the exploration of Bass' Straits and the north-west of Australia, and in the foundation of Hobarton; not to mention a fantastic march into the interior, when he pretended to take a French traveller beyond the track of any other white man.

Flinders was accompanied by Robert Brown, the botanist, the friend of the Hookers, and by other naturalists and artists sent out by Sir Joseph Banks, the great botanist and traveller, who had sailed with Cook, and was now President of the Royal Society. This then was the means of Jorgensen's introduction to Banks and Hooker; and when later he reached England after a long whaling voyage, he gratified Banks's philanthropic zeal by leaving in his care two Tahitians and two Maoris he had brought back with him.

Thereafter returning to his native land, he came in for the bombardment of Copenhagen in 1807. In accordance with the decree calling on all able-bodied Danes to fight, he was placed in command of a privateer, and early next spring ingeniously cut through the ice a month before any could expect it, and captured several English ships, but was himself captured off Flamborough Head by Captain Langford (not Longford, as his chronicler has it).

But he was never without friends, and besides Sir Joseph Banks, he was welcomed and aided by an official whom he had met near the British lines at Copenhagen.

Then came his great exploit. As a result of the war, Danish supplies were cut off from Iceland, and the Icelanders were starving. With the permission of the British Government a merchant named Phelps sent a ship from Liverpool with stores and provisions. For commander a man was needed who knew the Danish language and Danish ways. Jorgensen, a prisoner of war, volunteered apparently without official sanction, and took out the ship in mid-winter, a ' desperate enterprise ' signalised by a high rate of insurance.

The Danish governor at first refused permission to land any of the provisions, but he was overruled by the insistence of the hungry people. Trade was profitable ; on Jorgensen's return Mr. Phelps resolved to come in person on a second expedition, taking a larger ship, armed with ten guns. He was joined by Dr. William Jackson Hooker, then twenty-four years of age, who has left a full account of the affair in his ' Tour in Iceland.'

This time the Danish governor was still more stiff-necked ; finally Jorgensen landed with some sailors—on a Sunday—carried off the governor, who was quietly at home while the rest of the folk were at church, and proclaimed himself Protector without the shadow of resistance. The Icelanders welcomed the relief from an oppressive administration, and Jorgensen in his Autobiography, written in 1835 and 1838, records his ' satisfaction in knowing that the laws and regulations which I then made remain for the most part in force and undisturbed to this day.'

But after a brief nine weeks his protectorate was cut short by the arrival of a British cruiser, whose captain thought the honour of England involved by this unauthorised attack on a friendly government. Phelps and Jorgensen were warned off. The government itself was left in the hands of ' some of the most respectable of the inhabitants,' namely the Chief Justice and Sheriff of the Western County. The ex-governor in Phelps's ship, Jorgensen in another which he had commandeered, set sail to lay their respective cases before the British Government. It was early on this voyage that Phelps's ship caught fire, and crew and passengers were rescued by Jorgensen's personal energy and courage.

The upshot was disastrous to the nine weeks' king of Iceland. He was arrested a few weeks later on the charge of having broken his parole, ' although,' he asserts, ' I had never given one.' A year of prison and the hulks was the ruin of him. His prison companions, the vilest of gaol-birds, drew him into habits of drunkenness, and, more persistently ruinous, of gambling. After his release

neither interest in London, where he had political information to
sell, nor the quiet of a visit to Dr. Hooker at Halesworth, availed
to cure him. Time after time he gambled away his all. His debts
brought him back to prison in the Fleet; and when the Foreign
Office paid these off, and supplied him with funds for a secret mission
on the Continent, he gambled away these also. Boldness, however, never failed him. By a piece of bluff he secured a free passage
to Ostend, and once on the Continent his letters of credit became
available.

His adventures in Germany, Poland, and France were kaleido-
scopic. As occasion demanded, he employed the methods of Borrow
or of the hero of Koepenick. He witnessed Waterloo and entered
Paris after the Allies. At one time he ruffled it in the capital with
the best; at another he was stripped by the gamblers, and sneaked
away on foot for new adventures. He was rehabilitated by a Scotch
watchmaker in whose shop he had noted a chronometer by Jorgensen
the elder. He was introduced to Grand Dukes and to a greater
than these, Goethe. But although he did not fulfil his mission as
planned out, he brought back enough information to earn reward
from the Foreign Office on his return.

Again he broke his excellent resolve to make good use of his
money and emigrate to South America. Three years, from 1817
to 1820, he spent in gambling and dissipation. At the end of
this a third spell of prison awaited him. He was charged with
' converting ' his landlord's furniture, and sentenced to seven years'
transportation. But thanks to his influential friends, he was temporarily detained in England and was employed as an assistant in
the hospital of Newgate prison, obtaining medical knowledge that
was to stand him in good stead afterwards. Nor did he only attend
to the bodies of the convicts; he used to preach Sunday sermons
to his fellow prisoners. At the end of twenty months, however, it
was found that the offence for which he had been condemned had
been committed by his fellow-lodger, and he was set free on condition
of leaving the country within a month.

It was a fatal delay, for he was within reach of temptation.
He succumbed, gambled away all he possessed, outstayed his allotted
month, and on his belated way to the docks, was betrayed by an
old Newgate acquaintance for the sake of the reward, and for his
default was formally condemned to death, a sentence commuted
to transportation for life.

He managed to be reappointed to his old post in Newgate hospital,
and stayed there three years before the sentence was carried out in
1825. On board the convict ship he was made dispenser of the
hospital, and on the death of the surgeon—a stalwart of the calomel
school—took entire charge of the sick as far as Capetown, with the

greatest success. On May 4, 1826, he saw once more the fair city of Hobarton, the site for which he had helped to clear in the wilderness twenty-four years before.

Here he had wild adventures in the service of the Van Diemen's Land Company among blacks and bushrangers. Back in Hobart in 1827 with a ticket of leave, he turned his versatile talents to editing a paper; then was appointed to the constabulary and did service in the pursuit of bushrangers and blacks for two more years, till granted a free pardon. By this time he was, he tells us, entirely cured of his gambling propensities. Still he was unable to make good use of such moneys as came into his hands; he married a termagant wife, and, as we have seen, sank lower and lower till his death.[1]

At one time or another he published four books, on travel, religion, and the state of Van Diemen's Land; three more unpublished MSS. are no longer extant, though one, the History of the Black War in Van Diemen's Land, was used by James Bonwick in 'The Lost Tasmanian Race'; several other MSS., including romance and drama, are to be found in the British Museum, in addition to his many letters to W. J. Hooker, Dawson Turner, and Henry Jermyn.[2]

APPENDIX B

LIST OF WORKS BY THE LATE SIR JOSEPH HOOKER

Taken, with slight additions, from the Kew Bulletin, No. 1, 1912.

1837

Polytrichum semilamellatum, Grimmia laxifolia, Glyphocarpa Roylei, nn. spp. (Hook. Ic. Pl. 1837, vol. ii. t. 194.)

1840

Musci Indici; or list of Mosses collected in the East Indies by Dr. Wallich (by W. H. Harvey); to which are added those

[1] For a capital account of this romantic figure see *The Convict King*, by J. F. Hogan (Ward & Downey, 1891), to which I am much indebted.
[2] Henry Jermyn (1767–1820) was a Suffolk antiquary of Halesworth (J. D. Hooker's birthplace), who collected materials for a history of Suffolk in connexion with his friend D. E. Davy. Neither published, and their MSS. are in the British Museum.

collected by Dr. Royle in the northern part of India (by
J. D. H.). (Hook. Journ. Bot. 1840, vol. ii. pp. 1–21.)
Contributions towards a Flora of Van Diemen's Land, chiefly
from the collections of Ronald Gunn, Esq., and the late Mr.
Lawrence. (Hook. Journ. Bot. 1840, vol. ii. pp. 399–421.)
Entosthodon obtusifolius, E. Mathewsii, E. latifolius, nn. spp.
(Hook. Ic. Pl. 1840, vol. iii. t. 245); *Tridontium tasmannicum,*
n. sp. (t. 248); *Stackhousia flava,* n. sp. (t. 269); *Boronia nana,*
n. sp. (t. 270); *Stenopetalum incisifolium,* n. sp. (t. 276);
Baeckia thymifolia, B. prostrata, B. affinis, nn. spp. (t. 284);
Myriophyllum variaefolium, n. sp. (t. 289); *Goniocarpus ser-
pyllifolius,* n. sp. (t. 290); *Claytonia australasica,* n. sp. (t. 293);
Calandrinia calyptrata, n. sp. (t. 296); *Epilobium macranthum,*
n. sp. (t. 297); *Baeckia leptocaulis,* n. sp. (t. 298); *Milligania
cordifolia,* n. sp. (t. 299); *Caldasia argentea,* n. sp. (t. 300).

1841

Xanthosia dissecta, n. sp. (Hook. Ic. Pl. 1841, vol. iv. t. 302);
Hydrocotyle cordifolia, n. sp. (t. 303); *Didiscus humilis,* n. sp.
(t. 304); *Meionectes Brownii,* n. sp. (t. 306); *Didiscus pilosus*
(t. 307); *Leptospermum rupestre,* n. sp. (t. 308); *Baeckia
micrantha* (t. 309); *Tillaea macrantha,* n. sp. (t. 310); *Goniocarpus vernicosus,* n. sp. (t. 311); *Hydrocotyle tripartita* (t. 312).

1842

On the examination of some fossil wood from Macquarie Plains,
Tasmania. (Tasmanian Journ. Nat. Sci. 1842, vol. i. p. 24.)

1843

Notes on the Botany of H.M. discovery ships 'Erebus' and 'Terror,'
in the Antarctic Voyage; with some account of the Tussac
Grass of the Falkland Islands. (By W. J. Hooker, from
letters of J. D. H.) (Hook. Lond. Journ. Bot. 1843, vol. ii.
pp. 247–329.) Reprint. London, 1843.

1844

The Botany of the Antarctic Voyage of H.M. discovery ships
'Erebus' and 'Terror' in the years 1839–1843, under the
command of Captain Sir James Clark Ross—
   Part I. Flora antarctica. London, 1844–1847. 2 vols.,
xii + 574 pp., 198 pl. 4to.
   (Pp. 289–302 translated in Ann. Sci. Nat. ser. 3, Bot. 1846,
vol. v, pp. 193–225, pl. 5–9.)

Part II. Flora Novae-Zelandiae. Vol. i., Flowering Plants. London, 1853–1855. xxxix + 312 pp., pl. 1–70.—Vol. ii., Flowerless Plants. 1855. 378 pp., pl. 71–130. 4to.

Introductory Essay, pp. i–xxxix reprinted, London, 1853. (Analysis of the Introductory Essay, pp. ii–xxxvi by A. Gray in Amer. Journ. Sci. Arts, 1854, ser. 2, vol. xvii. pp. 241–252, 334–350.)

Part III. Flora Tasmaniae. Vol. i., Dicotyledones. London, 1855–1860. cxxviii + 18 + 359 pp., pl. 1–100.—Vol. ii., Monocotyledones and Acotyledones. 1860. 422 pp., pl. 101–200. 4to.

Introductory Essay, pp. i–cxxviii reprinted, London, 1859. (Pp. i–xxix, c–cv reprinted in Amer. Journ. Sci. Arts, 1860, vol. xxix. pp. 1–25, 305–326 ; pp. i–xxvi translated in Oesterr. Bot. Zeitschr. 1861, vol. xi. pp. 65–81, 118–128, 155–167.)

Some account of a new *Elaeodendron* from New Zealand. (Hook. Lond. Journ. Bot. 1844, vol. iii. pp. 228–230, pl. 8.)

Catalogue of the names of a Collection of Plants made by Mr. Wm. Stephenson, in New Zealand. (Hook. Lond. Journ. Bot. 1844, vol. iii. pp. 411–418.)

Hepaticae Antarcticae ; being characters and brief descriptions of the Hepaticae discovered in the southern circumpolar regions during the voyage of H.M. discovery ships ' Erebus ' and ' Terror.' (By J. D. H. and T. Taylor.) (Hook. Lond. Journ. Bot. 1844, vol. iii. pp. 454–480.)

Notes on the Cider Tree (*Eucalyptus Gunnii*). (Hook. Lond. Journ. Bot. 1844, vol. iii. pp. 496–501.)

Musci Antarctici ; being characters, with brief descriptions, of the new species of Mosses discovered during the voyage of H.M. discovery ships ' Erebus ' and ' Terror ' in the southern circumpolar regions, together with those of Tasmania and New Zealand. (By J. D. H. and W. Wilson.) (Hook. Lond. Journ. Bot. 1844, vol. iii. pp. 533–556.)

Hepaticae Novae Zelandiae et Tasmaniae ; being characters and brief descriptions of the Hepaticae discovered in the Islands of New Zealand and Van Diemen's Land, during the voyage of H.M. discovery ships ' Erebus ' and ' Terror,' together with those collected by R. C. Gunn and W. Colenso. (By J. D. H. and T. Taylor.) (Hook. Lond. Journ. Bot. 1844, vol. iii. pp. 556–582.)

Lichenes Antarctici ; being characters and brief descriptions of the new Lichens discovered in the southern circumpolar regions, Van Diemen's Land, and New Zealand, during the voyage of H.M. discovery ships ' Erebus ' and ' Terror.' (By J. D. H.

APPENDIX B 489

and T. Taylor.) (Hook. Lond. Journ. Bot. 1844, vol. iii. pp. 634–658.)

Lomaria Colensoi, n. sp. (Hook. Ic. Pl. 1844, vol. vii. t. 628) ; *Myrtus pedunculata*, n. sp. (t. 629) ; *Fagus fusca*, n. sp. tt. 630, 631) ; *Callixene parviflora*, n. sp. (t. 632) ; *Loranthus Colensoi*, n. sp. (t. 633) ; *Ranunculus macropus*, n. sp. (t. 634) ; *Gentiana bellidifolia*, n. sp. (t. 635) ; *G. Grisebachii*, n. sp. (t. 636) ; *Fagus Solandri*, n. sp. (t. 639) ; *Veronica nivea*, n. sp. (t. 640) ; *V. diffusa*, n. sp. (t. 645) ; *Fagus Menziesii*, n. sp. (t. 652) ; *F. cliffortioides*, n. sp. (t. 673) ; *Stellaria decipiens*, n. sp. (t. 680) ; *Epilobium confertifolium* (t. 685) ; *Cardamine corymbosa* (t. 686).

1845

Hepaticae Antarcticae, supplementum ; or specific characters, with brief descriptions, of some additional species of the Hepaticae of the Antarctic regions, New Zealand, and Tasmania, together with a few from the Atlantic Islands and New Holland. (By J. D. H. and T. Taylor.) (Hook. Lond. Journ. Bot. 1845, vol. iv. pp. 79–97.)

On the Huon Pine, and on *Microcachrys*, a new genus of *Coniferae* from Tasmania ; together with remarks upon the geographical distribution of that order in the Southern Hemisphere. (Hook. Lond. Journ. Bot. 1845, vol. iv. pp. 137–157, pl. 6.)

Algae Antarcticae, being characters and descriptions of the hitherto unpublished species of Algae, discovered in Lord Auckland's Group, Campbell's Island, Kerguelen's Land, Falkland Islands, Cape Horn, and other circumpolar regions, during the voyage of H.M. discovery ships, ' Erebus ' and ' Terror.' (By J. D. H. and W. H. Harvey.) (Hook. Lond. Journ. Bot. 1845, vol. iv. pp. 249–276, 293–298.)

Algae Novae Zelandiae ; being a catalogue of all the species of Algae yet recorded as inhabiting the shores of New Zealand, with characters and brief descriptions of the new species discovered during the voyage of H.M. discovery ships ' Erebus ' and ' Terror ' ; and of others communicated to Sir W. Hooker by Dr. Sinclair, the Rev. W. Colenso, and M. Raoul. (By J. D. H. and W. H. Harvey.) (Hook. Lond. Journ. Bot. 1845, vol. iv. pp. 521–551 ; vol. vii. pp. 443–445.)

On *Fitchia*, a new genus of arborescent *Compositae* (Trib. *Cichoraceae*), from Elizabeth Island (lat. 26°, long. 125° W.), in the South Pacific. (Hook. Lond. Journ. Bot. 1845, vol. iv. pp. 640–643, pl. 23, 24.)

Note on some marine animals, brought up by deep-sea dredging,

VOL. II 2 I

during the Antarctic Voyage of Captain Sir James C. Ross.
(Ann. Nat. Hist. 1845, vol. xvi. pp. 238–239.)
Aralia polaris. (Hook. Ic. Pl. 1845, vol. viii. t. 747.)

1846

Note on a fossil plant from the Fish River, South Africa. (Trans.
Geol. Soc. 1846, vol. vii. p. 227.)
Description of *Pleuropetalum,* a new genus of *Portulaceae,* from the
Galapagos Islands. (Hook. Lond. Journ. Bot. 1846, vol. v.
pp. 108–109, pl. 2.)
Description of a new genus of *Compositae* [*Scleroleima*], and a new
species of *Plantago* [*P. Gunnii*], from the mountains of Tasmania. (Hook. Lond. Journ. Bot. 1846, vol. v. pp. 444–447,
pl. 13, 14.)

1847

J. C. Ross, A voyage of discovery and research in the Southern
and Antarctic Regions, during the years 1839–43, vol. i. pp.
83–87, 144–149, 158–163 (cp. Appendix v. pp. 341–346);
vol. ii. pp. 5–8, 245–253, 261–277, 288–302. London, 1847.
Florae Tasmaniae Spicilegium; or Contributions towards a Flora
of Van Diemen's Land. (Hook. Lond. Journ. Bot. 1847,
vol. vi. pp. 106–125, 265–286, 461 [bis]–479 [bis].)
Botany of the Niger Expedition; notes on Madeira plants. (By
W. J. Hooker and J. D. H.). (Hook. Lond. Journ. Bot. 1847,
vol. vi. pp. 125–139.)
Description of a new species of *Lysipoma,* from the Andes of
Columbia. (Hook. Lond. Journ. Bot. 1847, vol. vi. pp. 286–
287, pl. 9a.)
Algae Tasmanicae; being a catalogue of the species of Algae collected on the shores of Tasmania, with characters of the new
species. (By J. D. H. and W. H. Harvey.) (Hook. Lond.
Journ. Bot. 1847, vol. vi. pp. 397–417.)

1848

On the diatomaceous vegetation of the Antarctic Ocean. (Brit.
Assoc. Rep. 1847 [1848], pt. 2, pp. 83–85.)
On the vegetation of the Carboniferous period, as compared with
that of the present day (Mem. Geol. Survey, 1848, vol. ii.
pp. 387–430; Edinburgh New Phil. Journ. 1848, vol. xlv.
pp. 362–369; vol. xlvi. pp. 73–78; pp. 398–400 reprinted in
Amer. Journ. Sci. Arts, 1849, vol. viii. pp. 131–133.)
On some peculiarities in the structure of *Stigmaria.* (Mem. Geol.
Survey, 1848, vol. ii. pp. 431–439.)

Remarks on the structure and affinities of some *Lepidostrobi*. (Mem. Geol. Survey, 1848, vol. ii. pp. 440–456.)

Observations made when following the Grand Trunk Road across the hills of Upper Bengal, Páras-Náth, &c. in the Soane Valley ; and on the Kymaor branch of the Vindhya hills. (Journ. As. Soc. Bengal, 1848, vol. xvii. pt. 2, pp. 355–411 ; translated in Berghaus, Zeitschr. für Erdk. 1849, vol. ix. pp. 230–242.) Reprint. Calcutta, 1849.

Botanical mission to India. (Hook. Lond. Journ. Bot. 1848, vol. vii. pp. 237–268, 297–321 ; Hook. Kew Journ. Bot. 1849, vol. i. pp. 1–14, 41–56, 81–89, 113–120, 129–136, 161–175, 226–233, 274–282, 301–308, 331–336, 337–344, 361–370 ; 1850, vol. ii. pp. 11–23, 52–59, 88–91, 112–118, 145–151, 161–173, 213–218, 244–249.)

Letters to A. von Humboldt, 1848–1849. (Translated in Berghaus, Zeitschr. für Erdk. vol. ix. p. 230 ; Berghaus, Geogr. Jahrb. vol. i.)

1849

The Rhododendrons of Sikkim-Himalaya. (Edited by W. J. Hooker.) London, 1849–1851. 14 + 7 pp., 30 pl. with descriptive text. fol.

Notes, chiefly botanical, made during an excursion from Darjiling to Tonglo. (Journ. As. Soc. Bengal, 1849, vol. xviii. pt. 1, pp. 419–446 ; Journ. Hort. Soc. 1852, vol. vii. pp. 1–23.) Reprint. Calcutta, 1849.

Flora nigritiana. (By J. D. H. and G. Bentham.) (W. J. Hooker, Niger Flora, pp. 199–577, pl. 17–50. London, 1849. 8vo.)

Enumeration of the Plants of the Galapagos Islands, with descriptions of the new species. (Proc. Linn. Soc. 1849, vol. i. pp. 276–279 ; Trans. Linn. Soc. 1851, vol. xx. pp. 163–234.)

Extract from a letter to Professor Wheatstone [on the temperature of the soil in Egypt, &c.]. (Brit. Assoc. Rep. 1848 [1849], pt. 2, pp. 17–19.)

1850

Letter from Churra Poonji, Khasiah Hills. (Gard. Chron. 1850, pp. 694, 710.)

Webb and Berthelot, Histoire Naturelle des Iles Canaries, vol. iii. pt. 3, 1836–1850, pp. 430–432.—*Balanophoreae*.

1851

A fourth excursion to the passes into Tibet by the Donkiah Lah. (Journ. Geogr. Soc. 1851, vol. xx. pp. 49–52, with map.)

492 APPENDICES

On the physical character of Sikkim-Himalaya; a letter to A. von
Humboldt, 1850. (Hook. Kew Journ. Bot. 1851, vol. iii.
pp. 21–31.) Reprint, with sketch-map.

On the vegetation of the Galapagos Archipelago, as compared with
that of some other tropical islands and of the continent of
America. (Proc. Linn. Soc. 1849, vol. i. pp. 313–314; Trans.
Linn. Soc. 1851, vol. xx. pp. 235–262.)

Report on substances as used as Food. (Report of Juries, Class III.,
Great Exhibition, London, 1851, pp. 123–162.)

1852

Description of a new species of *Amomum*, from tropical West Africa.
(Hook. Kew Journ. Bot. 1852, vol. iv. pp. 129–130, pl. 5;
Pharm. Journ. vol. xii. pp. 192–194.

On the climate and vegetation of the temperate and cold regions
of East Nepal and the Sikkim-Himalaya Mountains. (Journ.
Hort. Soc. 1852, vol. vii. pp. 69–131; Journ. Agric. Soc. India,
1854, vol. viii. pp. 35–65, 73–95.) Reprint. London, 1852.

Report of enquiry into the best mode of detecting vegetable substances mixed with Coffee for the purposes of Adulteration,
&c. (By J. Lindley and J. D. H.) London, 1852. 8 + 13 pp.,
3 + 4 col. pl. fol. Lithographed.

Luminous plants. (Gard. Chron. 1852, p. 86.)

Fagus Gunnii, n. sp. (Hook. Ic. Pl. 1852, vol. ix. t. 881); *Cardamine radicata*, n. sp. (t. 882); *Rhododendron Lowii*, n. sp.
(t. 883); *R. verticillatum* (t. 884); *R. rugosum*, n. sp. (t. 885);
R. acuminatum, n. sp. (t. 886); *R. ericoides*, n. sp. (t. 887);
Nepenthes villosa, n. sp. (t. 888); *Phyllocladus hypophylla*, n. sp.
(t. 889); *Rhododendron buxifolium*, n. sp. (t. 890); *Vaccinium
buxifolium*, n. sp. (t. 891); *V. coriaceum*, n. sp. (t. 892); *Leptospermum recurvum*, n. sp. (t. 893); *Diplycosia ciliolata*, n. sp.
(t. 894); *Drapetes ericoides*, n. sp. (t. 895); *Drimys piperita*,
n. sp. (t. 896); *Agalmyla tuberculata*, n. sp. (t. 897); *Leuco-
pogon lancifolius*, n. sp. (t. 898).

1853

On the distribution and organic contents of the 'Ludlow Bone
Bed,' in the districts of Woolhope and May Hill. With a
note on the seed-like bodies found in it. (By J. D. H. and
H. E. Strickland.) (Journ. Geol. Soc. 1853, vol. ix. pp. 8–12.)

On a new genus [*Milligania*] and some new species of Tasmanian
plants. (Hook. Kew Journ. Bot. 1853, vol. v. pp. 296–300,
pl. 7–9.)

Note on the occurrence of an eatable *Nostoc* in the Arctic Regions and

in the mountains of Central Asia. (Phytologist, 1853, vol. iv.
pp. 856–859 ; Proc. Linn. Soc. 1855, vol. ii. pp. 166–169.)
Lindley, The Vegetable Kingdom, ed. 3, 1853, pp. 88–90, 94.—
Balanophoraceae, Mystropetalinae.
Botanical Expedition to Oregon ; a review. (Hook. Kew Journ.
Bot., 1853, vol. v. pp. 315–317.)

1854

Himalayan Journals ; or notes of a naturalist in Bengal, the Sikkim
and Nepal Himalayas, the Khasia mountains, &c. London,
1854. Vol. i. xviii + 408 pp., 5 col. pl., 2 maps. Vol. ii.
xii + 487 pp., 7 col. pl.—Ed. 2. London, 1855. Vol. i.
xviii + 348 pp. Vol. ii. xii + 345 pp. — Another ed.
Minerva Library, London, 1891, 1 vol., xxxii + 574 pp., 13 pl.,
2 maps.—Re-issue. London, 1905. 606 pp. 8vo.
Notes on the fossil plants from Reading. (Journ. Geol. Soc. 1854,
vol. x. pp. 163–166.)
On a new species of *Volkmannia* (*V. Morrisii*). (Journ. Geol. Soc.
1854, vol. x. pp. 199–202.)
On the structure and affinities of *Trigonocarpon* (a fossil fruit
of the coal-measures). (Proc. Roy. Soc. 1854–55, vol. vii.
pp. 28–31 ; Ann. Nat. Hist. 1854, vol. xiv. pp. 209–212.)
On the functions and structure of the rostellum of *Listera ovata*.
(Phil. Trans. 1854, pp. 259–264 ; translated in Ann. Sci. Nat.
1855, ser. 4, Bot., vol. iii. pp. 85–90.)
On some species of *Amomum*, collected in Western Tropical Africa
by Dr. Daniell, Staff Surgeon. (Hook. Kew Journ. Bot. 1854,
vol. vi. pp. 289–297.) Reprint. London, 1854.
On *Maddenia* and *Diplarche*, new genera of Himalayan plants.
(By J. D. H. and T. Thomson.) (Hook. Kew Journ. Bot. 1854,
vol. vi. pp. 380–384, pl. 11–12.) Reprint. London, 1854.
Rhododendron anthopogon. (Gard. Chron. 1854, p. 182.)
On the possibility of impregnating ovules after the removal of the
stigma. (Gard. Chron. 1854, p. 629.)
Lomaria nigra (Hook. Ic. Pl. 1854, vol. x. t. 960) ; *Lycopodium
scariosum*, var. *decurrens* (t. 966) ; *Lomaria vulcanica* (t. 969) ;
Asplenium adiantoides, var. *Richardi* (t. 977) ; *A. adiantoides*,
var. *minus* (t. 983) ; *A. adiantoides*, var. *Colensoi* (t. 984) ;
Cyathea Cunninghami, n. sp. (t. 985).

1855

Flora indica : being a systematic account of the plants of British
India, together with observations on the structure and affinities
of their natural orders and genera. (By J. D. H. and T.

Thomson.)   Vol. i. [all published] xvi + 280 + 285 pp.,
2 maps. London, 1855. 8vo.

Illustrations of Himalayan Plants, chiefly . . . made for the
late J. F. Cathcart . . . the plates . . . by W. H. Fitch.
London, 1855. iv pp., 24 pl. with descriptive text. fol.

On the structure of certain Limestone nodules enclosed in seams of
Bituminous Coal, with a description of some Trigonocarpons
contained in them. (By J. D. H. and E. W. Binney.) (Phil.
Trans. 1855, pp. 149–156.)

On some minute seed vessels (*Carpolithes ovulum*, Brongniart) from
the Eocene beds of Lewisham. (Journ. Geol. Soc. 1855, vol.
xi. pp. 562–565.)

On some small seed-vessels (*Folliculites minutulus*, Bronn) from
the Bovey Tracey Coal. (Journ. Geol. Soc. 1855, vol. xi.
pp. 566–570.)

On *Hodgsonia*, Hook. fil. et Thoms., a new and remarkable genus
of *Cucurbitaceae*. (Proc. Linn. Soc. 1855, vol. ii. pp. 257–259.)

On some remarkable spherical exostoses developed on the roots of
various species of *Coniferae*. (Proc. Linn. Soc. 1855, vol. ii.
pp. 335*–336*.)

On *Decaisnea*, a remarkable new genus of the tribe *Lardizabaleae*.
(By J. D. H. and T. Thomson.) (Proc. Linn. Soc. 1855,
vol. ii. pp. 349–351.)

On *Enkyanthus himalaicus* and *Cassiope selaginoides*, two new
species of Himalayan *Ericaceae*. (By J. D. H. and T. Thomson.)
(Hook. Kew Journ. Bot. 1855, vol. vii. pp. 124–126, pl. 3, 4.)

On *Chortodes*, a subgenus of *Flagellaria*, from the Isle of Pines
(New Caledonia). (Hook. Kew Journ. Bot. 1855, vol. vii.
pp. 198–200, pl. 6.)

Longevity of seeds. (Gard. Chron. 1855, pp. 805–806.)

1856

On the structure and affinities of *Balanophoreae*. (Trans. Linn.
Soc. 1856, vol. xxii. pp. 1–68, pl. 1–16.)

On three new species of *Acrotrema*, from Ceylon. (Hook. Kew
Journ. Bot. 1856, vol. viii. pp. 241–243.)

Géographie Botanique Raisonnée . . . par M. Alph. de Candolle ;
a review (Hook. Kew Journ. Bot. 1856, vol. viii. pp. 54–64,
82–88, 112–121, 151–157, 181–191, 214–219, 248–256). Reprint.
London, 1866.

1857

On some Collections of Arctic Plants, chiefly made by Dr. Lyall,
Dr. Anderson, Herr Miertsching, and Mr. Rae, during the
Expeditions in search of Sir John Franklin, under Sir John

Richardson, Sir Edward Belcher, and Sir Robert M'Clure. (Journ. Linn. Soc., Bot., 1857, vol. i. pp. 114–124.)

On the botany of Raoul Island, one of the Kermadec group in the South Pacific Ocean. (Journ. Linn. Soc., Bot., 1857, vol. i. pp. 125–129.)

On the growth and composition of the ovarium of *Siphonodon celastrineus*, Griffith, especially with reference to the subject of its placentation. (Trans. Linn. Soc. 1857, vol. xxii. pp. 133–141, pl. 26.)

Descriptions of two new Dilleniaceous plants from New Caledonia and Tropical Australia. (Hook. Kew Journ. Bot. 1857, vol. ix. pp. 47–49, pl. 1, 2.)

On *Notospartium*, a new genus of *Leguminosae* from New Zealand. (Hook. Kew Journ. Bot. 1857, vol. ix. pp. 176–177, pl. 3.)

On *Bryocarpum*, a new genus of Himalayan *Primulaceae*. (By J. D. H. and T. Thomson.) (Hook. Kew Journ. Bot. 1857, vol. ix. pp. 199–200, pl. 5.)

On *Loxodiscus*, a new genus of *Sapindaceae*, from New Caledonia. (Hook. Kew Journ. Bot. 1857, vol. ix. pp. 200–201, pl. 6.)

On three new Indian *Scrophularineae*, with description of *Lancea*, gen. nov. (By J. D. H. and T. Thomson.) (Hook. Kew Journ. Bot. 1857, vol. ix. pp. 243–246, pl. 7, 8.)

On a new species of *Diapensia*, from the Eastern Himalaya. (Hook. Kew Journ. Bot. 1857, vol. ix. pp. 372–373, pl. 12.)

British North American Exploring Expedition [Additional Instructions]. (Hook. Kew Journ. Bot. 1857, vol. ix. pp. 216–219.)

1858

Enumeratio plantarum Zeylaniae : an enumeration of Ceylon plants with descriptions of the new and little-known genera and species, observations on their habitats, uses, native names, &c. (By G. H. K. Thwaites, assisted by J. D. H.) London, 1858–1864. viii + 483 pp. 8vo.

Praecursores ad Floram Indicam : being sketches of the natural families of Indian plants, with remarks on their distribution, structure, and affinities. (By J. D. H. and T. Thomson.) (Journ. Linn. Soc., Bot., 1858, vol. ii. pp. 1–29, 54–103, 163–180, pl. 2 ; 1860, vol. iv. pp. 106–157 ; 1861, vol. v. pp. 128–181.)

Cynoglossum nobile, n. sp. (Gard. Chron. 1858, p. 240.)

1859

On the origin and development of the pitchers of *Nepenthes*, with an account of some new Bornean plants of that genus. (Trans.

Linn. Soc. 1859, vol. xxii. pp. 415–424, pl. 69–74 ; translated in Ann. Sci. Nat., ser. 4, Bot., 1859, vol. xii. pp. 222–231.)

On a new genus of *Balanophoreae* (*Dactylanthus Taylori*) from New Zealand, and two new species of *Balanophora* (*B. Harlandi* and *B. Lowii*). (Trans. Linn. Soc. 1859, vol. xxii. pp. 425–427, pl. 75.)

1860

The monstrous *Begonia frigida* at Kew, in relation to Mr. Darwin's ' Theory of Natural Selection.' (Ann. Nat. Hist. 1860, vol. v. pp. 350–352.)

Vaccinium rugosum, n. sp. (By J. D. H. and T. Thomson.) (Gard. Chron. 1860, p. 384.)

On the species of *Cordyline* now in cultivation from New Zealand and Australia. (Gard. Chron. 1860, pp. 791–792 ; translated in Belgique Horticole 1861, vol. xi. pp. 66–70.)

1861

On *Fropiera*, a new Mauritian genus of calycifloral exogens, of doubtful affinity. (Journ. Linn. Soc., Bot., 1861, vol. v. pp. 1–2, pl. 1.)

On *Barteria*, a new genus of *Passifloreae*, from the Niger River. (Journ. Linn. Soc., Bot., 1861, vol. v. pp. 14–15, pl. 2.)

An account of the Plants collected by Dr. Walker in Greenland and Arctic America during the Expedition of Sir Francis M'Clintock, R.N., in the Yacht ' Fox.' (Journ. Linn. Soc., Bot., 1861, vol. v. pp. 79–88.)

Colonial Floras. (Nat. Hist. Review, 1861, pp. 255–266.)

1862

Genera Plantarum ad exemplaria imprimis in herbariis kewensibus servata definita. (By G. Bentham and J. D. H.) London, 1862–1883. 3 vols, 8vo. [For the joint and separate work of the authors, *see* Journ. Linn. Soc., Bot., 1883, vol. xx. pp. 304–308.]

Illustrations of the Floras of the Malayan Archipelago and of Tropical Africa. (Trans. Linn. Soc. 1862, vol. xxiii. pp. 155–172, pl. 20–28.)

Outlines of the distribution of Arctic plants. (Read June 21, 1860 ; Trans. Linn. Soc. 1862, vol. xxiii. pp. 251–348, with map ; pp. 251–276 and 281–309 reprinted in an abridged form in Admiralty Arctic Manual, London, 1875, pp. 197–238.)

On three Oaks of Palestine. (Trans. Linn. Soc. 1862, vol. xxiii. pp. 381–387, pl. 36–38.)

On the vegetation of Clarence Peak, Fernando Po ; with Descriptions of the Plants collected by Mr. Gustav Mann on the higher parts of that mountain. (Journ. Linn. Soc., Bot., 1862, vol. vi. pp. 1–23.)
On the Cedars of Lebanon, Taurus, Algeria, and India. (Nat. Hist. Review, 1862, pp. 11–18, pl. 1–3.)

1863

On *Welwitschia*, a new genus of *Gnetaceae*. (Trans. Linn. Soc. 1863, vol. xxiv. pp. 1–48, pl. 1–14 ; translated in Flora, 1863, pp. 459–464, 473–479, 489–496, 506–510, 513–520 ; and in Diario de Lisboa, Jun. 2, 1863.)
On a new *Heliconia* with the habits of a *Musa*, sent from New Grenada by Dr. A. Anthoine to the Royal Gardens, Kew. (Journ. Linn. Soc., Bot., 1863, vol. vii. pp. 68–69.)
Note on the embryo of *Ancistrocladus*. (By J. D. H. and G. Bentham.) (Journ. Linn. Soc., Bot., 1863, vol. vii. p. 111.)
The Botany of Syria and Palestine. (W. Smith's Dict. Bible, vol. ii., London, 1863.)
Enumeration of the Mountain Flowering Plants and Ferns . . . of the Cameroons Mountains, of Clarence Peak, Fernando Po, and of the Peak of San Thomé. (Burton : Abeokuta and the Cameroons Mountains, vol. ii. pp. 270–277. London, 1863.)
A. F. Henslow, Cotton and the want of it. London, 1863. 19 pp., 2 pl. 16mo. (Botanical description by J. D. H.)

1864

Handbook of the New Zealand Flora, &c. London, 1864–67. 15* + lxviii + 798 pp. 8vo.
On the Plants of the Temperate Regions of the Cameroons Mountains and Islands in the Bight of Benin ; collected by Mr. Gustav Mann, Government Botanist. (Journ. Linn. Soc., Bot., 1864, vol. vii. pp. 171–240, pl. 1 ; pp. 171–181 translated in Petermann, Mitteilungen, 1865, pp. 22–26.)
On the Genus *Euptelea*, Sieb. & Zucc. (By J. D. H. and T. Thomson.) (Journ. Linn. Soc., Bot., 1864, vol. vii. pp. 240–243, pl. 2.)
Note on the replacement of species in the Colonies and elsewhere. (Nat. Hist. Review, 1864, pp. 123–127.)
Epistephium Williamsii, n. sp. (Curtis's Bot. Mag. 1864, t. 5485.)

1865

Curtis's Botanical Magazine, comprising the Plants of the Royal Gardens of Kew and of other botanical establishments in

Great Britain; with suitable descriptions. Vols. xci.–cxxx.
London, 1865–1904, pl. 5486–7991. 8vo. (Vols. cxxix.–cxxx.
assisted by W. B. Hemsley.)

Catalogue of the plants distributed at the Royal Gardens, Kew
. . . from the Herbaria of Griffith, Falconer, and Helfer.
London, 1865, 37 pp. 8vo.

Description of a new genus (*Brandisia*) of *Scrophularineae* from
Martaban. (By J. D. H. and T. Thomson). (Journ. Linn.
Soc., Bot., 1865, vol. viii. pp. 11–12, pl. 4.)

On the identity of *Pinus Peuce*, Griseb., of Macedonia, with the
P. excelsa of the Himalaya Mountains. (Journ. Linn. Soc.,
Bot., 1865, vol. viii. pp. 145–147.)

Discovery of *Asplenium viride*, in New Brunswick. (Nat. Hist.
Review, 1865, p. 150.)

1866

Reports on the progress and condition of the Royal gardens at
Kew during the years 1865–1882. London, 1866–1884. 8vo.

Description of some new and remarkable species of *Aristolochia*
from Western Tropical Africa : *Aristolochia Goldieana, A.
triactina, A. Mannii.* (Trans. Linn. Soc. 1866, vol. xxv.
pp. 185–188, pl. 14.)

Lecture on Insular Floras, delivered before the British Association
for the Advancement of Science at Nottingham, Aug. 27, 1866.
(Gard. Chron. 1867, pp. 6–7, 27, 50–51, 75–76 ; Journ. Bot.
1867, vol. v. pp. 23–31 ; translated in Ann. Sci. Nat. 1866,
ser. 5, Bot., vol. vi. pp. 267–299.) Reprint. London, 1867.

1867

Hooker's Icones Plantarum ; or figures, with descriptive characters
and remarks, of new and rare plants, selected from the
Kew herbarium. Vols. xi.–xx., pt. 1. London, 1867–1890.
tt. 1001–1925. 8vo. (Editor and part author.)

Boott, Illustrations of the genus *Carex*. Part iv. London, 1867.
Pp. 127–233, pl. 412–600. fol. (Edited by J. D. H.)

Martius, Flora Brasiliensis, vol. xiv. pt. ii., 1867, pp. 1–76, pl. 1–22.
—*Rosaceae.*

On the struggle for existence amongst plants. (Popular Sci. Review,
1867, vol. vi. pp. 131–139.)

Begonia Veitchii, n. sp. (Gard. Chron. 1867, p. 734.)

1868

W. H. Harvey, The genera of South African plants. Ed. 2.
London, 1868. lii + 483 pp. (Edited by J. D. H.)

Oliver, Flora of Tropical Africa :—
    Vol. i., 1868, pp. 298–303.—*Impatiens.*
    Vol. ii., 1871, pp. 439–464, 521–580.—*Melastomaceae, Cucurbitaceae, Begoniaceae.*
On seeds and saplings of forest trees. (Canadian Naturalist, 1886, vol. iii. pp. 453–457.)

1869

Presidential Address to the British Association, Norwich, 1868. (Brit. Assoc. Rep. 1868 [1869], pp. 58–75.)
On the true *Fuchsia coccinea,* Aiton. (Journ. Linn. Soc., Bot., 1869, vol. x. pp. 458–461.)

1870

The Students' Flora of the British Islands. London, 1870. xx + 504 pp. Ed. 2. Ib., 1878. xx + 539 pp. Ed. 3. Ib., 1884. xxiii + 563 pp. 8vo.
Nepenthes. (Nature, 1870, vol. iii. pp. 147–148 ; Journ. Bot. 1871, vol. ix. pp. 49–50.)

1871

The ascent of the Great Atlas. (Proc. Roy. Geogr. Soc. 1871, vol. xv. pp. 212–221 ; Brit. Assoc. Rep. 1871 [1872], sect. rep., pp. 179–180.) Reprint. London, 1871.
The Admiralty Manual of Scientific Enquiry, ed. 4, article xv. Botany, pp. 377–384. (By W. J. Hooker, revised by J. D. H.) London, 1871.
Plants of the peninsula of Sinai (determined by D. Oliver). (E. H. Palmer's Desert of the Exodus. London, 1871.)

1872

The Flora of British India. (By J. D. H., assisted by various botanists.) London, 1872–1897. 7 vols. 8vo. Editor throughout ; author of the following Orders :—
    Vol. I., 1872–1875. xl + 740 pp.—*Ranunculaceae, Dilleniaceae, Magnoliaceae, Anonaceae, Menispermaceae, Berberideae, Nymphaeaceae, Papaveraceae, Fumariaceae* (with T. Thomson) ; *Cruciferae* (with T. Anderson) ; *Capparideae, Resedaceae, Violaceae, Bixineae, Pittosporeae* (with T. Thomson) ; *Caryophylleae* (with M. P. Edgeworth) ; *Lineae, Malpighiaceae ; Zygophylleae, Geraniaceae* (excl. *Balsamineae*) (with M. P. Edgeworth) ; *Balsamineae, Rutaceae, Chailletiaceae, Ilicineae.*
    Vol. II., 1876–1880. 792 + 1 pp.—*Sabiaceae, Anacardiaceae, Coriarieae, Moringeae, Connaraceae, Rosaceae.*

Vol. III., 1880–1882. 712 pp.—*Rubiaceae, Compositae, Primulaceae, Apocynaceae.*

Vol. IV., 1883–1885. 780 pp.—*Asclepiadaceae, Scrophulariaceae, Orobanchaceae, Selagineae, Labiatae, Plantagineae, Nyctagineae, Illecebraceae, Amarantaceae.*

Vol. V., 1886–1890. 910 pp.—*Chenopodiaceae, Phytolaccaceae, Polygonaceae, Podostemonaceae, Nepenthaceae, Cytinaceae, Aristolochiaceae, Piperaceae, Chloranthaceae, Myristiceae, Monimiaceae, Laurineae, Proteaceae, Thymelaeaceae, Elaeagnaceae, Loranthaceae, Santalaceae, Balanophoreae, Euphorbiaceae, Urticaceae* (excl. *Ficus* et *Artocarpus*), *Juglandeae, Myricaceae, Casuarineae, Cupuliferae, Salicineae, Ceratophylleae, Gnetaceae, Coniferae, Hydrocharideae, Burmanniaceae, Orchideae.*

Vol. VI., 1890–1894. 748 pp.—*Orchideae, Haemodoraceae, Irideae, Amaryllideae, Taccaceae, Dioscoreaceae, Roxburghiaceae, Liliaceae, Pontederiaceae, Philydraceae, Xyridaceae, Commelinaceae, Flagellarieae, Juncaceae* ; *Palmeae* (with O. Beccari) ; *Pandaneae, Typhaceae, Aroideae, Lemnaceae, Triurideae, Alismaceae, Naiadaceae, Eriocauleae.*

Vol. VII., 1896–1897. 842 pp.—*Gramineae* (with O. Stapf and J. S. Gamble).

Kew Gardens and the National Herbarium. (Nature, 1872, vol. vii. pp. 45–46, 103.)

1873

Le Maout and Decaisne, A General System of Botany, descriptive and analytical ; translated by Mrs. Hooker, with additions, appendix and synopsis of Orders by J. D. H. London, 1873. xii + 1066 pp. 4to.

De Candolle, Prodromus systematis naturalis, regni vegetabilis, vol. xvii. 1873, pp. 90–116.—*Nepenthaceae, Cytinaceae.*

Hololachne Shawiana, n. sp., *Apocynum Hendersonii*, n. sp. *Deyeuxia anthoxanthoides*, n. sp. (Henderson and Hume, Lahore to Yarkand, pp. 313, 327, 339, with plates. London, 1873.)

On *Melianthus Trimenianus*, Hk. f., and the affinities of *Greyia Sutherlandi.* (Journ. Bot. 1873, vol. xi. pp. 353–358, pl. 138.)

The production of Honeydew. (Entomologist, 1873, vol. vi. pp. 463–464.)

Potato Disease. Answers to Circular addressed to Cultivators of Potatoes in the counties of Ross, Inverness, Nairn and Moray. (By Col. J. A. Grant, with remarks by Prof. Church and J. D. H.) Inverness, 1873. 8vo.

1874

Address to the Depart. of Bot. and Zool. of the Brit. Assoc. at Belfast, Aug. 1874.—The carnivorous habits of plants. (Brit.

Assoc. Rep. 1874 [1875], pp. 102–116; Nature, 1874, vol. x.
pp. 366–372; translated in Revue Scientif. 1874, vol. vii.
pp. 481–489.) Reprint. London, 1874.
Notes on some plants from Smith Sound collected by Dr. Bessels.
(A. H. Markham, A Whaling Cruise to Baffin's Bay and the
Gulf of Boothia, p. 296. London, 1874. Reprinted in
Admiralty Arctic Manual, 1875, p. 321.)

1875

On the subalpine vegetation of Kilima Njaro, E. Africa. (Journ.
Linn. Soc., Bot., 1875, vol. xiv. pp. 141–146.)
On *Hydnora americana*, R. Br. (Journ. Linn. Soc., Bot., 1875,
vol. xiv. pp. 182–188.)
On the discovery of *Phylica arborea*, Thouars, a tree of Tristan
d'Acunha, in Amsterdam Island in the S. Indian Ocean; with
an enumeration of the phanerogams and vascular cryptogams
of that island and of St. Paul. (Journ. Linn. Soc., Bot., 1875,
vol. xiv. pp. 474–480.)
Observations on some Indian species of *Garcinia*. (Journ. Linn.
Soc., Bot., 1875, vol. xiv. pp. 484–486.)
Presidential address to the Royal Society, Nov. 1875.—The scientific
work of the year, &c. London, 1875. 25 pp. (Proc. Roy.
Soc. 1875 [1876], vol. xxiv. pp. 72–94.)
Instructions in Botany. (Admiralty Arctic Manual, 1875, Instructions, pp. 62–67.)

1876

Botany. (Macmillan & Co.'s Science Primers.) Ed. 1. London,
Feb. 1876; reprinted Nov. 1876. Ed. 2. 1877; reprinted
1878, 1879, 1880, 1881, June 1883, Oct. 1883, May 1884,
Oct. 1884, 1885. Ed. 3. 1886; reprinted 1887, 1888, 1890,
1892, 1894, 1897, 1900, 1904, 1909. 16mo. Portuguese trans-
lation from the second English edition. (By J. A. Henriques.)
Porto e Braga, 1877.
Evidences of ancient glaciers in central France. (Nature, 1876,
vol. xiii. pp. 31–32.)
Presidential address to the Royal Society, Nov. 1876.—The scientific
work of the year, &c. London, 1876. 27 pp. (Proc. Roy.
Soc. 1876 [1877], vol. xxv. pp. 339–362.

1877

Notes on the botany of the Rocky Mountains. (Nature, 1877,
vol. xvi. pp. 539–540; Amer. Journ. Sci. 1877, vol. xiv. pp.
505–509; Archives Sci. Phys. Nat. 1878, vol. lxiii. pp. 240–247.)

Presidential address to the Royal Society, Nov. 1877.—The scientific
work of the year, &c. London, 1877. 26 pp. (Proc. Roy.
Soc. 1877 [1878], vol. xxvi. pp. 427–446.)

1878

Journal of a tour in Marocco and the Great Atlas. (By J. D. H.
and John Ball.) London, 1878. xvi + 499 pp., 8 pl., 1 map.
8vo. [Includes the following appendices by J. D. H. :—
D. On some of the economic plants of Marocco, pp. 386–404 ;
E. On the Canarian Flora as compared with the Maroccan,
pp. 404–421 ; F. Comparison of the Maroccan Flora with that
of the Mountains of Tropical Africa, pp. 421–423.]
G. S. Nares, Narrative of a Voyage to the Polar Sea during 1875–76
in H.M. ships 'Alert' and 'Discovery,' vol. ii. appendix
xiv., Botany, pp. 301–310. London, 1878. 8vo.
The distribution of the North American Flora. (Proc. Roy. Instit.
1879, vol. viii. pp. 568–580 ; Gard. Chron. 1878, vol. x.
pp. 140–142, 216–217 ; translated in Ann. Sci. Nat. 1878,
ser. 6, Bot., vol. vi. pp. 318–339.) Reprint. London, 1879.
13 pp. 8vo.
Presidential address to the Royal Society, Nov. 1878.—The scientific
work of the year, &c. London, 1878. 28 pp. (Proc. Roy.
Soc. 1878 [1879], vol. xxviii. pp. 43–63 ; Nature, 1878, vol.
xix. pp. 109–113, 132–135.)

1879

Observations on the botany of Kerguelen Island. (Phil. Trans.
1879, vol. clxviii. pp. 9–23, pl. 1–2.)
Epipactis helleborine. (Bot. Gaz. 1879, vol. iv. p. 225.)

1880

On the discovery of a variety of cedar of Lebanon on the mountains
of Cyprus ; with letter thereupon by Sir Samuel Baker, F.R.S.
(Journ. Linn. Soc., Bot., 1880, vol. xvii. pp. 517–519.)
W. Smith, A Dictionary of the Bible, ed. 5. London, 1880. 8vo.
(Botanical articles by J. D. H.)

1881

Presidential address to the Geogr. Sect. of the Brit. Assoc. at York,
Sept. 1, 1881.—On geographical distribution. (Brit. Assoc.
Rep. 1881 [1882], pp. 727–738 ; Nature, 1881, vol. xxiv.
pp. 443–448.)

Notes on arctic plants. (A. H. Markham, A Polar Reconnaissance, App. A. London, 1881.)

Drosera spatulata. (Gard. Chron. 1881, vol. xvi. p. 852.)

Begonia socotrana, Jasminum gracillimum, Nepenthes Northiana, nn. spp. (Gard. Chron. 1881, vol. xv. pp. 8–9 ; vol. xvi. p. 717.)

The Compass Plant. (Gard. Chron. 1881, vol. xv. p. 74. Reprinted from Bot. Mag. t. 6534.)

1882

The vegetation of the Rocky Mountain region and a comparison with that of other parts of the world. (By J. D. H. and Asa Gray.) (Bull. U.S. Survey 1882, vol. vi. pp. 1–62.)

On *Dyera,* a new genus of rubber-producing plants belonging to the natural order *Apocynaceae,* from the Malayan Archipelago. (Journ. Linn. Soc., Bot., 1882, vol. xix. pp. 291–293.)

On some undescribed and imperfectly known Indian Species of *Primula* and *Androsace.* (By George Watt, revised by J. H. D.) (Journ. Linn. Soc., Bot., 1882, vol. xx. pp. 1–18, pl. 1–18.)

1884

Royal Gardens, Kew. Official Guide to the Museums of Economic Botany. No. 1. Dicotyledon and Gymnosperms. London, 1883 [1884]. 153 pp. 8vo.

Tropical African Mountain Flora. (Nature, 1884, vol. xxx. p. 635.)

Notes on the Flora of Parasnath. (By C. B. Clarke, with an introductory note by J. D. H.) (Journ. Linn. Soc., Bot., 1884, vol. xxi. pp. 252–255.)

1885

Royal Gardens, Kew. Official Guide to the Royal Botanic Gardens and Arboretum. Ed. ' 29 ' [*i.e.* 30]. London, 1885. 184 pp. 8vo.

1886

The Admiralty Manual of Scientific Enquiry, ed. 5, Article xiv. Botany, pp. 418–432. London, 1886.

On the *Castilloa elastica* of Cervantes, and some allied Rubberyielding Plants. (Trans. Linn. Soc., Bot., 1886, ser. 2, vol. ii. pp. 209–215, pl. 27–28.)

The Himalayan Larch. (Gard. Chron. 1886, vol. xxv. p. 718, fig. 157.)

The Himalayan Silver Fir (*Abies Webbiana*). (Gard. Chron. 1886, vol. xxv. p. 788, figs. 174, 175.)

The Himalayan Hemlock Spruce (*Tsuga brunoniana*). (Gard. Chron. 1886, vol. xxvi. p. 72.)

The Botany of the Rocky Mountain Region ; a review. (Nature, 1886, vol. xxxiii. pp. 433–435.)

A Sketch of the Flora of South Africa ; a review. (Nature, 1886, vol. xxxiv. pp. 77–79.)

1887

Bentham, Handbook of the British Flora. Editions 5–8. London, 1887–1908. 8vo. (Revised by J. D. H.)

On *Hydrothrix*, a new genus of *Pontederiaceae*. (Ann. Bot. 1887, vol. i. pp. 89–94, pl. 7.)

Anniversary dinner of the Royal Society, 1887. Reply to the Toast . . . ' The Medallists,' Gloucester, 1887. 14 pp. 8vo.

1888

The Royal Horticultural Society. (Gard. Chron. 1888, vol. iii. p. 171.)

1889

Pachytheca. (Ann. Bot. 1889, vol. iii. pp. 135–140, pl. 8.)

1890

Indian *Orchideae*. (Hook. Ic. Pl. 1890, vol. xxi. tt. 2001–2050 ; 1891, tt. 2051–2075 ; 1892, tt. 2076–2100 ; 1892, vol. xxii. tt. 2101–2125 ; 1893, tt. 2126–2175 ; 1894, tt. 2176–2200 ; 1894, vol. xxiv. tt. 2317–2322, 2334, 2335.)

Eulogium on Robert Brown, delivered 1888 (Proc. Linn. Soc. 1890, pp. 54–67.)

1892

Juncus nematocaulon, n. sp. (Hook. Ic. Pl. 1892, vol. xxiii. t. 2234) ; *J. sikkimensis*, n. sp. (t. 2235).

1893

Index kewensis plantarum phanerogamarum . . . ad annum 1885 . . . sumptibus beati C. R. Darwin, ductu et consilio J. D. Hooker, confecit B. D. Jackson. Oxford, 1893–1895. 2 vols. 4to.

1895

A Century of Indian Orchids. (Ann. Bot. Gard. Calcutta, 1895, vol. v. pt. 1, pp. 1–68, pl. 1–101.)

David Lyall, M.D.; an obituary notice. (Journ. Bot. 1895, vol. xxxiii. pp. 209–211.)

1896

Journal of the Right Hon. Sir Joseph Banks during Captain Cook's first voyage. (Edited by J. D. H.) London. 1896. lii. + 466 pp., 2 portraits, 4 maps. 8vo.
Ischnochloa Falconeri, n. gen. et sp. (Hook. Ic. Pl. 1896, vol. xxv. t. 2466.)

1898

A handbook to the Flora of Ceylon, containing descriptions of all the species of flowering plants indigenous to the island, and notes on their history, distribution, and uses. (By H. Trimen, continued by J. D. H.) Part iv. London, 1898. iii. + 384 pp. Part v. 1900. 477 pp., 2 maps. 8vo. Plates 76–100. 1898. 4to.

1901

[Speech at] the opening of the new Botanical Department at the Glasgow University. (Ann. Bot. 1901, vol. xv. pp. 551–555.)
Rev. William Colenso. 1811–1899; obituary notice. (Year-book Roy. Soc. Lond. 1901, pp. 191–194; Proc. Roy. Soc. Lond. 1904, vol. lxxv. pp. 57–60.)

1902

A sketch of the life and labours of Sir William Jackson Hooker, with portrait [and bibliography]. (Ann. Bot. 1902, vol. xvi. pp. ix–ccxxi.)

1904

The Imperial Gazetteer of India, ed. 3, vol. i. chap. 4, Botany, pp. 157–212. Oxford, 1907.—Advance issue, entitled 'A sketch of the flora of British India.' London, 1904. 55 pp. 8vo. Introduction etc. reprinted in Journ. Bot. 1904, vol. xlii. pp. 221–227, with portrait of author.—Reprint. Oxford, 1906. 60 pp. 8vo.
An epitome of the British Indian species of *Impatiens*. (Rec. Bot. Surv. India, 1904–1906, vol. iv. pp. 1–58, and index.)
On the species of *Impatiens* in the Wallichian herbarium of the Linnean Society. (Journ. Linn. Soc., Bot., 1904, vol. xxxvii. pp. 22–32.)

1906

Premature shedding of leaves of Scots pine. (Gard. Chron. 1906, vol. xl. p. 278.)
George Bentham ; a review. (Kew Bull. 1906, pp. 187–188.)

1908

Les espèces du genre *Impatiens* dans l'herbier du Muséum de Paris. (Nouv. Arch. Mus. Hist. Nat. Paris, 1908, ser. 4, vol. x. pp. 233–272, pl. 2–6.)
Asiatic species of *Impatiens*. (Hook. Ic. Pl. 1908, vol. xxix. tt. 2851–2875 ; 1910, vol. xxx. tt. 2301–2325 ; 1911, tt. 2951–2975.)

1909

On some species of *Impatiens* from Indo-China and the Malayan Peninsula. (Kew Bull. 1909, pp. 1–12.)
A review of the known Philippine Islands species of *Impatiens*. (Kew Bull. 1909, pp. 282–289.)
Impatiens Hawkeri. (Curtis's Bot. Mag. 1909, t. 8247.)

1910

New *Impatiens* from China. (Kew Bull. 1910, pp. 269–275.)
Indian species of *Impatiens*. Generis *Impatiens* species indicae novae et minus rite cognitae a cl. A. Meebold detectae. (Kew Bull. 1910, pp. 291–300.)
Impatiens Hubertii, I. orthosepala, I. Winkleri, nn. spp. (Kew. Bull. 1910, pp. 74–76.)

1911

Lecomte, Flore générale de l'Indo-Chine, 1911, vol. i. pp. 611–629.—*Balsaminaceae*.
On the *Balsaminaceae* of the State of Chitral. (Kew Bull. 1911, pp. 209–211.)
On some species of *Impatiens* from the Malayan Peninsula : II. (Kew Bull. 1911, pp. 249–250, with plate.)
Indian species of *Impatiens*. On some Western Peninsular Indian *Balsamineae* collected by Mr. A. Meebold. (Kew Bull. 1911, pp. 353–356.)
Impatiens Herzogii. (Curtis's Bot. Mag. 1911, t. 8396.)

APPENDIX C

LIST OF DEGREES, APPOINTMENTS, SOCIETIES, AND HONOURS

1839	Glasgow	M.D.	Degree
1839	Edinburgh	Collegium Regium Chirurgorum Civitatis Edinensis	Diploma
1839	Chatham	Assistant-Surgeon of H.M.S. *Erebus*	Appointment
1842	London	The Linnean Society . .	Fellow
1843–45	Edinburgh	Assistant to Professor Graham, Professor of Botany	Appointment
1844	Woolwich	Assistant-Surgeon of H.M. Yacht *William and Mary*	Appointment
1845	Breslau	Caesareae Leopoldino-Carolinae Academiae Naturae Curiosorum Universitatis Vratislaviensis sub Cognomine Graham	Member
1846	London	Geological Survey of the United Kingdom	Botanist
1847	London	The Royal Society . . .	Fellow
1847	London	Botanical Exploring Expedition to East India	Appointment
1848	Calcutta	The Asiatic Society of Bengal	Honorary Member
1850	Florence	Imperiale e Reale Museo di Fisica e Storia Naturale di Firenzè	Silver Medal
1851	London	The Athenæum Club (under Rule 2)	Member
1851	London	The Exhibition of the Works of Industry of All Nations	Juror and Medal
1851	Paris	Société Nationale et Centrale d'Agriculture	Foreign Correspondent
1852	Paris	Société de Géographie . .	Diploma of Honourable Mention
1852	Haarlem	Société Royale Hollandaise des Sciences	Member

1852	Munich	Academia Literarum et Scientiarum Regia Boica, Monachii	Corresponding Member
1853	St. Petersburg	La Comité Scientifique du Ministère des Domains de l'Etat	Corresponding Member
1853	Dürkheim	Die Pollichia, ein Naturwissenschaftlicher Verein der Bayerischen Pfalz	Honorary Member
1853	Cherbourg	Société des Sciences Naturelles de Cherbourg	Corresponding Member
1854	London	The Royal Society . .	Royal Medal
1854	Berlin	Regia Scientiarum Academia Borussica	Member
1855	London	Assistant Director of the Royal Gardens, Kew	Appointment
1857	Ratisbon	Regia Societas Botanica Ratisbonensis	Member
1857	Vienna	Kaiserliche Königliche Geographische Gesellschaft, Wien	Corresponding Member
1858	Dublin	Dublin University Zoological and Botanical Association	Corresponding Member
1859	St. Petersburg	Société Russe d'Horticulture	Member
1859	St. Petersburg	Imperialis Academia Scientiarum Petropolitana	Corresponding Member
1859	Erlangen	Societas Physico - Medica Erlangensis	Corresponding Member
1860	London	Scientific Expedition to Syria and Palestine	Botanist
1861	Edinburgh	Societas Medica Edinburgena	Honorary Member
1862	Stockholm	Regia Scientiarum Academia Svecica	Foreign Member
1862	Königsberg	Die Ostpreussische Physikalisch - Ökonomische Gesellschaft zu Königsberg	Foreign Member
1863	Dublin	Natural History Society of Dublin	Corresponding Member
1863	New Zealand	Philosophical Institute of Canterbury, New Zealand	Honorary Member

1864	Newcastle	Natural History Society of Northumberland, Durham, and Newcastle-on-Tyne	Honorary Member
1865	Munich	Academia Literarum et Scientiarum Regia Boica, Monachii	Foreign Member
1865	London	Societas Londinensis pro Scientia Horticulturale	Honorary Member
1865	Vienna	K. K. Gartenbaugesellschaft in Wien	Member
1865	Upsala	Regia Scientiarum Societas Upsaliensis	Member
1865	Göttingen	Die Königliche Gesellschaft der Wissenschaften zu Göttingen	Foreign Member
1865	New Zealand	New Zealand Industrial Exhibition, Dunedin	Silver Medal
1865	London	Director of the Royal Gardens, Kew	Appointment
1866	Edinburgh	Societas Botanica Edinensis	Honorary Fellow
1866	Boston	American Academy of Arts and Sciences	Foreign Hon. Member
1866	London	Royal Geographical Society	Fellow
1866	Georgetown	Royal Agricultural and Commercial Society of British Guiana	Honorary Member
1866	Cambridge	Philosophical Society of Cambridge	Member
1866	Cambridge	LL.D.	Honorary Degree
1866	Oxford	D.C.L.	Honorary Degree
1866	Paris	Institut Impérial de France Académie des Sciences	Corresponding Member
1866	Lancashire	Todmorden Botanical Society	Member
1867	Paris	Exposition Universelle à Paris	Juror and Medal
1867	Antwerp	La Société d'Horticulture d'Anvers	Honorary Member
1867	Copenhagen	Det Kongelige Danske Videnskabernes Selskab, Kjöbenhavn	Foreign Member

1867	Helsingfors	Societas Pro Fauna et Flora Fennica	Honorary Member
1868	Vienna	K. K. Landwirthschafts Gesellschaft in Wien	Corresponding Member
1868	Norwich	British Association for the Advancement of Science	President
1868	London	Societas Regia Medico-Chirurgica Londinensis	Honorary Fellow
1868	Natal	Natural History Association of Natal	Honorary Member
1868	London	The Ethnological Society	Fellow
1869	Philadelphia	The American Philosophical Society of Philadelphia	Member
1869	Caracas	La Sociedad de Ciencias Fisicas y Naturales de Caracas	Honorary Member
1869	Dublin	Royal Irish Academy . .	Honorary Member
1869	St. Petersburg	Two Jasper Cups from the Russian Emperor	Gift
1869	London	The Most Honourable Order of the Bath	Companion (Civil)
1870	San Francisco	The Academy of Sciences	Honorary Member
1870	Paris	Société Impériale Zoologique d'Acclimatation	Honorary Member
1870	Florence	Società Geografica Italiana	Hon. Member
1870	Lancashire and Cheshire	Historic Society of Lancashire and Cheshire	Honorary Member
1870	Liverpool	The Literary and Philosophical Society	Honorary Member
1870	Dublin	The Royal Dublin Society	Honorary Member
1871	New Zealand	The New Zealand Institute	Honorary Member
1871	Montpellier	Société d'Horticulture et d'Histoire Naturelle de L'Hérault	Corresponding Member
1872	Moscow	Imperial Society of Botany	Member (?)
1872	Frankfurt	Die Senkenbergische Naturforschende Gesellschaft	Corresponding Member
1872	Brussels	Académie Royale des Sciences, des Lettres et des Beaux Arts de Belgique	Associate

1873	Rio de Janeiro	Sociedad Vellosiana do Rio de Janeiro	Corresponding Member
1873	London	The Royal Society . . .	President
1873	Munich	K. B. Academie der Wissenschaften	Foreign Member
1873	London	University College . .	Life Governor
1873	Brazil	The Imperial Order of the Roze	Commander
1873	Philadelphia	The Academy of Natural Sciences	Corresponding Member
1873	Bologna	Academia Scientiarum Instituti Bononiensis	Corresponding Member
1873	Stockholm	The Royal Swedish Order of the Polar Star (Riddar Nordstjerne Orden)	Knighthood (Riddar)
1873	Nimègue	La Société Botanique Neerlandaise à Leyde	Corresponding Member
1874	London	Royal Botanic Society, Regent's Park	Honorary Member
1874	Norwich	The Norfolk and Norwich Naturalists' Society	Honorary Member
1874	Glasgow	The Philosophical Society of Glasgow	Honorary Member
1874	Yokohama	The Asiatic Society of Japan	Honorary Member
1875	Florence	Società Toscana d'Orticultura	Honorary Member
1875	Watford	The Natural History Club, Watford	Honorary Member
1875	London	The Royal Institution of Great Britain	Member
1875	Paris	L'Académie de Médecine	Foreign Associate
1875	London	The University of London	Member of the Senate
1875	Russia	The Society of Naturalists of the Imperial Kazan University	Honorary Member
1875	Rome	Reale Accademia dei Lincei	Member
1876	Palermo	Academia Panormitana Scientiarum ac Litterarum	Honorary Member
1876	Boston	The Massachusetts Horticultural Society	Corresponding Member

1876	London	The Royal Institution	Manager and Vice-President
1876	London	Science and Art Department of the Committee of Council of Education, S. Kensington Museum	Vice-President
1876	Rome	Regia Lynceorum Academia	Hon. Foreign Member
1876	Norwich	The Medico - Chirurgical Society of Norwich	Honorary Member
1876	Lisbon	L'Académie Royale des Sciences	Corresponding Foreign Member
1877	Iowa	The Davenport Academy of Natural Sciences	Honorary Member
1877	Boston	The Society of Natural History	Honorary Member
1878	London	The Most Exalted Order of the Star of India	Knight Commander
1878	Toulouse	Académie des Sciences, Inscriptions et Belles Lettres de Toulouse	Foreign Associate
1878	London	The Pharmaceutical Society of Great Britain	Honorary Member
1878	Hamburg	Der Gartenbau-Verein für Hamburg, Altona und Umgegend	Honorary Member
1878	Dublin	Trinity College, Dublin (Doctor Utriusque Juris), J.U.D.	Degree
1878	Breslau	Die Schlesische Gesellschaft für Vaterländische Cultur	Honorary Member
1878	Brussels	Société Royale de Botanique de Belgique	Associate
1879	London	The Club	Member
1879	Paris	Universal Exhibition . .	Member of Commission
1879	New York	The New York Academy of Sciences	Honorary Member
1879	Sydney	The Sydney International Exhibition	Member of Commission and Medal
1879	Berlin	Der Verein für Beförderung des Gartenbaues	Honorary Member

1879	France	Société d'Horticulture d'Angers	Honorary and Corresponding Member
1880	London	The Royal Society . .	Portrait
1880	Sydney	The Royal Society for New South Wales	Honorary Member
1880	Lausanne	La Société des Sciences Naturelles du Canton de Vaud	Honorary Member
1881	Melbourne	Melbourne International Exhibition	Silver Medal
1881	York	The British Association for the Advancement of Science	President of Geographical Section
1881	New York	The Syracuse Botanical Club	Honorary Member
1881	London	The International Medical Congress	Vice-President
1881	Vienna	K.K. Geographische Gesellschaft in Wien	Honorary Member
1882	Edinburgh	Societas Regia Edinensis	Hon. Fellow
1882	Milan	Società Crittogamologica Italiana	Honorary Member
1882	Bath	Royal Literary and Scientific Institution	Vice-President
1883	London	The Association for the Advancement of Medicine by Research	Member
1883	Rome	Reale Accademia dei Lincei	Foreign Member
1883	Washington	National Academy of Sciences of the United States of America	Foreign Associate
1883	London	Royal Geographical Society	Founder's Medal
1883	London	Linnean Society (on completion of ' Genera Plantarum ')	Congratulations
1883	London	Society for the Encouragement of Arts, Manufactures, and Commerce	Albert Medal
1883	Berlin	Die Deutsche Botanische Gesellschaft	Honorary Member
1883	London	The Worshipful Company of Salters	Honorary Freedom

1884	Edinburgh	International Forestry Exhibition	Special Diploma
1884	Shrewsbury	Caradoc Field Club . .	Hon. Member
1884	London	Royal Institution of Great Britain	Manager
1884	London	British Association for the Advancement of Science (Canada Meeting)	Vice-President
1884	Edinburgh	The University of Edinburgh LL.D.	Honorary Degree
1884	London	International Health Exhibition	Bronze Medal
1884	Sydney	The Royal Society of New South Wales	Clarke Memorial Medal
1885	Turin	Regia Taurinensis Academia	Corresponding Member
1885	Amsterdam	Regia Academia Disciplinarum Nederlandica	Member
1885	Calcutta	Agricultural and Horticultural Society of India	Honorary Member
1885	London	Colonial and Indian Exhibition	Medal
1886	France	La Comité de l'Association pour la Protection des Plantes	Member
1886	Boston	The Appalachian Mountain Club	Corresponding Member
1886	Boston	Massachusetts Horticultural Society	Diploma
1886	Shettleston, N.B.	Scottish Society of Literature and Art	Honorary Member
1886	Modena	Società dei Naturalisti in Modena	Honorary Member
1887	London	The Royal Society . . .	Copley Medal
1887	Paris	La Société de Géographie	Foreign Corresponding Member
1887	Texas	Trinity Historical Society, Dallas, Texas	Non-Resident Member
1887	Ghent	Société Royale d'Agriculture et de Botanique de Gand	Honorary Member
1887	Glasgow	Natural History Society of Glasgow	Honorary Member

1888	London	The Linnean Society . .	Portrait
1888	London	The Linnean Society . .	Centenary Medal
1888	London	The Italian Exhibition .	Honorary Member of Committee
1888	Woking	Gordon Boys' Home . .	Member of Council
1888	Surrey	Surrey Archæological Society	Member
1888	Bologna	The University of Bologna	Honorary Doctor
1889	Adelaide	Jubilee International Exhibition, Adelaide	Medal
1889	New York	Torrey Botanical Club, Columbia College, New York	Honorary Member
1889	London	The Alpine Club . . .	Hon. Member
1890	Antwerp	Exposition Internationale de Géographie Botanique commerciale et industrielle, Anvers	Honorary Member
1891	Copenhagen	La Société Botanique de Copenhagen	Honorary Member
1891	Budapest	Magyar Tudomanyos Akademia (the Hungarian Academy of Sciences)	Foreign Member
1891	London	The Royal Statistical Society	Fellow
1891	London	Royal Naval Exhibition .	Diploma
1892	London	The Royal Society . .	Darwin Medal
1892	Manchester	The Manchester Literary and Philosophical Society	Honorary Member
1893	Königsberg	Physikalisch-Ökonomische Gesellschaft	Foreign Member
1893	Berlin	Gesellschaft für Erdkunde zu Berlin	Honorary Member
1893	Paris	Académie Internationale de Géographie Botanique	Free Member and Medal
1894	Geneva	Société de Physique et d'Histoire Naturelle	Honorary Member
1894	Boston	The Appalachian Mountain Club	Honorary Member
1895	Moscow	Societas Caesarea Naturae Curiosorum Mosquensis	Honorary Member

1895	Edinburgh	The Scottish Natural History Society	Honorary Fellow
1896	Rome	Società Italiana Delle Scienze	Foreign Member
1897	Amsterdam	Nederlandsche Maatschappij Tuinbouw en Plantkunde	Honorary Member
1897	London	Royal Horticultural Society	Victoria Medal of Honour
1897	London	The Most Exalted Order of the Star of India	Knight Grand Commander
1897	Manchester	The Manchester Literary and Philosophical Society	Wilde Medal
1897	London	The Salters' Company . .	Gift (on his 80th birthday)
1897	London	The Linnean Society (completion of Flora of British India)	Commemorative Medal
1898	Manchester	The Manchester Literary and Philosophical Society	Wilde Gold Medal
1900	Halle	Die Kaiserliche Leopoldinisch - Carolinische Deutsche Akademie der Naturforscher	Cothenius Medal
1900	Paris	Institut de France, Académie des Sciences	Foreign Member
1900	Dominica	Agricultural Society of Dominica	Honorary Member
1902	Berlin	Der Königliche Preussische Orden ' Pour Le Mérite ' für Wissenschaften und Künste	Order
1904	Berlin	Academia Scientiarum Borussica	Foreign Member
1905	London	British Science Guild . .	Vice-President
1907	Munich	Regia Scientiarum Academia Bavarica	Address on his 90th birthday
1907	New York	Academy of Sciences . .	Honorary Member

| 1907 | Upsala | Regia Academia Scientiarum Suecica (the Linnean Bicentenary) | Commemorative Gold Medal |
| 1908 | London | The Linnean Society (50th Anniversary of joint papers by Darwin and Wallace) | Darwin - Wallace Medal (Silver) |

INDEX

ABRAHAM'S Oak, i. 533

Abyssinia, no link with Cape and Australian temperate floras, i. 462

Aconite, ii. 282

Acquired characters, ii. 123

Acropera, ii. 105

Adams, J. C., on Norwich Address, ii. 120 and *note*

Adamson, early friend of J. D. H., i. 156

Adenanthera pavonina, ii. 107

Agassiz, A., x Club guest, i. 544

Agassiz, L., quoted, ii. 117; and species-mongering, 473, 474, 475

Airlie, Lord, character and military study, ii. 372

Airy, Sir G., as P.R.S., i. 543, ii. 132, 133; order of Polar Star, 186 *sq.*

Aiton, W., house in Kew Gardens, i. 341, 345, 348

Albani, Mme., ii. 156

Albatross' eggs, a practical joke, i. 104

Algae, Antarctic, i. 173, 175, 183, 189, 190, Darwin's disputed specimen, 176; how replaced on British coast, 212

Allman, G. J., i. 197 and *note*

Almond, Rev. Mr., i. 88

America, the future of botany in, i. 475; democracy and politics, ii. 39–45; visit to, 205–17, cp. 261; N. Amer. botany, 216, 220 *sq.*, 232, 426; a universal peacemaker, 325

Amnothea communis, i. 57

Amsterdam Island, i. 83

Amundsen, i. 55

Anatomy, botanical, advance of, ii. 419 *sq.*

Anderson, Dr. Thomas, i. 406 *n.*, ii. 1 *n.*

 Letters to: 'Lionising,' i. 406; W. J. H. and recognition of Kew

Herbarium, 419; reviews of the 'Origin,' 515; the *Quarterly* article, 520; over-work and society, 536, 537; Henslow's death: *Servate animam æquam*, 537

 Papyrus, ii. 1; Cinchona, 2; at Darjiling, *ib.*; rival sites, *ib.*, and theories, *ib.*; frosts, *ib.*; alleged species of, *ib.*; a German cultivator, 3; nomenclature, 3; John Scott, 3; orchids: seeds from Sikkim, 8; Indian duplicates, 9; Calcutta Botanical Garden, 9, 10; Colonial Floras and Flora Indica, 13; arrangements for, 13; Latin or English, 13; delay about Flora Indica, 13 *sq.*; further delay, 14; initiative desired from Calcutta, 14; comes from India Council, 15; delayed by other work, 15; pressure of general work, 16; progress depends on his return from India, 16 *bis*; microscope work, 25; on Henslow, 61 *bis*.

Anderson, William, i. 76 and *note*, 100

Andraea, i. 74, 83

Andropogon, Hackel on, 284, 285

Angiopteris, i. 468

Angiosperms, primitive type of, ii. 22 *sq.*, 25

Anogeissus, ii. 390

Anopterus glandulosus, i. 106

Antarctic, the, botanical barrenness, i. 55, 111, 114, 163, beyond the Arctic, 55, 82; plants common to, 140; zoological interest, 55–60, 67–70, 122; deep sea life, *ib.*; collections wasted, 56; diatoms discovered, 55–6, 58–60; botanical generalisations, 66, 74, 75; relative proportion of plants, 76 and *note*, 79 *sq.*; some strictly Antarctic

INDEX

539

J. D. H. as to exam. papers, 388 ;
a dried plant examination scheme,
389 ; pioneer of practical elementary teaching, 394 ; his unfinished book of village botany
used by Prof. Oliver, 391 ; criticisms of MS., 393–8 ; his botanical
diagrams, 392 ; English botanical
names, 394 *sq.*, 428 *n.*, 479 ;
travel with, 434–5 ; sends Galapagos plants, 436 ; and the
' Origin,' i. 511 ; resists unfair
attacks on, 512 *sq.* ; partial acceptance of, 520 ; at Oxford, 1860,
523, 526 ; death, ii. 60, i. 537 ;
character, ii. 60 *sq.*, cp. 123 ;
faculties inherited from, 366
Letter from : Fair hearing for
Darwin, i. 512 *sq.*
Letters to : Value of a botanical
career, i. 374 ; botany at Oxford,
383 ; value of botany in medical
education, 388 ; handbooks and
examinations, 389 ; his botanical
diagrams, 392 ; criticises his MS.,
393 *bis*, 395, 396 ; aid for Linnean
Journal, 410 ; Nelumbiaceae, 423,
a paradox, 424 ; unnecessary questions, 429 ; 430 ; to Paris, 1855,
434 ; senseless attacks on Darwin,
514 *bis* ; a Darwinian botany
book, 535
Henslow, L., i. 396
Heredity, as illustrated by the Hooker
family, ii. 308, 366 ; in literary
criticism, 366
Herkomer, Sir H., R.A., ii. 342
Hermaphroditism, perpetual, i. 452
Hermite Island, i. 53, 66 ; visited,
128, 133, 134–8 ; a plant from, ii.
358
Herschel, Sir J., i. 147 *n.* ; and
Humboldt, 186 ; ii. 56 ; unfair
quotation of, 127 ; Lyell's letters
to, 202 ; cameo of, 436
Heysham, Dr., ii. 429
Higher and Lower types of plants, i.
444, 445 *sq.*, 480 ; conifers, 451, 460,
463 ; gymnosperms, 481 ; Australian
the highest, 463 ; old world plants
have competitive superiority, *ib.*,
note, 494 ; relation to ' progressive
development,' 507
Himalaya, early hope of exploring,
i. 167 ; objective, 251 ; difficulties,
251 *sq.*, 264–72 ; first expedition,

274–84 ; second, 289–319 ; captivity, 290, 312 *sq.* ; obstruction, 290,
292–4, 295 *sq.*, overcome, 297 *sq.* ;
the snow-line, 300 ; geography of,
revolutionised, 327 *sq.*
Mr. Elwes in, ii. 125 ; travellers
in, 266 ; glacial period in, 320
sq. ; geological rhyme, 321 ; im-
pressiveness of view from Darjeeling, 365
Himalayan Botany, a parallel with
the Antarctic, i. 258 ; general
features of, *ib.* 259 ; rhododendrons,
q.v. ; map to show distribution,
259 ; a lichen as at Cockburn I.,
55, 305 ; tardy advance of the
flora, 307 ; hermaphrodite heads
of maize, *ib.*
ii. 248, 426 ; and the Caucasus,
363 ; effect on, of elevation of
the chain, 392
Himalayan Journals, no intention of
writing Indian travels, i. 255 ; impressions quoted, 281 *sq.* ; arrest
of Campbell, 313 ; work on, 341 ;
published, 363 *sq.* ; dedicated to
Darwin, 363 ; a pole-star of his
life, *ib.* ; delight in practical
philosophy, 364 ; E.I.C. treatment
of, *ib.*
Military appreciation of, ii. 183 ;
useful to his son, 374, and to a
mining prospector, 471 ; position
in Geographical Science, 412 ; Dr.
Freshfield's appreciation, 452
Hinds, R. B., i. 438 and *note*
Hirst, T. A., i. 539 *n.* ; in the *x* Club,
i. 539, 541 ; his minute on the
Atom, 543, 546 ; on Magee's
sermon, ii. 119
History, enjoys as a boy, i. 28–29 ;
the burden of modern education,
ii. 252, 329
Hodgson, Brian H., i. 247 ; friendship,
248 *sq.* ; an evening with, 261 ;
description of, 262 ; unable to
join first expedition, 266 ; hopeless
about, 272 ; help from, 274, 285,
291, during captivity, 318
Visited, ii. 203 ; travels, 266 ;
authority on Buddhism, 433
Letter from : Old age and
friendship, ii. 257
Letters to : Science workers in
1862, ii. 31 ; Bishop Colenso, 57,
57–8 ; Wedgwood ware, 78 ; work

six hours' sleep, 220; in India, 239, 260, 323, cp. ii. 198, 357, 461, 466
Sources of general knowledge, 24, 36; schooldays, 22–4; college, 24–9, 31, M.D. 38 *sq.*; gains from Latin, but not from Moral Philosophy, 22; prizes, 28; early work in botany, 22, 24, 25 *bis*, 32, a slight on which he resents, 35, and entomology, 24, 25 *bis*, 26; tastes and acquirements, 28 *sq.*; love of history, *ib.*; public speaking, hindrance to, 29 *sq.*, 194, ii. 309; early critical herbarium work, 30, 40; religious observance, 33, 46, 106; first attendance at Brit. Assoc., Newcastle, 34–6; is careful about money, 27, 32, 33; unique botanical training, 37; aided by his father's position, *ib.*, 40, 68, 220
Antarctic Expedition: medical service necessary, 38, though little required, 45, 57, and disliked by H., ii. 439, 457; appointment, i. 41–4 *sq.*; equipment, 45–8; his father's gifts, esp. a watch, 46–7; activity as naturalist, 55 *sq.*, 68 *sq.*; zoological work in the intervals of botany, 55–60, esp. on the Diatoms, 58 *sq.*; discouraged by his father, 63, 161, 262; temporary only, 60, 114, 161, 262; drawing, scientific and artistic, under difficulties at sea (*see* Drawing); collections, early, 60, 64, 72; later improve, 65, 139; suggest generalisations, *see under* Collections; his opportunities and advice from his father, 64 *sq.*; his natural seriousness, 53, 106, 107, 120, 161; needless warnings, 53, 106; friendship with fellow officers, 67 *sq.*, 92 *sq.*, and with Ross, 68 *sq.*; reads 'Voyage of the *Beagle*' in proof, 66 and *note*, 136; visualising power in making comparisons, 67, 87, 90, 93; enthusiastic imagination, 74; home ties, 74, 156; home-like friendships, 107; happiness, independent of circumstance, 79, 94, lies in work, 157; no tedium for a naturalist at sea, 73; modest ambitions, 72, 79, 83, 113, 114, 143, 164–6
Antarctic work, limits of, 82, 85; is critical of systems, 84, 132;

ignorant of tropical plants, 85; Antarctic Journal, 86, *s.v.* Journal; and letters, 87; contributions to Ross' 'Voyage,' 86 *n.*, 139, 173; the Scottish Highlands, a standard of beauty, 87, 90, 93, 94, 135; at Madeira 87 *sq.*, Teneriffe, 91, Cape Verde I., 91 *sq.*, St. Paul's I., 95, Trinidad, 95 *sq.*, St. Helena, 96 *sq.*, Cape Town, 99, Kerguelen I., 99 *sq.*; interest in economic botany, 88, and ideas on gardening, 97; return for hospitality, 93; keenness of observation, 97 *sq.*; records a practical joke, 104; Tasmania, 105, 120, affection for, 107 *sq.*, festivities, 119 *sq.*; first voyage to the South, 109 *sq.*; Sydney, 120–3; New Zealand, 124; second voyage to the ice, 124 *sq.*; ideas on the ice, 127 *sq.*; the Falklands, 127, 128–34; general reading, 131, and botanical books, *ib. sq.*; grouping of plants, 132, theory and practice, *ib.*; delight in finding strictly Antarctic Orders, 133; in Fuegia, 134 *sq.*; third voyage to the ice, 139 *sq.*; depression of, to all but Ross and Hooker, 140, 141; official secrecy about the voyage, 141, broken by Prince Albert, 142–5; preliminary botanical account, 146; Cape Town visited, 147–51
Personal: Limited circle of friends, 143, 154, 156; music and art, 153–4; bereavements, 154 *sq.*; difficulties in correspondence, 154 *sq.*, 157 *sq.*; happiness in work, 157; community of interest with his father, 157, 159, 160–1, 166; plans to aid Kew and W. J. H., 159–61, 206; prospects, afloat or ashore, 162 *sq.*; ready to become a botanical hermit, *ib.*; succession to his father, 166; would like further travel, 167; work at Kew, 169, and on Flora Antarctica, 168–76; Govt. aid, 170; vicissitudes of, 189, 196, 199, 206 *sq.*, 214; arranging the Herbarium, 'stable occupation,' 170; affection for animals, *ib.*; received into Linn. Soc., 171; rivalry with French Antarctic botanists, 174; good and bad species, 174–5; openings

7; statue at Exeter, 8 *n.*; ii. 472 and *note*; visited, *ib.*

Hooker, Richard Symonds, his name and skull, ii. 266; a budding doctor, 357, 457 and *note*; reading with, 434

Hooker, Robert H., presents statue of Richard Hooker to Exeter, i. 8 *n.*; ii. 472

Hooker, Thomas, Rev., founds the American branch of the family, i. 8; and American democracy, ii. 323

Hooker, William Dawson, i. 18, 20; delicacy, 20; boyish quickness, 21–23; tastes, 23, 25, 27; at College, 22, 24, 25; his first book, 22; death, 154 *sq.*

Hooker, William Henslow, ii. 192, 196, 268; illness of, 430; sends an ivory walking-stick, 462

Hooker, Sir W. J., position, scientific, i. 3, and official, 37; birth, 8; career and discoveries, 9 *sq.*; travel, 10; the Hooker-Turner-Paget alliance, 10, 16, and with D. Turner, 16; the brewery and financial loss, 10; at Glasgow, 11; his teaching and influence, 11–14, 136; botanical excursions, 13; works, 14 *sq.*, a parallel with his son's, 15; personality and home life, 20, cp. 67; K. H., 27; aid to his son, 37, 38–40, 67, 68, 166, 176, 215, appreciated by J. D. H., 220; criticises *Erebus* officers, 45, 67; botany and 'dissipation,' zoological and other, 63 *sq.*, 161, 262; collections, advice on, 64, 101, delight in good, 65, would like to join in, 101 *n.*; preliminary account of Antarctic botany, 146; advises shortening of Continental trip, 178; friends in Paris, 180; Herbarium and Library, more useful than Brit. Mus., 192, as gift to his son, 192, first offered to Kew, 215; estimate of J. D. H., 206; small aid from Govt., 210; official residence offered, at a price, 346, then freely, with a lien on the Herbarium and Library, 346–7; modesty, though in need of an Assistant, 349; unselfish love of science, 376; opposed to Sunday opening, 377; trustee of Fielding Herbarium, 382; science

at Oxford, 383; grows careless as editor, 411; claims overlooked, 418–20; identity of several species of Juniper, 472

ii. 3, 10 *sqq.*; his strenuous age, 45, character, 64; death, 47, 64, 67, account of, 68; his Library and Herbarium, 47–8; portrait, 178; the Bot. Mag., 242 *n.*; natural inheritance from, 307 *sq.*; connexion with Jorgensen, 346–7–8, 483, 484, 485, 486; visit to Iceland, 347 and *note*; quotes 'Loves of the Plants' in his lectures, 354; Memoir of, 379–82; originator of economic botany, 410; a maker of Kew, 417; J. D. H.'s devotion to his memory, 148, 382, 410, 430, 444, 473, 474; his systematic work, 421, and its aim, *ib.*; and the Linnean Society, 430; works to the last, 444

Letters from: Botanical excursion, i. 23; Latin: the Boy's Own Book, 24; useful knowledge of botany, 25; unprofitable travel, 25; College work, 27, 31; visit to D. Turner, *ib.*; Joseph's botanical progress, 32; Antarctic, preparations for, 38, 39, 40; officers of the *Erebus*, 45; Fuegian collections, 139; prospect of succession, 166; J. D. H. strengthened by the voyage, 168; (to D. T.) on Joseph's engagement, 220

Letters to: Interview with Ross, i. 41–44, with R. G. Commission, 44; Govt. and Science, 45; medical duties, *ib.*; work at sea, 57, 60, 60–1; as naturalist *de facto*, 68, 70; aid from Ross, 68 *sq.*, 70; botany at sea, 71, 72; no tedium, 73; St. Helena plants, 71, unsatisfactory, 72; uprooted seaweeds, 73 and *note*; Kerguelen's Land, its plants, 77, 78; affection for, 79; described, 100–4; the Falklands and Cryptogams, 79, and Geog. Distribution, 81; an Antarctic Flora, 82, and local floras, *ib.*; Antarctic mosses and their grouping, 83 *sq.*; Tasmania, 106, 107, 107 *sq.*; first Polar voyage, 109 *sqq.* and 110 *n.*; only marine zoology possible so

Reeve, Lovell (publisher of Fl. Ant., &c.), i. 171; shirks botany, 370; 509; ii. 260
'Reign of Law,' by F. T. Palgrave, ii. 119, 120; by the Duke of Argyll, 114
Reimers, i. 178
Religion, training in, i. 19 *sq.*, 33, 46, 106; in State education, ii. 326 *n.*, 338; of pure reason, 337; the Greek Church in Russia, 87
Religious views, on standing godfather, i. 323 *n.*, ii. 59; liberal and anticlerical, ii. 54–9, op. Henslow, 61, 66; and scientific differences, 55 *sq.*; the function of anthropomorphism, 113; scripture chronology, 118; Natural Theology, 118; 83, 114. *See also s.v.* Providence
Renaissance, the new, i. 1
Reputation, scientific, how best to attain, i. 56; indifference to ordinary form of, 83
Rheumatic fever, attack of, i. 91
Reunion, ii. 11
Rhodes, 'superb,' i. 529
Rhododendron Book, i. 255, 325 *n.*, 326 and *note*; appreciated, *ib.*; work on, 341; published, 355; enquirers should read, 430; edits Cathcart's, 355
Rhododendrons, i. 254, 256, 257; seed, 285, 287; loftiest shrub, 325; acclimatised, ii. 343; 'wearing the willow,' 444; *R. anthopogon*, ii. 471, *Campbelliae*, i. 254, *nivale*, described, 325 and *note*, *Thomsoni*, on his memorials, ii. 481
Rich, Anthony, ii. 235
Richard, Achille, i. 160 and *note*, 423
Richardson, Sir John, i. 32 and *note*; description of, 34; helps J. D. H. qualify for the Antarctic, 37, 38; discriminating aid at Haslar, 39, 40; encourages H.'s zoological work, 56; medallion of, ii. 477
Richardson, Lady, i. 32 *n.*
    Letter from: Naval experience, i. 152
Richmond, 'lackadaisical' portrait of J. D. H., ii. 72
Rigby, Edward, M.D., i. 17, 18; family, 17, 18, 19
Rio de Janeiro, i. 53, 85, 93
Ripon, Lord, ii. 169

Roberts, Lord, ii. 373
Robertson, Archibald, i. 45 and *note*
Rodriguez, ii. 138
Romanes, G. J., at Kew, ii. 230 and *note*, 231; Darwin obituary, 259
Rogers, Rev. W., i. 544 *n.*
Rome, ii. 252–3
Rosen, Robert and Theophile de, marry Rigby sisters, i. 17, 18; H. visits, ii. 86
Ross, Sir J. C., i. 6; friendship with the Hookers, 37–8, 67; ii. 439 *bis*; interview with, 41 *sq.*; 44; Arctic experience, 50; relations with Wilkes and d'Urville, 51 *sq.*; Antarctic voyages, 52; discoveries, 54; and good fortune, ii. 362; and his officers, i. 67; relations with J. D. H., 67, 68–72, 101 *bis*; share in zoological work, 68, 69, 113; magnetic, 99, 105 *n.*; strictness, 106, relaxed, 125; view of the collision, 126; his book, 85; appreciated in India, 244; H.'s contributions to, 86 *n.*, 139, 171, 173, 245, a sort of humbug! *ib.*; impressed by the Barrier, 118; interest in prolonging the expedition, 140; receives Geog. Soc. Gold Medal, 141; keeps the strict Admiralty rule about letters and collections, 141 *sq.*; mitigation of, 141; secrecy overreaches itself, 145; his letter to Prince Albert, 145; motto for, 146; H.'s future career, 164 *sq.*; dedication to, of Fl. Ant., Pt. I., 171, *q.v.*; amends account of fossil tree, 172; 189
    A fellow Linnean, ii. 429; care as a navigator, 442 *sq.*; coldly received on his return, 443; at Lady Franklin's, 456; portrait, 477
    Letters to: Deep-sea life, i. 122 *n.*; Antarctic notes, 172; the Geol. Survey, and continuance of Fl. Ant., 208; duties at, 209; India and Fl. Ant., 216 *sq.*
Rosse, Earl of, i. 350, ii. 56
Roxburgh, i. 473, ii. 281, 283 and *note*
Royal Companies' Islands, i. 83
Royal Society, speech at Anniversary, 1887, i. 3, 5; urges Antarctic exploration, i. 49; instructions to

illustrates Bot. Mag., ii. 243 n.,
481 n. ; acknowledgments to, Pre-
face, vii
Smith, Robertson, x Club guest,
i. 544
Smyrna, comfort of Europeans in-
ferior to India, i. 529
Snowdon, storm on, ii. 203
Solander, Dr., i. 10 n., 139 ; Wedgwood cameo of, ii. 437
' Solomon Grundy,' i. 197–8
Sonder, O. W., i. 468 and note ;
ii. 15
South Georgia, i. 83
South Shetlands, i. 53, 139
Spach, i. 472
Spain, botanising in, i. 433, ii. 91
Sparshall, Mr., i. 4
Species, on naming, i. 56, 83 ;
difficulties in Galapagos collections,
169, 443 ; first discoverers of, 173 ;
changes in nomenclature, 174, 190 ;
-making, 174 ; quot homines, 176 ;
and varieties, 190, 221 ; fixity of,
shaken by wide knowledge, 366–8,
abandoned in Tasmanian Essay,
353, 504–9 ; no common standard
of differentiation, 367, for peculiar
species, 438, 443, and varieties,
456 ; reduced by careful examination, 422 ; an argument for
reduction, 444 ; validity of, 441 ;
mundane, labour of establishing,
442 ; botanists differ as to, 443 ;
shaken to their foundations by
intermediates, 447, cp. 449 ; are
very few, 447 ; Bentham begins
to ' lump,' 453 ; Decaisne also
follows H.'s lead, ib. ; objective
and subjective, 455, 478–9, 485 ;
personal idiosyncrasy in placing,
458, as new or varieties, ib., 467 ;
the two aims of systematists, 454 ;
a definition of, 466 ; in improving
classification H. reduces species,
466 sq. ; many specimens required
for determination, 466, but prove
many species to be varieties, 467
(cp. ii. 286) ; ' swimming in
synonymy,' ib. ; created by extinction of intermediates, 470, 505 ;
and habit, 472, 475 ; number of
known, 473 ; synonymy, 473 ;
domestic varieties ranked as, 474 ;
fixity of, and an open mind, ib.,
507, 508 sq. ; mutability of,

adopted in Tasmanian Essay, 481
sq., 484 ; transitional forms not
found at once, 497 ; influence of
external conditions, 498 ; ' species '
the coin of science, 505 ; extent of
mutability, 506 ; centrifugal variation, ib. ; regulation of, ib. ; reversion, ib. ; bigoted idea of the term,
508 and note
    Limits of, contrast between
Gray's and Hooker's manuals,
ii. 235, 236 n. ; merged by great
numbers of specimens, 286 (cp. i.
467) ; founded on single specimens,
397, 401 ; mutability of, 421 ; the
struggle over, 427 ; how regarded
by the two Hookers, 421 ; H.'s
view in relation to Darwin, 427
' Species Filicum,' work on, i. 169
Spence, William, i. 98 and note, 30 ;
ii. 429
Spencer, Herbert, i. 526 n. ; quoted
by Dr. Draper, 526 ; in the x
Club, 538, 539 ; metaphysics, 543 ;
the guardian of order, 543, 544 ;
' the battle of the ladies,' 544
    Power to appreciate Pangenesis,
ii. 110 ; the unknowable, 119,
quoted, ib., 120 ; a forgotten
opinion, 194 ; scientific reason for
variation, 306 ; on acquired habits,
348 ; reads, 434 ; his works and
non-scientific readers, 454 ; his
educational ideas, ' for bachelors'
children,' ib.
Spottiswoode, W., i. 540 n. ; in the
x Club, i. 540, 541, 542, 545
    A social centre, ii. 82, 192 ;
Magee's sermon, 119 ; liberality,
136 ; as Treas. R. S., 135, 139 ;
researches, 141 ; aid in Ayrton
affair, 171 ; Darwin's funeral, 259 ;
death while P.R.S., 263, i. 545
Sprengel, K., i. 131 and note
Stanley, Dean, ii. 56
Stanley, Lord, i. 174
Stapf, O., ii. 286 and note, 287, 288
bis, 290
Steenbock, Count, ii. 188
Stephens, J. F. (1792–1852), F.L.S.,
i. 26. He published ' Illustrations
of British Entomology,' 1827–
37
Steuarts, early friends of J. D. H.,
i. 156
Steudel (cp. Index Kewensis), mul-

END OF VOL. II.

AT THE BALLANTYNE PRESS
PRINTED BY SPOTTISWOODE, BALLANTYNE AND CO. LTD.
COLCHESTER, LONDON AND ETON, ENGLAND

Printed in the United States
By Bookmasters